Cambridge Studies in Biological and Evolutionary Anthropology 40

Shaping Primate Evolution

Form, Function, and Behavior

Shaping Primate Evolution is an edited collection of state-of-the-art papers about how biological form is described in primate biology, and the consequences of form for function and behavior. The contributors are highly regarded, internationally recognized scholars in the field of quantitative primate evolutionary morphology. Each chapter elaborates upon the analysis of the form–function–behavior triad in a unique and compelling way. This book is distinctive not only in the diversity of the topics discussed, but also in the range of levels of biological organization that are addressed, from cellular morphometrics to the evolution of primate ecology. The book is dedicated to Charles E. Oxnard, whose influential pioneering work on innovative metric and analytic techniques has gone hand-in-hand with meticulous comparative functional analyses of primate anatomy. Through the marriage of theory with analytical applications, this volume will be an important reference work for all those interested in primate functional morphology.

FRED ANAPOL is a professor in the Department of Anthropology (adjunct in Biological Sciences) and the Director of the Center for Forensic Science at the University of Wisconsin–Milwaukee, where he teaches skeletal biology, primate variation and evolution, and forensic anthropology. His research focuses on evolutionary and developmental morphology and physiology of the neuromusculoskeletal system of mammals, especially primates.

REBECCA Z. GERMAN is a professor in Biological Sciences at the University of Cincinnati. Her research interests involve functional and evolutionary morphology, evolutionary developmental biology, and biostatistics. She has worked extensively on mammals and, in particular, on marsupials; work which has taken her regularly to Australia, including a term as senior Fulbright Fellow at the University of Western Australia.

NINA G. JABLONSKI is the Irvine Chair and Curator of Anthropology at the California Academy of Sciences. She is an evolutionary anthropologist with broad interests in primate and human evolution. She is the author of numerous publications, including several edited volumes on the biology of Old World monkeys and on the relationship between environmental change and primate evolution. Her recent research also embraces the controversial topics of the evolution of human bipedalism and of human skin coloration. Alongside several journal appointments, she is also Series Editor for *Cambridge Studies in Biological and Evolutionary Anthropology*.

Cambridge Studies in Biological and Evolutionary Anthropology

Also available in the series

21 *Bioarchaeology* Clark S. Larsen 0 521 65834 9 (paperback)
22 *Comparative Primate Socioecology* P. C. Lee (ed.) 0 521 59336 0
23 *Patterns of Human Growth*, second edition Barry Bogin 0 521 56438 7 (paperback)
24 *Migration and Colonisation in Human Microevolution* Alan Fix 0 521 59206 2
25 *Human Growth in the Past* Robert D. Hoppa & Charles M. FitzGerald (eds.) 0 521 63153 X
26 *Human Paleobiology* Robert B. Eckhardt 0 521 45160 4
27 *Mountain Gorillas* Martha M. Robbins, Pascale Sicotte & Kelly J. Stewart (eds.) 0 521 76004 7
28 *Evolution and Genetics of Latin American Populations* Francisco M. Salzano & Maria C. Bortolini 0 521 65275 8
29 *Primates Face to Face* Agustín Fuentes & Linda D. Wolfe (eds.) 0 521 79109 X
30 *Human Biology of Pastoral Populations* William R. Leonard & Michael H. Crawford (eds.) 0 521 78016 0
31 *Paleodemography* Robert D. Hoppa & James W. Vaupely (eds.) 0 521 80065 5
32 *Primate Dentition* Daris R. Swindler 0 521 65289 8
33 *The Primate Fossil Record* Walter C. Hartwig (ed.) 0 521 66315 6
34 *Gorilla Biology* Andrea B. Taylor & Michele L. Goldsmith (eds.) 0 521 79281 9
35 *Human Biologists in the Archives* D. Ann Herring & Alan C. Swedlund (eds.) 0 521 80104 4
36 *Human Senescence and Life Span* Douglas E. Crews 0 521 57173 1
37 *Patterns of Growth and Development in the Genus* Homo Jennifer L. Thompson, Gail E. Krovitz & Andrew J. Nelson (eds.) 0 521 82272 6
38 *Neanderthals and Modern Humans – An Ecological and Evolutionary Perspective* Clive Finlayson 0 521 82087 1
39 *Methods in Human Growth Research* Roland C. Hauspie, Noel Cameron and Luciano Molinari (eds) 0 521 82050 2

Shaping Primate Evolution

Form, Function, and Behavior

EDITED BY

FRED ANAPOL
University of Wisconsin–Milwaukee

REBECCA Z. GERMAN
University of Cincinnati

NINA G. JABLONSKI
California Academy of Sciences

CAMBRIDGE UNIVERSITY PRESS
Cambridge, New York, Melbourne, Madrid, Cape Town, Singapore,
São Paulo, Delhi, Dubai, Tokyo, Mexico City

Cambridge University Press
The Edinburgh Building, Cambridge CB2 8RU, UK

Published in the United States of America by Cambridge University Press, New York

www.cambridge.org
Information on this title: www.cambridge.org/9780521143417

First published 2004
First paperback printing 2010

A catalogue record for this publication is available from the British Library

Library of Congress Cataloguing in Publication data

Shaping primate evolution / edited by Fred Anapol, Rebecca Z. German &
Nina G. Jablonski.
p. cm. – (Cambridge studies in biological and evolutionary
anthropology ; 40)
ISBN 0 521 81107 4
1. Primates – Evolution. 2. Primates – Morphology. I. Anapol, Fred
Charles. II. German, Rebecca Z. III. Jablonski, Nina G. IV. Series.
QL737.P9S453 2004
599.8′138–dc22 2003060226

ISBN 978-0-521-81107-1 Hardback
ISBN 978-0-521-14341-7 Paperback

Dedication

For Charles Oxnard – mentor, inspiration, friend:

He was a scholar, and a ripe and good one;

Exceeding wise, fair-spoken, and persuading.

King Henry VIII. Act iv, Scene 2

Contents

Contributors

Gene H. Albrecht, Department of Cell and Neurobiology, Keck School of Medicine, University of Southern California, Los Angeles, CA 90033, USA

Fred Anapol, Department of Anthropology, University of Wisconsin–Milwaukee, P.O. Box 413, Sabin Hall, Milwaukee, WI 53201, USA

Fred L. Bookstein, Michigan Center for Biological Information, University of Michigan, Ann Arbor, MI 48109, USA; and Institute of Anthropology, University of Vienna, Austria

Matt Cartmill, Department of Biological Anthropology and Anatomy, Duke University Medical Center, P.O. Box 3170, Durham, NC 27710, USA

George Chaplin, Department of Anthropology, California Academy of Sciences, Golden Gate Park, San Francisco, CA 94118-4599, USA

Robin Huw Crompton, Department of Human Anatomy and Cell Biology, University of Liverpool, Ashton Street, Liverpool L69 3GE, UK

Brigitte Demes, Department of Anatomical Sciences, School of Medicine, Health Sciences Center, Stony Brook University, Stony Brook, NY 11794-8081, USA

Willem de Winter, Leiden Experts on Advanced Pharmacokinetics & Pharmacodynamics, Archimedesweg 31, 2333 CM, Leiden, The Netherlands

John G. Fleagle, Department of Anatomical Sciences, School of Medicine, Health Sciences Center, Stony Brook University, Stony Brook, NY 11794-8081, USA

Bruce R. Gelvin, Department of Anthropology, California State University, Northridge, CA 91330, USA

Rebecca Z. German, Department of Biological Sciences, University of Cincinnati, Cincinnati, OH 45221-0006, USA

J. Patrick Gray, Department of Anthropology, University of Wisconsin–Milwaukee, P.O. Box 413, Sabin Hall, Milwaukee, WI 53201, USA

Walter Stalker Greaves, Department of Oral Biology, College of Dentistry, University of Illinois at Chicago, 801 South Paulina St., Chicago, IL 60612, USA

Colin P. Groves, School of Archaeology and Anthropology, Australian National University, Canberra, ACT 0200, Australia

Michael M. Günther, Department of Human Anatomy and Cell Biology, University of Liverpool, Ashton Street, Liverpool L69 3GE, UK

William L. Hylander, Department of Biological Anthropology and Anatomy, Duke University Medical Center, P.O. Box 3170, Durham, NC 27710, USA

Nina G. Jablonski, Department of Anthropology, California Academy of Sciences, Golden Gate Park, San Francisco, CA 94118-4599, USA

Kirk R. Johnson, Department of Biological Anthropology and Anatomy, Duke University Medical Center, P.O. Box 3170, Durham, NC 27710, USA

Françoise K. Jouffroy, CNRS-UMR-8570, Laboratoire d'Anatomie Comparée, Muséum National d'Histoire Naturelle, 55 rue Buffon, F-75005 Paris, France; and Department of Anatomical Sciences, School of Medicine, Health Sciences Center, Stony Brook University, Stony Brook, NY 11794-8081, USA

D. Casey Kerrigan, Department of Physical Medicine and Rehabilitation, School of Medicine, University of Virginia, Charlottesville, VA 22908-1007 USA

Robert S. Kidd, School of Science, Food and Horticulture, The University of Western Sydney, Locked Bag 1797, Penrith South, NSW, 1797 Australia

Yu Li, Department of Anatomy, University of Bristol, University Walk, Bristol BS8 1TD, UK

Peter W. Lucas, Department of Anatomy, University of Hong Kong, 21 Sassoon Road, Hong Kong

Monique F. Médina, CNRS-UMR-8570, Laboratoire d'Anatomie Comparée, Muséum National d'Histoire Naturelle, 55 rue Buffon, F-75005 Paris, France

Joseph M. A. Miller, Department of Pathology and Laboratory Medicine, Geffen School of Medicine, University of California, Los Angeles, CA 90095, USA

Paul O'Higgins, Hull York Medical School and Department of Biology, The University of York, Heslington, York, YO10 5DD, UK

Charles E. Oxnard, School of Anatomy and Human Biology, University of Western Australia, 35 Stirling Highway, Crawley, WA, 6009 Australia

Ruliang L. Pan, School of Anatomy and Human Biology, University of Western Australia, 35 Stirling Highway, Crawley, WA, 6009 Australia

Matthew J. Ravosa, Department of Cell and Molecular Biology, Northwestern University Medical School, 303 East Chicago Avenue, Chicago, IL 60611, USA

Kaye E. Reed, Institute of Human Origins, Department of Anthropology, Arizona State University, Box 874101, Tempe, AZ, 85287-4101, USA

F. James Rohlf, Ecology and Evolution Department, Stony Brook University, Stony Brook, NY 11794-5245, USA

Callum F. Ross, Department of Anatomical Sciences, School of Medicine, Health Sciences Center, Stony Brook University, Stony Brook, NY 11794-8081, USA

Russell Savage, Department of Human Anatomy and Cell Biology, University of Liverpool, Ashton Street, Liverpool L69 3GE, UK

Nazima Shahnoor, Department of Anthropology, University of Wisconsin–Milwaukee, P.O. Box 413, Sabin Hall, Milwaukee, WI 53201, USA

Jack T. Stern, Jr., Department of Anatomical Sciences, School of Medicine, Health Sciences Center, Stony Brook University, Stony Brook, NY 11794-8081, USA

Christopher J. Vinyard, Department of Biological Anthropology and Anatomy, Duke University Medical Center, P.O. Box 3170, Durham, NC 27710, USA

Christine E. Wall, Department of Biological Anthropology and Anatomy, Duke University Medical Center, P.O. Box 3170, Durham, NC 27710, USA

Weijie Wang, Department of Human Anatomy and Cell Biology, University of Liverpool, Ashton Street, Liverpool L69 3GE, UK

Preface: shaping primate evolution

The last half-century has witnessed a dramatic improvement in our understanding of the relationship between form and function in biology. This phenomenon has been fueled by innovations in many fields, from molecular biology to mechanical engineering and multivariate statistics. The importance of "big thinkers" has also been critical – people who can understand and synthesize information from diverse technical fields and then apply and integrate such information in the context of organismal biology. The theoretical and substantive innovations brought about by "big thinkers" such as Charles E. Oxnard have transformed the last half-century of zoology and anthropology.

Charles Oxnard's fascination with size and form in primates shaped not only his career, but also the careers of students and colleagues. This volume is a tribute to how that fascination has influenced numerous others. Many of the highly regarded, internationally known contributors to this volume participated in a recent symposium (American Association of Physical Anthropologists, April 2000) honoring Professor Oxnard for his many and various contributions to the study of primate evolutionary morphology. Each contributor has been influenced strongly by Professor Oxnard and each contribution elaborates on the analysis of the form–function–behavior triad in a unique and compelling way.

This book is diverse both in the topics covered and in the range of levels of biological organization that are addressed, from the cellular level (Jouffroy and Médina) to the evolution of primate ecology (Fleagle and Reed). Part I, *Craniofacial form and variation*, is the most topically focused section of the book, exploring the theoretical and practical implications of choosing the appropriate dataset for addressing questions regarding heterochrony in mammals (German), interspecific scaling in African (O'Higgins and Pan) and Asian (Pan and Groves) leaf-eating monkeys, and interpretation of the early hominid fossil record (Miller, Albrecht, and Gelvin).

Part II, *Organ structure, function, and behavior*, includes a diverse collection of papers that examine the architectural (Anapol, Shahnoor, and Gray) and cytochemical (Jouffroy and Médina) underpinnings of muscle morphology, the hominoid foot and the interpretation of early bipedal locomotion (Kidd), how the mechanical properties of plants effect selection for tooth shape (Lucas), and the correlation between structural proportions of the primate brain and biomathematical studies of primate limbs (De Winter).

In Part III, *In vivo organismal verification of functional models*, two of the more intriguing questions of primate evolution, the functional implications of symphyseal fusion in the mandible (Hylander, Vinyard, Ravosa, Ross, Wall, and Johnson), and the controversial issue of division of labor between the fore and hind limbs of primates (Li, Crompton, Wang, Savage, and Gunther), are addressed through direct measurements of behavior in living animals.

In Part IV, *Theoretical models in evolutionary morphology*, human bipedality is addressed in terms of the viability of several theories concerning the origins of human bipedality within the context of human anatomy (Jablonski and Chaplin), and modeled as an inverted pendulum of varying length (Stern, Demes, and Kerrigan). The mammalian jaw also is modeled by considering skull shape and masticatory muscle orientation (Greaves).

Part V, *Primate diversity and evolution*, addresses the relationship between where primates live and their phylogenetic relationship (Fleagle and Kaye), and the taxonomic position of *Daubentonia madagascarensis* (Groves). In the last two chapters, the evolution of Oxnardian morphometric interpretation is seen both through the eyes of two unsurpassed contributors to the measurement of shape using quantitative morphometrics (Bookstein and Rohlf) and through the eyes and mind of Oxnard himself.

It is no coincidence that those on the cutting edge of primate functional morphology are associated with Professor Oxnard, through academic "descent," collegial association, or both. Professor Oxnard arguably has been the most influential scholar in the burgeoning field of primate morphometrics for the past 50 years. His unique and vast contributions include not only his meticulous comparative functional analyses of primate anatomy, but also his pioneering of innovative metric and analytical techniques that has resulted in the transposition of phylogenetic reconstruction from the qualitative to the quantitative arena.

Of course, no pre-mortem volume evolving from the life's work of Charles Oxnard would be complete without his own running commentary. Consequently, each section of the book is introduced within the historical context of Professor Oxnard's musings from his own career.

Finally, we would like to express our gratitude to, among others, the many generous and highly accomplished contributors to this volume, the many anonymous (and who shall remain so) reviewers of each chapter, Hugh Brazier, for his sterling job of copy-editing, and Joseph Morneau for his excellent clerical assistance.

Fred Anapol
Rebecca Z. German
Nina G. Jablonski

1 Charles Oxnard: an appreciation

MATT CARTMILL
Duke University

In an extraordinary scientific career extending across the whole second half of the twentieth century and into the twenty-first, Charles Oxnard has placed his unique stamp on nearly every aspect of biological anthropology. He has profoundly influenced the growth and direction of our discipline all around the world, beginning in Europe and moving westward through North America to Asia and Australia. His research accomplishments have been almost as global as his residence patterns. When we think of his work as a whole, we tend to think first of his morphometric work – his lifelong quest for finding reliable ways of taking huge numbers of data or complicated shapes, and crunching them into simpler functions that reveal a small number of underlying patterns reflecting diet, or locomotor behavior, or phylogeny. And most of us think mainly of the works in which Oxnard has applied these approaches to the study of primate and human evolution. But a glance at Oxnard's long bibliography shows an amazingly diverse span of other work, from classical comparative studies of primate anatomy down through studies of growth and development, bone biology, and vitamin B12 metabolism in primates, to the patterns and causes of sexual dimorphism, lower back pain, and osteoporosis in aging. In this introductory chapter, I intend only to sketch briefly the story of what I take to be the central theme in Charles Oxnard's career as a scientist – namely, primate biometrics and its implications for human evolution.

To appreciate the importance of Oxnard's work in this field, we need to look back across the historical landscape of paleoanthropology to 1958, when Oxnard was beginning his graduate studies at the University of Birmingham under Solly Zuckerman. Zuckerman had come to Birmingham in 1945 from Oxford, where he had worked under Le Gros Clark. The two men had quarreled (Oxnard, 1997), and Zuckerman had taken the job at Birmingham to start a program that would strike out in new directions to remedy what he thought of as the unscientific character of primate biology.

Shaping Primate Evolution, ed. F. Anapol, R. Z. German, and N. G. Jablonski. Published by Cambridge University Press.

When Oxnard began his graduate work, Birmingham was in many ways an exciting place to be. Zuckerman and his colleagues were innovative and intellectually lively. They were out on the forefront of the methodological revolution in biometry that the digital computer was beginning to make possible. But they were also manning an embattled outpost of an increasingly unpopular school of thought about human phylogeny – namely, the idea, stemming ultimately from Henry Fairfield Osborn, that the human lineage had been separate from all other mammals throughout most of the Cenozoic. Zuckerman believed that "it was reasonable to infer from the available evidence that man . . . had begun his independent evolution as far back as the Oligocene" (Zuckerman, 1954, p. 349). Throughout his career, Zuckerman was convinced that none of the australopithecines could possibly be a human ancestor. In fact, he thought that *Australopithecus* was more likely to be ancestral to modern gorillas and chimpanzees (p. 396).

All this had seemed more plausible back in the 1930s, when Zuckerman had first begun publishing on human evolution. Most experts then thought that both the cercopithecoid and gibbon lineages were represented in the Fayum Oligocene. It was not much of a stretch to think that the hominid lineage might have been around at the same time. And up to the end of the Second World War, *Australopithecus* was generally dismissed as an aberrant ape that showed a few interesting convergences with real hominids like *Eoanthropus*, *Pithecanthropus*, and *Homo* (Gregory, 1949).

But by the time Oxnard arrived at Birmingham, the *Zeitgeist* had started leaning in the other direction. The postcranial fossils from South Africa that had come to light over the preceding decade had shown that *Australopithecus* was distinctively human-like in some respects, especially in the lower limb. By the early 1950s, many of the other leading experts in this field had identified *Australopithecus* as something very close to the long-sought missing link between man and his simian ancestors (Dart, 1940, 1948, 1949; Le Gros Clark, 1947, 1952; Gregory, 1949; Broom *et al.*, 1950; Washburn, 1951). Zuckerman and his colleagues set to work to refute this thesis, and immediately got into trouble.

A 1950 paper by Ashton and Zuckerman compared australopithecine teeth with those of *Homo* and living apes, and concluded that "in their metrical attributes, [they] are more ape-like than human." But the paper contained a mathematical error; and when this was pointed out and corrected (Yates and Healey, 1951; Ashton and Zuckerman, 1951), the australopithecines looked far less ape-like. Undaunted, Zuckerman argued in 1954 that the sagittal crest seen in some *Paranthropus* skulls proved that this form must have had a flaring, gorilla-like nuchal crest. "The implication," he wrote, "is thus clear that *Paranthropus* carried its head on its vertebral column far more in the manner of a gorilla than of a man" (Zuckerman, 1954, p. 390). And Zuckerman spoke with some contempt of the ignorance of those who could possibly think otherwise.

But in 1959, the *Zinjanthropus* find proved that an australopithecine could in fact have a high sagittal crest without having a gorilla-like nuchal shelf (Leakey, 1959; cf. Holloway, 1962; Tobias, 1967, pp. 23–25). Zuckerman and his ideas were again discredited, and they grew increasingly peripheral to the mainstream of paleoanthropology. As Zuckerman put it sarcastically in 1966 (p. 92), "the anatomical findings which my colleagues and I have reported have been consistently out-of-step . . . in the context of generally accepted views about the australopithecines . . . It is something of a record for an active team of research workers whose strength has seldom been below four, never to have produced an acceptable finding in some 15 years of assiduous study!"

Zuckerman had his own fixed ideas about human evolution. But I think that he was correct in saying that the conventional wisdom of the time rested on some intellectual fashions that had no empirical basis. In the 1960s, it was regarded as enlightened and virtuous to obliterate taxonomic distinctions. Splitting was out of style and lumping was in, and there was a general effort to interpret fossil taxa as direct ancestors of living ones. During this period, Neanderthals joined the human species, *Pithecanthropus* and *Sinanthropus* joined the human genus, and *Australopithecus* and *Ramapithecus* were welcomed into the family of man. As Zuckerman insisted, the leading lights of paleoanthropology in the 1960s (e.g., Mayr, 1951; Simons, 1961, 1963; Simpson, 1963; Brace, 1964; Simons and Pilbeam, 1965; Buettner-Janusch, 1966) shared a general wish to draw simple lines of descent through as many fossils as possible, especially where hominids were concerned; and this agenda was blinding them to some unwelcome facts about the australopithecines.

Oxnard received his Ph.D. in 1962 and went on working at Birmingham, collaborating with Ashton on a series of classic studies on comparative morphometrics and locomotor adaptations in the primate forelimb. After joining the University of Chicago faculty in 1966, Charles set to work painstakingly and systematically to uncover those unwelcome facts about the australopithecines, using innovative methods borrowed in part from other fields of science.

In 1968, he demonstrated that the fragmentary Sterkfontein scapula showed clear signs of having a glenoid and a spine that were markedly cranial in orientation, like those of apes but unlike those of human beings. Oxnard concluded that "the fragment was almost as well adapted for suspension of the body by the limbs as is the corresponding part of the present-day gibbon" (Oxnard, 1968a, p. 215).

That same year, Oxnard (1968b) reported on the supposed *Homo habilis* clavicle from Olduvai, and inferred from its ape-like axial twist that this fossil as well must have had a cranially directed glenoid. In two 1969 papers,

Oxnard put together these two girdle fragments and compared them with extant primates in a canonical analysis. He found that the composite form was specifically orangutan-like (Oxnard, 1969a, 1969b), and suggested that "the presumed common ancestor of man and the African apes may well have been an animal that lived in trees and which used its shoulder in a manner reminiscent of that of the orangutan" (Oxnard, 1969b, p. 94).

In 1972, Oxnard published a generalized distance analysis of the fossil hominid tali from Olduvai and Kromdraai, which showed them to be *sui generis*, differing both from later species of *Homo* and from the African apes. He further suggested that *Homo habilis* was probably not generically different from *Australopithecus* (Oxnard, 1972, p. 8).

In his book *Form and Pattern in Human Evolution*, Oxnard (1973a) showed using experimental stress analysis that the phalanges of the Olduvai hand functioned best in a suspensory posture, while those of *Homo* and *Pan* did not. He concluded that the Olduvai hand was orang-like in function. In a paper that same year with Zuckerman and three Birmingham co-authors, Oxnard concluded that the pelvis of *Australopithecus* was human-like in its weight-bearing features, but ape-like in muscle vectors. It was probably a biped, but not a human-like biped (Zuckerman *et al.*, 1973). Perhaps most tellingly, Oxnard (1973b) noted that in all known australopithecines, the hind-limb articular surfaces were much smaller than those of the forelimb.

Oxnard summarized all these studies in his 1975 book, *Uniqueness and Diversity in Human Evolution* (Oxnard, 1975a). In a review article of the same year in *Nature*, he concluded:

> the fossils have ankle, hand, and shoulder bones patterned somewhat after those of the orang-utan . . . we can only surmise that perhaps, as the orang-utan, the fossils had ankles, hands, and shoulders adapted for climbing. Because they have pelves that have articular relationships parallel to those of man, we may guess that . . . they stood and moved upright with a vertical load distribution. But . . . the muscular features of the pelvis are positioned in a way more like those of the great apes . . . [and] they have relatively small articular surfaces in the hindlimb as compared with the forelimb . . . They may have been bipedal in a way that is no longer seen, but have retained abilities for climbing, and perhaps minor arboreal acrobatics such as might be found in an intermediately sized ape-like creature.
>
> (Oxnard, 1975b, p. 394)

These words from over 25 years ago sound extraordinarily fresh and up to date. Essentially similar conclusions about the persistent arboreal habits and nonhuman bipedality of *Australopithecus* have since been urged upon us by studies of later finds from East Africa and the Transvaal (Stern and Susman, 1983; Susman *et al.*, 1984; Clarke and Tobias, 1995; McHenry and Berger,

1998). There is now increasing doubt about the inclusion of *habilis* and *rudolfensis* in the genus *Homo* (Wood and Collard, 1999; cf. Wolpoff, 1996, p. 387), and increasing evidence for the antiquity of the *erectus/ergaster* lineage alongside those of the better-known australopithecines (Larick and Ciochon, 1996; Kimbel *et al.*, 1996; Gabunia *et al.*, 2000).

All these issues are of course still debated. But on every one of these points, the current consensus has largely shifted to Oxnard's view of things. Just as Oxnard predicted in his 1984 magnum opus *The Order of Man* (pp. 331–332), new fossils and new investigations have borne out the results of his biometric work. Most of the rest of us are just now catching up with the positions that Charles Oxnard established over 25 years ago. That fact has not been sufficiently appreciated, and this seems like the right place to point it out.

I have touched only on one sector of Oxnard's research, and I have said nothing whatever about other facets of his extraordinary career: his work as an editor, or the long list of honors and awards that he has received, or his 22-year service to three Universities as a chair, a dean, and a mentor of students and faculty. But because I had the good luck to be Charles's first American graduate student, I want to say something about him as a teacher. I learned a lot of things from Charles during my years at Chicago. I learned the importance of anatomical detail, and the rigors and constraints of anatomical description. I learned to doubt my measurements, and I learned how to remove those doubts. I learned the ideals of scientific methodology that Charles had absorbed at Birmingham. All these things were valuable. But I could have learned them from other people, or elsewhere. What I gained uniquely from Charles, and could not have gained from anyone else, was an attitude.

More than any other scientist I have ever known, Charles Oxnard positively fizzes with what I can only call boyish enthusiasm for everything he does. Ever since I have known him, Charles has radiated love for his work and his knowledge and his profession. He gives you the feeling that doing research is such sheer, unadulterated fun that it's something of a scandal that the government pays people to do it. I never miss an opportunity to hear Charles give a paper, even on some point where I think he is wrong, because the infectious energy and enthusiasm that he communicates is a more precious gift than empirical certainty.

Through his life in science, Charles Oxnard has given us some rich and valuable presents: his methodologies, his research findings, his prescient analyses, his example as a rigorous investigator, his service as a member of our discipline and our university communities. But he has given us nothing so rich and valuable as the example he has set us of a lasting, unshakable, disinterested joy in trying to figure out how the world works. The contributors to this volume, whose work and careers have all been guided and encouraged by Charles's science and mentorship and example, hope that we can return some partial reflection of that joy to him in this celebratory anthology.

References

Ashton, E. H. and Zuckerman, S. (1950). Some quantitative dental characters of fossil anthropoids. *Phil. Trans. Roy. Soc. B*, **234**, 485–520.

(1951). Statistical methods in anthropology. *Nature*, **168**, 1116–1117.

Brace, C. L. (1964). The fate of the "classic" Neanderthals: a consideration of hominid catastrophism. *Curr. Anthropol.*, **5**, 3–43.

Broom, R., Robinson, J. T., and Schepers, G. W. H. (1950). Sterkfontein ape-man, *Plesianthropus*. *Mem. Transvaal Mus.*, **4**, 1–118.

Buettner-Janusch, J. (1966). *Origins of Man: Physical Anthropology*. New York, NY: Wiley.

Clarke, R. J. and Tobias, P. V. (1995). Sterkfontein Member 2 foot bones of the oldest South African hominid. *Science*, **269**, 521–524.

Dart, R. A. (1940). The status of *Australopithecus*. *Amer. J. Phys. Anthropol.*, **26**, 167–186.

(1948). A (?) Promethean *Australopithecus* from Makapansgat Valley. *Nature*, **162**, 175–176.

(1949). The predatory implemental technique of *Australopithecus*. *Amer. J. Phys. Anthropol.* n.s., **7**, 1–38.

Gabunia, L., Vekua, A., Lordkipanidze, D., *et al.* (2000). Earliest Pleistocene hominid cranial remains from Dmanisi, Republic of Georgia: taxonomy, geological setting, and age. *Science*, **288**, 1019–25.

Gregory, W. K. (1949). The bearing of the Australopithecinae upon the problem of man's place in nature. *Amer. J. Phys. Anthropol.* n.s., **7**, 485–512.

Holloway, R. L. (1962). A note on sagittal cresting. *Amer. J. Phys. Anthropol.*, **20**, 527–530.

Kimbel, W. H., Walter, R. C., Johanson, D. C., *et al.* (1996). Late Pliocene *Homo* and Oldowan tools from the Hadar Formation (Kada Hadar Member), Ethiopia. *J. Hum. Evol.*, **31**, 549–561.

Larick, R. and Ciochon, R. L. (1996). The African emergence and early Asian dispersals of the genus *Homo*. *Amer. Sci.*, **84**, 538–551.

Leakey, L. S. B. (1959). A new fossil skull from Olduvai, *Nature*, **184**, 491–493.

Le Gros Clark, W. E. (1947). Observations on the anatomy of the fossil Australopithecinae. *J. Anat.*, **81**, 300–333.

(1952). Hominid characters of the australopithecine dentition. *J. Royal Anthropol. Inst.*, **80**, 37–54.

Mayr, E. (1951). Taxonomic categories in fossil hominids. *Cold Spring Harbor Symp. Quant. Biol.*, **15**, 109–117.

McHenry, H. M. and Berger, L. R. (1998). Body proportions in *Australopithecus afarensis* and *A. africanus* and the origin of the genus *Homo*. *J. Hum. Evol.*, **35**,1–22.

Oxnard, C. E. (1968a). A note on the fragmentary Sterkfontein scapula. *Amer. J. Phys. Anthropol.*, **28**, 213–218.

(1968b). A note on the Olduvai clavicular fragment. *Amer. J. Phys. Anthropol.*, **29**, 429–432.
(1969a). Evolution of the human shoulder: some possible pathways. *Amer. J. Phys. Anthropol.*, **30**, 319–332.
(1969b). Mathematics, shape, and function: a study in primate anatomy. *Amer. Sci.*, **57**, 75–96.
(1972). Some African fossil foot bones: a note on the interpolation of fossils into a matrix of extant species. *Amer. J. Phys. Anthropol.*, **37**, 3–12.
(1973a). *Form and Pattern in Human Evolution: Some Mathematical, Physical, and Engineering Approaches*. Chicago, IL: University of Chicago Press.
(1973b). Functional inferences from morphometrics: problems posed by uniqueness and diversity among the primates. *Syst. Zool.*, **22**, 409–424.
(1975a). *Uniqueness and Diversity in Human Evolution: Morphometric Studies of Australopithecines*. Chicago, IL: University of Chicago Press.
(1975b). The place of the australopithecines in human evolution: grounds for doubt? *Nature*, **258**, 389–395.
(1984). *The Order of Man: a Biomathematical Anatomy of the Primates*. New Haven, CT: Yale University Press.
(1997). Zuckerman, Solly (Lord). In: *History of Physical Anthropology*, ed. F. Spencer. New York, NY: Garland. pp. 1130–1131.
Simons, E. L. (1961). The phyletic position of *Ramapithecus*. *Postilla Yale Peabody Mus.*, **57**, 1–9.
(1963). Some fallacies in the study of hominid phylogeny. *Science*, **141**, 879–889.
Simons, E. L. and Pilbeam, D. R. (1965). Preliminary revision of the Dryopithecinae (Pongidae, Anthropoidea). *Folia Primatol.*, **3**, 81–152.
Simpson, G. G. (1963). The meaning of taxonomic statements. In: *Classification and Human Evolution*, ed. S. L. Washburn. Chicago, IL: Aldine. pp. 1–31.
Stern, J. T., Jr. and Susman, R. L. (1983). Locomotor anatomy of *Australopithecus afarensis*. *Amer. J. Phys. Anthropol.*, **60**, 279–317.
Susman, R. L., Stern, J. T., Jr., and Jungers, W. L. (1984). Arboreality and bipedality in the Hadar hominids. *Folia primatol.*, **43**, 113–156.
Tobias, P. V. (1967). *Olduvai Gorge, vol. 2. The Cranium and Maxillary Dentition of Australopithecus (Zinjanthropus) boisei*. Cambridge: Cambridge University Press.
Washburn, S. L. (1951). The analysis of primate evolution, with particular reference to the origin of man. *Cold Spring Harbor Symp. Quant. Biol.*, **15**, 67–78.
Wolpoff, M. H. (1996). *Human Evolution*. New York, NY: McGraw-Hill.
Wood, B. A. and Collard, M. (1999). The human genus. *Science*, **284**, 65–71.
Yates, F. and Healey, M. J. R. (1951). Statistical methods in anthropology. *Nature*, **168**, 1116.
Zuckerman, S. (1954). Correlation of change in the evolution of higher primates. In: *Evolution as a Process*, ed. J. Huxley, A. C. Hardy, and F. B. Ford. Reprinted 1963. New York, NY: Collier. pp. 347–401.
(1966). Myths and methods in anatomy. *J. Roy. Coll. Surgeons Edinburgh*, **11**, 87–114.
Zuckerman, S., Ashton, E. H., Flinn, R. M., Oxnard, C. E., and Spence, T. F. (1973). Some locomotor features of the pelvic girdle in primates. *Symp. Zool. Soc. Lond.*, **33**, 71–165.

Part I
Craniofacial form and variation

The study of craniofacial form and variation has always been one of the most important areas for those interested in shaping primate evolution. Skulls were the most frequently collected specimens in museums. Skull parts, especially teeth, are most frequently found in the fossil record. Skulls and teeth are easily examined in the living. The bones of the face allow some estimation of how their owners appeared. Appearance and change in appearance as produced by medical and dental technologies have profound effects upon individual well-being. All these are good reasons why this is one of the most critical of anatomical regions.

At the same time, however, skulls, faces, jaws, and teeth are the most complex region of the body. More, perhaps, than in any other region, do a number of completely different functions have to be integrated in its structure. The genetics underlying cranium, face, jaw, and teeth are even now not well known and clearly far more complicated than the postcranium. The development and growth of the head depends upon complex mechanisms and processes, many of which have only been elucidated in the last two decades. In evolutionary terms, the "head problem" in chordates, reflecting at the same time both very ancient and very recent elements, has always been more difficult to understand than, say, the equivalent trunk problem or limb problem (which problems do not even rate quotation marks).

As a result, by far the best-known studies of this region have been carried out over the years by established workers. Beginning students, however, have also often been captivated by the range of these problems and frequently want to work on the skull. Most of the current students at the University of Western Australia are so challenged – and I am sure that it is also so in most other laboratories. There is an "alas, poor Yorick, I knew him well Horatio" influence upon us.

My own first investigations (for an Honors Bachelors degree) were in the same vein. I set out to study in one year (!) the comparative anatomy of the cranial nerves in mammals. This was reduced within the first week of reading to the cranial nerves in primates, and shortly after confined to the fifth cranial nerve. My first paper was limited to its maxillary division but concentrated upon its infraorbital and zygomatico-temporal branches in 21 specimens of seven species of Ceboidea!

Perhaps chastened by the complexities of the cranial nerves, my own doctoral studies, in contrast, were on the shoulder. I reasoned that the cranium, a region of such great complexity, should not be the entrée for a novice, especially not a novice who was looking for a new way to gain a handle on functional adaptation. This was perhaps the best decision I ever made. Function in the shoulder was, relative to the skull, rather simple – there was a chance that I could understand it, even given the lack of functional knowledge and technology in those days. The form of the scapula was primarily two-dimensional, rather than the complex three-dimensional skull. The bone was almost totally suspended by muscle and so there was every chance that its form would mirror muscular activity and little else. Instead of being mired in a morass of complexity, it seemed possible that I might actually produce a modus operandi of my own for studying functional adaptation.

However, it was my hope, even then, that if I could only work out a methodology for a relatively simple area, I might eventually work out how to tackle a complex one. And this has indeed occurred. But it has taken me 30 years to start studying the cranium, face, jaws and teeth, and the complexes of functional adaptations, and the developmental mechanisms and evolutionary relationships that underlie them. Most of my students and colleagues have beaten me to it and some of them have written the sections below. Thus, to me, the head came last. But in any book on shaping primate evolution, the head must come first.

Charles Oxnard

2 *The ontogeny of sexual dimorphism: the implications of longitudinal vs. cross-sectional data for studying heterochrony in mammals*

REBECCA Z. GERMAN
University of Cincinnati

Introduction

As the name suggests, studies of sexual dimorphism began with a focus on morphological differences between the sexes (Darwin 1871). Current use of the term "dimorphism" and current studies of sexual dimorphism have expanded to include ecological, behavioral, and physiological differences between the sexes (Harvey and Clutton-Brock, 1985). Charles Oxnard (1987) brought his unique quantitative perspective to the investigation of sexual dimorphism, showing that studies, particularly quantitative studies, of differences between the sexes in morphology are meaningful and not outdated. His work has provided inspiration for this chapter, which examines the role that data and analysis play in understanding evolution. As Oxnard identified multiple dimorphisms among taxa along morphological axes, this study examines heterochronic variation among taxa to show that different ontogenetic trajectories produced analogous multiple dimorphisms. Crucial to Oxnard's work, and to the results presented here, are matches among question, data, and method.

Studies of sexual dimorphism and growth

Most research addressing questions of growth and sexual dimorphism examines the ontogeny of that dimorphism, focusing on how growth produces adult differences (see German and Stewart, 2002 for review). A slight shift in focus

Shaping Primate Evolution, ed. F. Anapol, R. Z. German, and N. G. Jablonski. Published by Cambridge University Press.

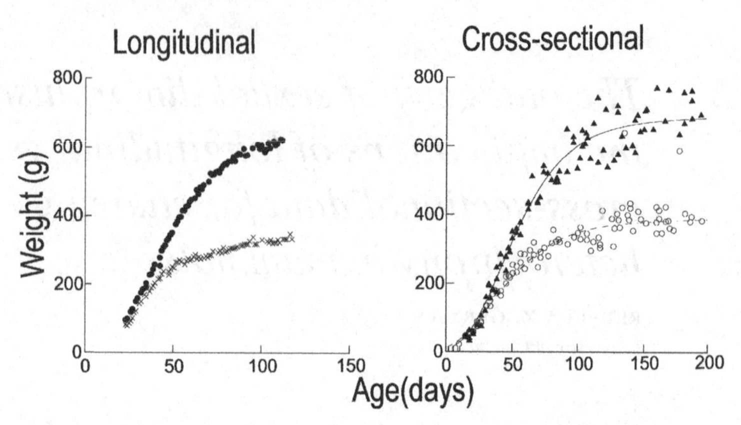

Figure 2.1. Longitudinal and cross-sectional growth curves for weight of male and female rats. In the longitudinal data, there are two individuals, with approximately 100 points each. In the cross-sectional data, there are approximately 100 individuals of each sex. Although the average growth curves are nearly identical, variation in the cross-sectional data is much higher, because it contains inter-individual variation. Data are from Stewart and German (1999) and Miller and German (1999).

to the sexual dimorphism of ontogeny, or the way the sexes grow, will generate alternative questions, centering on the differences in growth itself (e.g., Watts and Gavan, 1982; Glassman *et al.*, 1984; Coelho, 1985; Watts, 1986).

Growth in mammals has a number of distinguishing characteristics that make comparisons of growth between two groups difficult, whether they are sexes or species. Mammalian growth is nonlinear. Traditional heterochrony studies compare *the* rate of growth of a particular feature (Alberch *et al.*, 1979). In a mammalian study, group A may grow faster initially but decelerate their growth early on, while group B, by virtue of maintaining an initially slower rate for a longer duration, may be growing faster at a later point in development. Furthermore, across mammalian species, growth does not contain distinct homologous events, such as instars or metamorphosis, to facilitate a qualitative analysis. Finally, a closer examination of the ways in which growth data are collected suggests the importance of linking the design, including data and analysis, to the question being asked.

Longitudinal vs. cross-sectional data

Data used in growth studies fall into two broad classes, with possible and potential intermediates (Cock, 1966; Diggle *et al.*, 2002). Longitudinal data are those collected on a single individual over a span of time (Fig. 2.1). If the unit of analysis is an individual organism, then for any variable each unit will contain multiple data points to produce a single growth curve. Cross-sectional data are collected for a number of individuals of different ages, or a cross-section of a population (Fig. 2.1). If the organism is the unit of analysis, each unit is a data point. Thus, in Figure 2.1, the longitudinal curves and cross-sectional curves contain an equal number of points, although the longitudinal curves are two individuals, while the cross-sectional curves are approximately 200. There are several advantages to using cross-sectional data, especially the efficiency and relative low cost of collecting such data, making very large datasets possible. Furthermore, there are many situations in which the collection of longitudinal data is prohibitively difficult, such as in studies of endangered species or studies where the species is the unit of analysis, and thus must rely on museum collections. Finally longitudinal data are impossible to collect for studies involving fossil taxa. In all of these cases the practical advantages of cross-sectional data are far more significant than any advantage of using longitudinal data.

There are, however, differences between longitudinal data and cross-sectional data that have implications for answering certain scientific questions. These differences relate to situations where variance both in time or age and in response (size or shape) is critical to the biologic question being asked. For example, within a sex, longitudinal curves have only a small amount of variance around the main signal (Fig. 2.1). The cross-sectional curve for each sex is highly variable, reflecting the inter-individual variation that is not present in longitudinal data. Although these are obvious results, this variation has implications for the power and significance of results in some analyses.

One of the key implications of this variation arises with the quantification of growth in comparative studies among multiple groups. A cross-sectional curve is essentially a sample taken from hypothetical or potential longitudinal curves (Fig. 2.2). As such it falls within the envelope defined by the set of possible longitudinal curves. However, the cross-sectional curve hides or misrepresents many of the important quantitative aspects of a growth curve. Because heterochronic studies, particularly those of mammals, rely on estimates of rate and timing of growth, calculations of rates taken from primary growth data (size or shape as a function of time) are critical to results and conclusions. The rates of growth for the set of longitudinal curves and the cross-sectional curve from Figure 2.2 are very different (Fig. 2.3). Smoothing the cross-sectional data either prior to or after calculating growth rates does not improve this result.

The problem with cross-sectional data and estimation of growth rates and timing can be seen in an analysis of human growth. Figure 2.4 shows longitudinal data for a human male and female, and similar cross-sectional data. The cross-sectional curve clearly lacks the detail of the pubertal growth spurt, although seeing the longitudinal curves makes it possible to discern the same

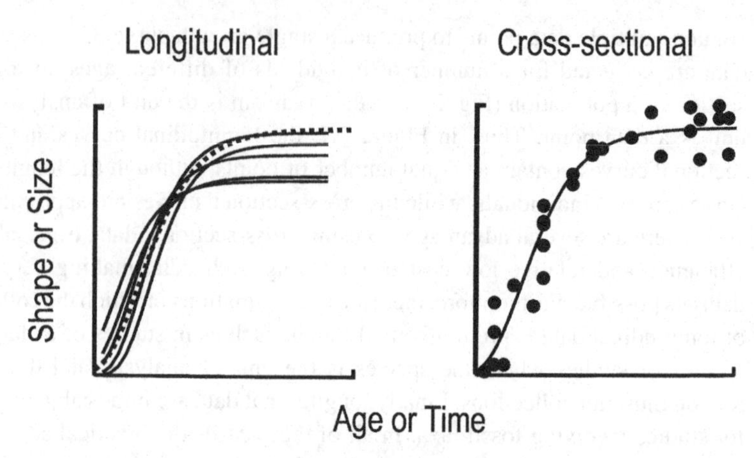

Figure 2.2. A cross-sectional sample taken from a set of longitudinal growth curves. The cross-sectional curve falls within the envelope of longitudinal curves.

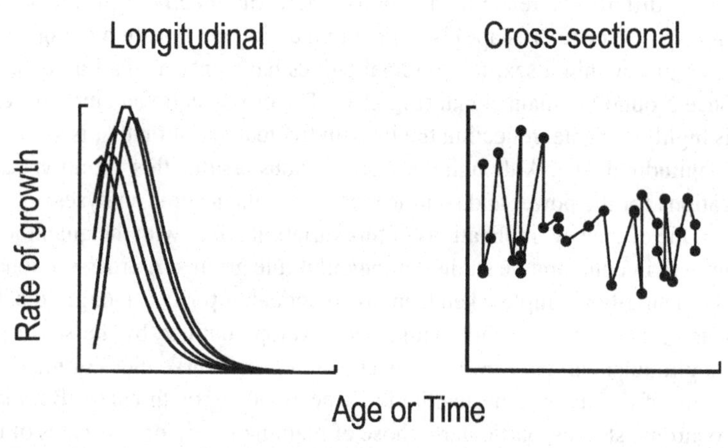

Figure 2.3. Rates of growth from longitudinal and cross-sectional curves. The rates of growth are calculated directly from the original data in Figure 2.2, as the change in size over a unit time. The cross-sectional curve data are calculated from the individual points in Figure 2.2. The longitudinal curves show similar patterns of rate change over time, although individual curves are offset along the time axis. The curve based on cross-sectional data does not have a discernible pattern, even though the original data that generated this curve are well within the envelope of variation defined by the longitudinal curves (Fig. 2.2). The "tail" of the cross-sectional data is a reflection of the variation in adult size in cross-sectional data.

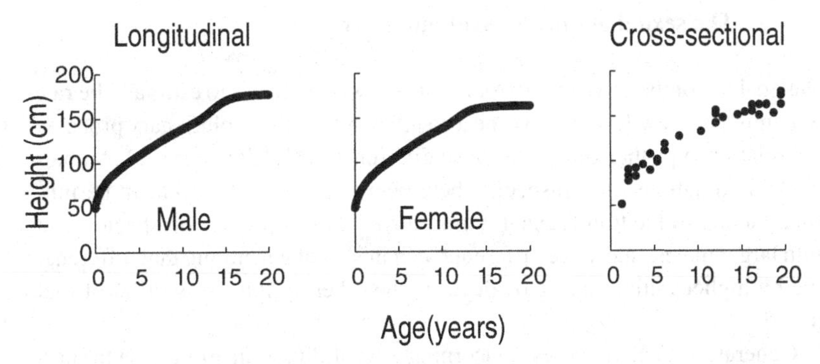

Figure 2.4. Longitudinal and cross-sectional human growth curves. Adolescent growth spurts can be seen as small bumps in the longitudinal curves. In males the growth spurt occurs at approximately 13–15 years, whereas in the female it occurs somewhat earlier, 11–13 years. No growth spurt is easily visible in the cross-sectional data.

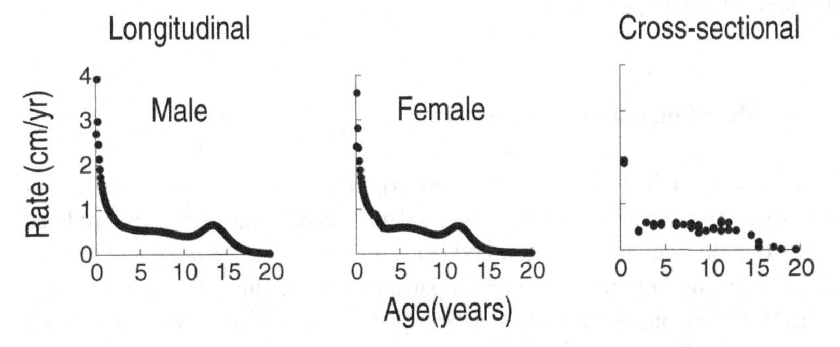

Figure 2.5. Rates of growth for the human growth curves shown in Figure 2.4. The growth spurts are easily discerned in the longitudinal curves, but not at all in the cross-sectional curve. The maximum rate of growth is also lower in the cross-sectional curve. The end or slowing of growth is much more distinct in the longitudinal curves than it is in the cross-sectional curves.

pattern in the cross-sectional data. The differences between rates of growth, shown in Figure 2.5, are even more striking. It is possible to estimate the timing of the human pubertal spurt from longitudinal curves (Watts and Gavan, 1982; Marubini and Milani, 1985; Tanner, 1986; Jolicoeur *et al.*, 1988), and the difference between males and females is obvious. However, the cross-sectional curve has lost this detail, not only the extent or value of the spurt, but also its timing.

The sexual dimorphism of ontogeny

The goal of comparative or heterochronic growth studies is to estimate the rates and timing of growth, and to use these to infer and test the evolutionary processes and relationships that generated those differences (McKinney and McNamara, 1991). Estimations of intraspecific heterochronic sexual dimorphism permit a finer parsing of the trait "sexual dimorphism." For example, two species, each with larger males, may have different mechanisms of generating that difference: one by higher initial male growth rates, the other by longer duration of male growth.

Generation of hypotheses concerning sexual dimorphism in growth, however, depends on data where measuring the rates and the timing of rates is possible. Such estimates may not be possible with cross-sectional data. Testing of hypotheses is dependent on the sample size and the amount of variation in the data. Even with very densely collected data, the among-group variation will be much larger, making any kind of inference as to group, in this case sex, differences nearly impossible, or requiring unrealistic sample sizes to achieve sufficient statistical power.

Modeling longitudinal growth

Mammalian growth is nonlinear, with an exponential phase that levels into an asymptote. The methods historically used to quantify these curves predates modern computer methods (Zeger and Harlow, 1987). The accurate and precise estimation of human growth curves in particular is the subject of considerable effort (McCance and Widdowson, 1986; Watts, 1986; Jolicouer *et al.*, 1988). There is extensive justification for different models within the family of exponential curves (Laird *et al.*, 1965; Koops, 1986; Jolicouer and Pirlot, 1988). The example below uses Gompertz curves (Maunz and German, 1996, 1997), a flexible and general form of the curve. What all of these have in common, besides a theoretical justification based on growth, are parameters that can be interpreted in a biological framework.

The Gompertz curve (Fig. 2.6) is a three-parameter curve that can be expressed in two different but equivalent algebraic forms:

$$y = A\mathrm{e}^{-b\mathrm{e}^{-kt}} \tag{2.1}$$

$$y = w\mathrm{e}^{(I/k(1-\mathrm{e}^{-kt}))} \tag{2.2}$$

In these equations, y is the variable being measured in units of size or shape, e is the base of the natural logarithm, A is the asymptote or maximum value of

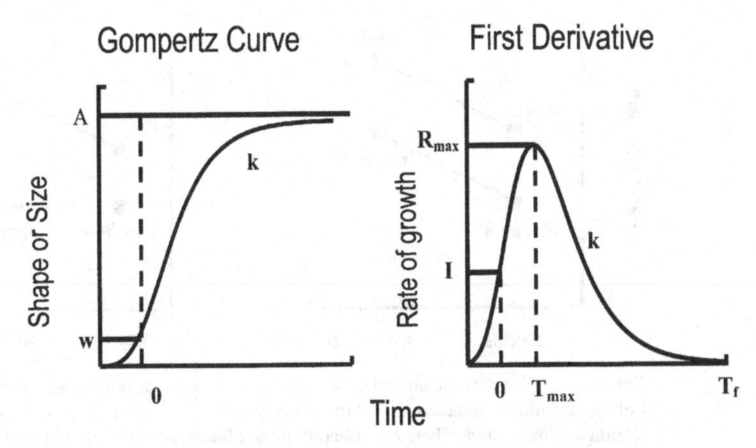

Figure 2.6. Gompertz curve for size vs. age, and the first derivative showing rate vs. age. The parameters, which have biological interpretations, are described in the text.

y, b is a parameter describing the delay in initial growth, I is the instantaneous growth rate at time equals zero and is equivalent to $b \times k$, w is the value of y at birth (time zero), and k is the rate of decay of the rate of growth or a measure of deceleration. One set of parameters may be calculated from a nonlinear regression, and the other set from the relationships:

$$w = A\mathrm{e}^{-b} \tag{2.3}$$

$$I = bk \tag{2.4}$$

The first derivative of the Gompertz equation provides the rate of growth through time, R_{max}, and an estimate of the time at which it occurs, T_{max}.

$$\mathrm{d}y/\mathrm{d}t = Abk\mathrm{e}^{-b\mathrm{e}^{-kt}}\mathrm{e}^{-kt} \tag{2.5}$$

The duration of growth, T_f, cannot be directly calculated. Instead, the time at which the rate of growth reaches a small percent of the maximum (1% or 5%), is used as T_f. These parameters permit the description of growth of any size or shape measurement (y).

Analysis procedures

Each measurement of size or shape is fitted with a Gompertz curve for each individual over the interval of time for which data exist. Thus the individual is the unit of analysis. This will generate a set of parameters (A, b, I, w, k, T_f, R_{max}) for each individual and measurement combination. These parameters are the dependent variables in a series of ANOVAs. The categorical

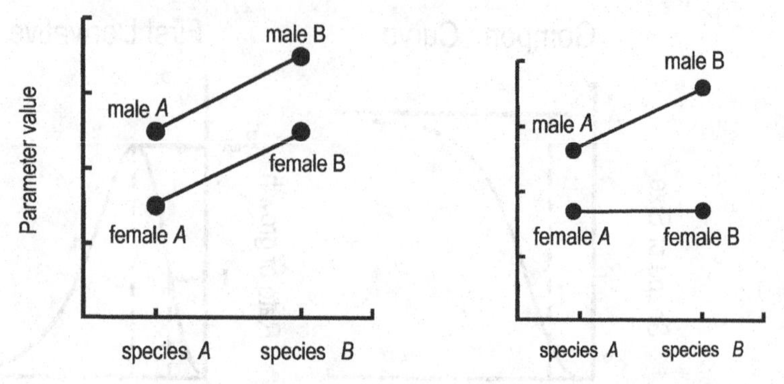

Figure 2.7. Non-significant and significant interaction terms in two-way ANOVA. Left: no significant interaction, and the effect of sex is the same in both species. Right: significant interaction, where the effect of the sex factor, sexual dimorphism, is different in the two species. A full model ANOVA partitions variation into the main factors, sex and species here, and then tests any residual variation in the interaction term.

predictor/independent variables are the groups being tested for differences. For studies of sexual dimorphism, sex is one fixed factor, and if the question involves species differences, species is another. The total number of dependent variables is the number of growth parameters (seven in the list above) times the number of size/shape variables in the analyses (two in the case of the example described later, skull length and weight). These analyses thus answer the question: are there differences between sexes (across species), and differences among species (across sexes) in the rates and timing of growth, measured by the Gompertz parameters?

Using a full linear model, including an interaction term, permits testing of an important biological hypothesis for heterochronic studies. The interaction term between the sex factor and the species factor tests the specific hypothesis, that for at least one species the sexual dimorphism in the growth parameter being analyzed is different from the sexual dimorphism in another species. For example, consider the parameter T_f, the duration of growth. A full model partitions the variation into three components, a species component, a sex component, and an interaction. Thus, if in our data males of all species are larger than females, and one species is larger than another, we would not be surprised to find that the sex factor and the species factor are significant, i.e., that larger animals grow for a statistically significant longer period of time (T_f). If difference between sexes in the duration of growth is proportional in the two species, i.e., one species can have a longer duration than the other, but the difference between sexes is what is expected for all species in the study, then there is no interaction (Fig. 2.7). A significant interaction indicates a significant difference in sexual dimorphism between the two species for, in this case (Fig. 2.7, second panel), the differences between males and females are no longer proportional. Species B is bigger, and males are bigger, but the sexual dimorphism is not equivalent between these two species. Thus, the interaction tests not if one species is growing differently, or if one sex is growing differently, but if the differences between sexes are consistent across species.

Sexual dimorphism in the growth of two species of marsupial

Are the differences in the timing and rates of growth between males and females the same in two closely related species? This question requires longitudinal data as outlined above. Cross-sectional data does not have the temporal resolution to address this question, given the changes in rate, and the significant difference in size between the species. I took data from a larger study (Maunz and German, 1996, 1997) to specifically quantify and analyze the heterochronic mechanisms producing sexual dimorphism in two closely related species. The two species were the short-tailed opossum (*Monodelphis domestica*), native to South America (adult males 80–150 g, adult females 60–100 g) and *Didelphis virginiana*, found primarily in the eastern half of North America (males 3000–6000 g, females 2000–4000 g). These two species are not sister taxa (Kirsch *et al.*, 1993; Patton *et al.*, 1996). They do, however, represent one of the smallest and the largest species of the family Didelphidae. Phylogenetic analysis of mitochondrial cytochrome b suggests that *Monodelphis domestica* is the more basal taxon of the two species (Patton *et al.*, 1996).

Following a 14-day internal gestation period, infants of both species were born at approximately the same size (0.01–0.015 g) and exhibited the same overall level of development (Russell, 1982). Each animal was weighed daily and radiographed three times a week beginning at approximately 40 days postnatal (earliest possible to avoid infant mortality), ending at 350 days (*M. domestica*) or 450 days (*D. virginiana*). All work was done under IACUC approval (91-05-27-01). The study included a total of 43 animals (*M. domestica* 12 males and 10 females, *D. virginiana* 11 males and 10 females). Following the procedures outlined above, a Gompertz curve was fitted to each individual for several variables; this study includes body weight (g) and total skull length (mm).

The two species were clearly different in adult size, and both were sexually dimorphic for these two measurements. The *Didelphis* were always the larger species, and the males were always the larger sex. In terms of growth, the two species were significantly different in all heterochronic parameters except initial instantaneous rate of growth (I) for body weight (Table 2.1). There was sexual dimorphism in adult size of skull length and body weight. For

Table 2.1. *Summary of F-ratios from two-way analysis of variance (ANOVA) testing for sexual dimorphism within and between species on two measurements of overall body size*

	w	I	k	T_f	A
Total skull length					
Species	21.346*	47.545*	18.471*	28.183*	770.01*
Sex	2.476	2.948	16.942*	9.604*	7.55*
Interaction	0.001	1.287	9.818*	5.816*	0.74
Log of weight					
Species	25.218*	2.904	27.504*	62.711*	2656.19*
Sex	0.643	5.135*	12.352*	2.555	16.976*
Interaction	4.934*	11.006*	10.527*	13.142*	5.52*

Degree of freedom for the mean square was 1 for all of the terms. Error term degrees of freedom ranged from 35 to 44. Asterisk (*) represents $P < 0.001$; no symbol represents $P > 0.05$. See text for explanation of Gompertz parameters.

body weight, the rate of decay of growth, and initial rate of growth were sexually dimorphic. For skull length, the rate of decay and the duration of growth were sexually dimorphic. All interactions were significant for the parameters of body weight, indicating differences in the patterns of growth generating sexual dimorphism for each species (Table 2.1). For skull length, k and T_f, parameters that describe the end of growth, had significant interactions showing that there were differences in the patterns of growth producing the same degree of adult sexual dimorphism (seen by an insignificant interaction term in the parameter A). Nonparallel lines in Figure 2.8 graphically represent the significant interactions. Males are larger than females in each species, but the way in which they grow to become larger is not the same in the two species. In particular, the rate of deceleration in male *Didelphis* is proportionally smaller (Fig. 2.8).

These results strongly suggest that the heterochronic processes generating adult sexual dimorphism in these two species are not homologous. Although both species are dimorphic, the intraspecific growth differences between males and females are not the same in the two species. It is highly unlikely that phylogenetic inertia is responsible for sexual dimorphism in the more derived species. Instead, external selective forces may have altered the growth mechanisms producing sexual dimorphism in at least one of the two species over evolutionary time. Whether these differences are directly due to sexual selection requires more information on the remainder of the clade, including the primitive reproductive state and an analysis of the sexual dimorphism of the outgroup species. Studies such as these are scarce but provide some of the most compelling

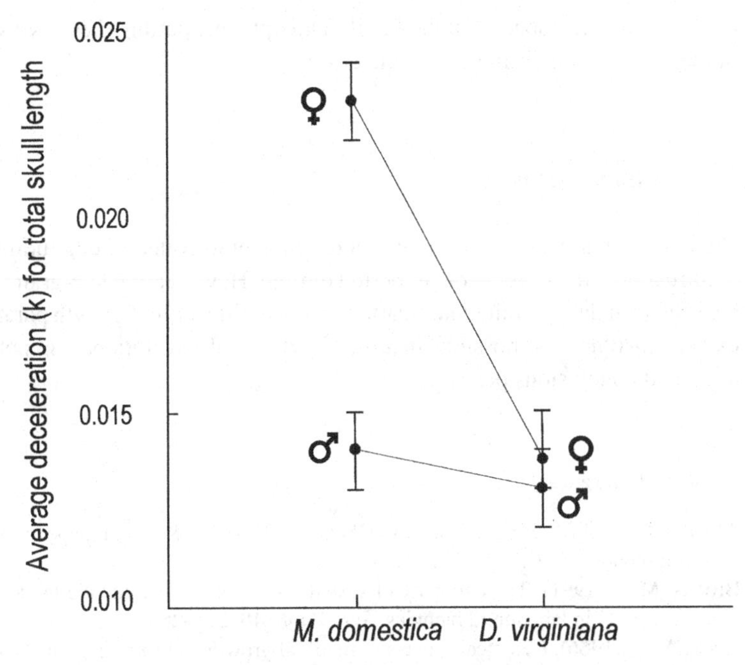

Figure 2.8. Different patterns of sexual dimorphism in one heterochronic parameter, *k* (deceleration of the growth rate), for the two species of marsupial. Each point represents the cell mean for one of the four treatment groups (species * sex). The bars around each point are the upper and lower values of the 95% confidence interval around each mean. The difference between the sexes for *k* is significant for *M. domestica* but not for *D. virginiana*. Since *k* values are different between sexes in one species but not in the other, different heterochronic mechanisms are responsible for the adult dimorphisms within each species. If the sexual dimorphism was the same in both species for the *k* parameter, the two lines drawn connecting the same sex of each species would be parallel.

evidence for the evolutionary mechanisms of sexual dimorphism (Shea, 1986; Brooks, 1991).

This lack of homology in sexual dimorphism is an explicit vindication of Oxnard's (1987) hypotheses on the existence of multiple "sexual dimorphisms" in primates. In an exploratory analysis, Oxnard generated significant hypotheses that sexual dimorphism was not a single entity across anatomical regions. He points out the implications of these differences for understanding patterns of primate evolution. The results presented here address the differences in growth that could generate the patterns seen in adult sexual dimorphism, but still with the ultimate goal of understanding the evolution of sexual dimorphism. These results, and those of Oxnard (1987), suggest the importance of further studies. A complete study of the evolution of the sexual dimorphism of growth requires results from other species in the family Didelphidae, putting these two species in the context of their monophyletic clade.

Conclusions

Cross-sectional data are easier and more efficient to collect. Longitudinal data require more effort over a longer period of time. However, the fine-grained time basis of such data permit identification and quantification of growth parameters that are not otherwise possible. In turn, such data make evolutionary hypotheses, tests, and conclusions possible.

References

Alberch, P., Gould, S. J., Oster, G. F., and Wake, D. B. (1979). Size and shape in ontogeny and phylogeny. *Paleobiol.*, **5**, 296–317.

Brooks, M. J. (1991). The ontogeny of sexual dimorphism: quantitative models and a case study in labrisomid blennies. *Syst. Zool.*, **40**, 271–283.

Cock, A. G. (1966). Genetical aspects of metrical growth and form in animals. *Q. Rev. Biol.*, **41**,131–190.

Coelho, A. M. (1985). Baboon dimorphism: growth in weight, length and adiposity from birth to 8 years of age. In: *Nonhuman Primate Models for Human Growth and Development*, ed. E. Watts. New York, NY: Alan R. Liss.

Darwin, C. (1871). *The Descent of Man and Selection in Relation to Sex*. London: Murray.

Diggle, P., Heagerty, P., Liang, K.-Y., and Zeger, S. (2002). *Analysis of Longitudinal Data*. 2nd edn. Oxford: Oxford University Press.

German, R. Z. and Stewart, S. A. (2002). Sexual dimorphism and ontogeny in primates. In: *Human Evolution Through Developmental Change*, ed. N. Minugh-Purvis and K. J. McNamara. Baltimore, MD: Johns Hopkins University Press.

Glassman, D. M., Coelho, A. M., Carey, K. D. and Bramblett, C. A. (1984). Weight growth in savannah baboons: a longitudinal study from birth to adulthood. *Growth*, **48**, 425–433.

Harvey, P. H. and Clutton-Brock, T. H. (1985). Life history variation in primates. *Evolution*, **39**, 559–581.

Jolicouer, P. and Pirlot, P. (1988). Asymptotic growth and complex allometry of the brain and body in the white rat. *Growth Devel. Age*, **52**, 3–10.

Jolicoeur, P., Pontier, J., Pernin, M-O., and Sempe, M. (1988). A lifetime asymptotic growth curve for human height. *Biometrics*, **44**, 995–1003.

Kirsch, J. A. W., Bleiweiss, R. E., Dickerman, A. W., and Reig, O. A. (1993). DNA/DNA hybridization studies of carnivorous marsupials. III. Relationships among species of *Didelphis* (Didelphidae). *J. Mammal. Evol.*, **1**, 75–97.

Koops, W. J. (1986). Multiphasic growth curve analysis. *Growth*, **50**, 169–177.

Laird, A. K., Tyler, S. A., and Barton, A. D. (1965). Dynamics of normal growth. *Growth*, **29**, 233–248.

Marubini, E. and Milani, S. (1985). Approaches to the analysis of longitudinal data. In: *Human Growth*, ed. F. Falkner and J. M. Tanner, 2nd edn. New York, NY: Plenum Press.

Maunz, M and German, R. Z. (1996). Sexual dimorphism and craniofacial heterochrony in the short-tailed opossum (*Monodelphis domestica*). *J. Mammal.*, **77**, 992–1005.

Maunz, M. and German, R. Z. (1997). Ontogeny and limb bone scaling in two new world marsupials: *Monodelphis domestica* and *Didelphis virginiana*. *J. Morphol.*, **231**, 117–130.

McCance, R. A. and Widdowson, E. M. (1986). Glimpses of comparative growth and development. In: *Human Growth*, ed. F. Falkner and J. M. Tanner. 2nd edn, New York, NY: Plenum Press.

McKinney, M. L. and McNamara, K. (1991). *Heterochrony: the Evolution of Ontogeny*. New York, NY: Plenum Press.

Miller, J. P and German, R. Z. (1999). Protein malnutrition affects the growth trajectories of the craniofacial skeleton but not final adult size in rats. *J. Nutrition*, **129**, 2061–2069.

Oxnard, C. E. (1987). *Fossils, Teeth and Sex*. Hong Kong: Hong Kong University Press.

Patton, J. L., dos Reis, S. F., and da Silva, M. N. F. (1996). Relationships among didelphid marsupials based on cytochrome b gene. *J. Mammal. Evol.*, **3**, 3–29.

Russell, E. M. (1982). Patterns of parental care and parental investment in marsupials. *Biol. Rev.*, **52**, 423–486.

Shea, B. T. (1986). Ontogenetic approaches to sexual dimorphism in anthropoids. In: *Sexual Dimorphism in Living and Fossil Primates*, ed. M. Pickford and B. Chiarelli. Florence: Sedicesimo.

Stewart, S. A. and German, R. Z. (1999). Sexual dimorphism and ontogenetic allometry of soft tissues in *Rattus norvegicus*. *J. Morph.*, **242**, 57–66.

Tanner, J. M. (1986). Use and abuse of growth standards. In: *Human Growth*, ed. F. Falkner and J. M. Tanner. 2nd edn, New York, NY: Plenum Press.

Watts, E. S. (1986). Evolution of the human growth curve. In: *Human Growth*, ed. F. Falkner and J. M. Tanner. 2nd edn. New York, NY: Plenum Press.

Watts, E. S. and Gavan, J. A. (1982). Postnatal growth of nonhuman primates: the problem of the adolescent spurt. *Hum. Biol.*, **54**, 53–70.

Zeger, S. L. and Harlow, S. D. (1987). Mathematical models from laws of growth to tools for biologic analysis: fifty years of *Growth*. *Growth*, **51**, 1–21.

3 *Advances in the analysis of form and pattern: facial growth in African colobines*

PAUL O'HIGGINS
University of York

RULIANG L. PAN
University of Western Australia

Introduction

One of Oxnard's enduring contributions to primatology has been the demonstration of the value of multivariate morphometric methods in providing insights into patterns of morphological variation and their ontogenetic, functional, ecological, and phylogenetic correlates. Oxnard's studies have spanned the whole range of primates and a very broad spectrum of anatomy from the skeleton to the soft tissues of muscle and brain. He was one of the first to emphasize the importance of ontogeny in providing an interpretive framework for adult anatomy (Oxnard, 1984, Fig. 2.4). The work and ideas presented in this chapter continue this morphometric tradition. Thus we present a study of ontogenetic and adult variation in the facial skeleton of one group of primates using geometric morphometric methods. While these methods are becoming widely used, they may still be unfamiliar to many readers. The methods section therefore presents an outline summary of some relevant geometric morphometric tools before detailing the specific analyses employed in this chapter.

The study examines patterns of facial variation and ontogeny within and between three African colobine taxa. The ontogenetic origins of diversity in facial skeletal morphology of this group of monkeys have never been studied and yet they are interesting since, unlike the Asian colobines, they form a closely related, possibly monophyletic group (subtribe Colobina: Delson, 1994). This is suggested by the fact that they share quite a number of similarities including a vestigial thumb and shortened midtarsals together with postcranial and dental

Shaping Primate Evolution, ed. F. Anapol, R. Z. German, and N. G. Jablonski. Published by Cambridge University Press.

features (Szalay and Delson, 1979; Strasser and Delson, 1987; Delson, 1994). The divergence between African and Asian colobines likely occurred about 10 million years ago (Delson, 1994; Stewart and Disotell, 1998).

There is some uncertainty with regard to the phylogeny and taxonomy of the Colobina. The black and white, red, and olive colobus monkeys were regarded by Napier and Napier (1967) as one genus (*Colobus*), with three subgenera, *C.* [*Colobus*], *C.* [*Piliocolobus*], and *C.* [*Procolobus*] respectively. An alternative view is that the red and olive colobus monkeys are more closely related to each other in terms of morphology and behavior than either of them is to the black and white colobus. Thus several authors have favored the black and white colobus being assigned to one genus, *Colobus*, with the red and olive colobus allocated to another genus, *Procolobus*, whilst recognizing their differences by separating them into two subgenera, *P. (Piliocolobus)* and *P. (Procolobus)* (Brandon-Jones, 1984; Strasser and Delson, 1987; Groves, 1989; Oates *et al.*, 1994). However others (Fleagle, 1988; Nowak, 1991; Groves, 2001), noting differentiation between these colobines in terms of behavior and ecology, regard them as three different genera, *Colobus*, *Piliocolobus*, and *Procolobus*. We follow this last scheme in this chapter for the simple reason that it avoids confusion when discussing the red and olive colobus monkeys. Bear in mind, however, that most classifications recognize greater similarities between red and olive colobus than between either of these and the black and white colobus.

These taxonomies have relied on morphological and behavioral data. Thus characters in *Procolobus* and *Piliocolobus*, such as the discontinuous callosites, the perineal organ, and a four-chambered stomach, have been regarded as derived features, while a large larynx, sub-hyoid sac, and lack of female swellings have been considered derived features of *Colobus* (Strasser and Delson, 1987). There are insufficient genetic and molecular data to resolve issues of African colobine relationships at the generic level since most studies have focused on broader issues of the relationships between these and other colobines or anthropoids (Hewett-Emmett *et al.*, 1976; Stewart and Disotell, 1998), or on the relationships between species in the same genus, based on chromosomal banding patterns (Disotell, 1996), mitochondrial DNA (Collura *et al.*, 1996) and lysozymes (Yang, 1998).

It is not an aim of this study to resolve any taxonomic disputes but rather to provide further data on African colobine morphological diversity. In particular the study aims to establish the extent to which differences in facial form within (dimorphism) and between black and white (*Colobus guereza*), red (*Piliocolobus badius*), and olive (*Procolobus verus*) varieties might be accounted for by patterning and ontogenetic scaling differences. As such, this study follows the model established in earlier studies of humans (O'Higgins and Strand Vidarsdottir, 1999), and other primate groups (O'Higgins and

Jones, 1998; Cobb and O'Higgins, 2000; O'Higgins, 2000; O'Higgins *et al.*, 2001; Collard and O'Higgins, 2001; O'Higgins and Collard, 2002).

In particular this set of studies of Colobina investigates two things: first, the extent to which sexual dimorphism and inter-taxon adult shape differences in the face might be explained in terms of ontogenetic or static allometric scaling; second, the extent to which variations in adult facial morphology and postnatal facial ontogenetic allometry support the generally accepted view that the red and olive colobus are more closely related to each other than either is to the black and white colobus. This latter analysis is undertaken cautiously since there is evidence that facial morphology and growth correlate poorly with phylogeny in some primate groups and at lower taxonomic levels (Collard and Wood, 2000, 2001; Collard and O'Higgins, 2001; O'Higgins and Collard, 2002).

Materials and methods

The sample of African colobines we have used is from the collections of the Natural History Museum, London and comprises 46 black and white colobus (adult: 12 male, 10 female; immature, from partial deciduous to nearly complete adult dentition: 9 male, 9 female, 6 unknown sex), 31 red colobus (adult: 8 male, 6 female; immature: 6 male, 9 female, 2 unknown sex), 42 olive colobus (adult: 9 male, 10 female; immature: 3 male, 13 female, 7 unknown sex). Note that details of dental eruption and so of maturation of each specimen are not presented, because our analyses examine patterns of shape variability and size-and-shape relationships (allometry) rather than growth (size vs. age) or development (shape vs. age). In the main, this is because dental data for these taxa only provide a crude estimate of age.

From each face 31 three-dimensional landmark coordinates were taken (Table 3.1) using a Microscribe digitizer (Immersion Corporation, San Jose, California) and submitted to geometric morphometric analyses directed at examining patterns of intra- and interspecific variation in the size and shape of the landmark configuration.

Statistical shape analysis

The geometric morphometric methods we employ are particularly useful in the context of the current study because they preserve geometry at all stages of analysis and so allow visualization of analytical results as well as statistical analysis. Further they partition "size" from "shape" and so facilitate the study of allometry. We first provide a brief review of these before detailing the analyses used in this study.

Table 3.1. *Definitions of landmarks*

Number	Definition: based on anatomical orientation of the face
1 &19	Most lateral point on zygomatico-frontal suture on orbital rim
2 &20	Most supero-lateral point on supraorbital rim
3 & 21	Uppermost point on orbital aperture
4 & 22	Zygomatico-frontal suture at the lateral aspect of the orbital aperture
5 & 23	Fronto-lacrimal suture at medial orbital margin
6 & 24	Zygomatico-maxillary suture at inferior orbital margin
7 & 25	Superior root of zygomatic arch
8 & 26	Inferior root of zygomatic arch
9 & 27	Zygomatico-maxillary suture at root of zygomatic arch
10 & 28	Most posterior point on maxillary alveolus
11 & 29	Deepest point in maxillary fossa
12 & 30	Maxillary–premaxillary suture at alveolar margin
13 & 31	Nearest point to maxillary–premaxillary suture on nasal aperture
14	Upper margin of supraorbital rim in the midline
15	Naso-frontal suture in the midline
16	Tip of nasal bones in the midline
17	Premaxillary suture at the inferior margin of the nasal aperture in the midline
18	Premaxillary suture at alveolar margin

Size of the landmark configuration

Centroid size (the square root of the sum of squared Euclidean distances from each landmark to the centroid) is used in the current analyses because it reflects the general scale of the landmark configuration and so is reasonable in terms of the scaling questions we address. Importantly it is also the natural choice in geometric morphometric studies since scaling by this measure has desirable statistical properties (Dryden and Mardia, 1998).

Shape of the landmark configuration: registration

Landmark configurations vary among specimens in shape and in location, rotation, and scale ("registration": see, for example, Bookstein, 1978). The term "form" refers to combined size and shape. In order to examine variations in the shape of landmark configurations it is therefore necessary to remove scaling and registration differences from raw coordinate data. There are, however, many ways in which configurations can be registered, and different registrations will generate different impressions of shape differences. Further, in studies of biological structures it is nearly always impossible to carry out registration in an unequivocal way (i.e., in a way that exactly mimics the "biological truth": Lele, 1993). The reason for this is simply that most anatomical structures do not have a "natural" register with each other in which there is one fixed center of growth, away from which all other parts move (e.g., in skulls growth occurs at diverse centers such that, in some sense, all bones move away from each other). Thus, the biological reality of apparent directions and magnitudes of relative landmark displacements resulting from superimposition must be interpreted with caution. In the absence of unequivocal registration there have been in the past some notable attempts to standardize registration and carry out multivariate analyses (e.g., Creel and Preuschoft, 1971; Corruccini, 1988). However, such studies remained somewhat tentative with regard to the reliance that could be placed on their findings until a detailed understanding of shape spaces was achieved (see Dryden and Mardia, 1998; Rohlf 1999, 2000a, 2000b).

In geometric morphometric analyses the registration method is chosen because of the statistical properties of the resulting shape space (see below) rather than to represent some a priori model (and so, registration) of transformation between forms. The consequence is that if there exist stable regions for registration these can be discovered through analysis (and used for subsequent registration if desired), thus enabling testing of models of morphogenesis. Registration is through generalized Procrustes analysis (GPA: Goodall, 1991) which has desirable outcomes. Thus, each shape (scaled form; scaling being such that each landmark configuration has centroid size = 1) can be represented as a point in a "shape space," Kendall's shape space (Kendall, 1984), that has the desirable property that small independent isotropic distributions (iid) of landmarks result in an isotropic distribution of points representing specimens in the shape space (e.g., Kent, 1994; Dryden and Mardia, 1998, p. 137). Conversely, deviations from iid will generally lead to a non-isotropic distribution of specimens in the shape space. Such distributions can be of biological interest since the principal directions of variation might be related to interesting biological variables such as age, "size," sex, species, etc. Although different registrations will lead to different distributions of specimens in the resulting shape space, if variations are small then all reasonable registrations lead to approximately similar results (Kent, 1994).

Kendall's shape space and the tangent space

Statistical inference in Kendall's shape space is not straightforward since it is non-Euclidean (i.e., it is curved). For triangles the space is equivalent to the surface of a sphere of unit diameter, but when the number of landmarks (k) is greater than the number of dimensions (m) and there are three or more dimensions the space is high-dimensional and more complex. In general the dimensionality of the shape space is given by $km - m - m(m - 1)/2 - 1$. Thus for two dimensions the space has $2k - 4$ and for three dimensions $3k - 7$ dimensions. Because the space is non-Euclidean, great care is needed in carrying out statistical analyses. One approach that is particularly appealing, since it naturally allows the study of allometry, is to carry out principal components analysis (PCA) in the tangent space to Kendall's shape space (Dryden and Mardia, 1993; Kent, 1994). For triangles we take the scatter of points on the spherical shape space representing variation within our sample and project it into a Euclidean tangent plane, as a cartographer might project a map from a globe onto a flat sheet of paper. The coordinates of the points representing specimens are no longer given in terms of the sphere, but rather as coordinates in the plane. As long as the projection has not resulted in excess distortion (as might occur if the projection encompasses a large proportion of the sphere) we can carry out useful analyses in this plane. For higher dimensions the tangent space to the shape space can be imagined as a space of $km - m - m(m - 1)/2 - 1$ dimensions. Procrustes tangent coordinates can be estimated using the partial Procrustes tangent space projection given by Dryden and Mardia (1993).

Visualization of shape variations

One advantage of geometric morphometric methods is that it is possible to produce visualizations of variations in shape represented by the principal components by simply "warping" ("morphing") the mean shape along each PC of interest using the eigenvectors (loadings) to "shift" landmarks (Kent, 1994).

A further approach to visualizing variations between landmark configurations dispenses with the issue of registration by describing differences between pairs of landmark configurations in terms of deformations (stretchings and contractions of the configurations to make them fit each other exactly) rather than as absolute differences in landmark locations between configurations following registration. The best-known such representation of deformations in biology is as a "transformation grid" (Thompson, 1917). For 3-D data, Cartesian transformation grids are readily calculated from triplets of thin-plate splines (TPS: Bookstein, 1989; Marcus *et al.*, 1996; Dryden and Mardia, 1998) in which the grids are derived from a smooth mathematical function applied to the whole landmark configuration. The grids derived from TPS indicate how the space surrounding a reference shape might be deformed into that surrounding a target shape such that landmarks in the reference map exactly into those of the target. The thin-plate spline ensures that this deformation involves minimum bending. The statistical and graphical models of shape transformations that result from these approaches are readily interpretable and highly visual (e.g., Bookstein, 1978, 1989; O'Higgins and Dryden, 1992, 1993; Marcus *et al.*, 1996).

Shape and its correlates

The approach of PCA of GPA coordinates (i.e., landmark coordinates after GPA) readily leads to the testing of differences in mean shape related to species or sex. Any such apparent differences on a single principal component (PC) can be assessed for significance using *t*-tests of PC scores. If the significance of joint differences on more than one PC is to be assessed then this can be achieved using standard mutivariate techniques such as Hotelling's T^2 as long as the number of PCs used in the computation is considerably smaller than the number of specimens in the comparison. An alternative when sample size is small relative to the dimensionality of the shape space is to utilize a permutation test (Good, 1993) in which the true difference between means (Procrustes distance) is compared with the distribution of differences between means obtained by randomly permuting group membership many times.

In studies where ontogenetic changes in size are the major source of variations in shape it is likely that the first few principal components will adequately represent the relationship between size and shape (allometry). It can be readily explored by examining plots and correlations or regressions of PC scores vs. centroid size for the significant principal components. The extent to which allometric shape variations (i.e., the shape changes associated with size) differ between taxa can be assessed by computing the angles between size-related shape vectors. If, as in this study, the first PC (computed after joint GPA) adequately represents the size-related shape vector in each taxon then differences in ontogenetic allometric trajectories between pairs of taxa can be simply expressed as the angle between first PCs. The significance (i.e., of the difference from zero) of this angle can be assessed by carrying out a permutation test (Good, 1993). This is performed by randomly permuting group membership and recalculating the angle (permuted angle) between PC1s many times. The significance is estimated as the proportion of times the true angle is exceeded by the permuted angles.

Details of specific analyses used in the present study

The first analysis investigates patterns of adult sexual dimorphism within and between the three species. Thus the landmarks were submitted to GPA and sex mean shapes and centroid sizes were computed. These were used to estimate Procrustes distances and centroid size differences between sex means in each species. The significance of the former was assessed by permutation tests (1000 iterations) and of the latter by *t*-tests.

The second analysis examines facial size and shape differences between pooled sex samples of the adults of each species. The significance of any apparent difference between species is assessed as above by *t*-tests and permutation tests. Since the numbers of adults of either sex are not identical within each species and adult sample size is small the findings of this part of the study should be regarded as approximate. The extent to which adult intra- and inter-taxon differences in facial shape might be explained in terms of static and evolutionary

Table 3.2. *Adult sexual dimorphism in facial size and shape amongst African colobines*

Species	Centroid size difference (cm: M − F means)	Procrustes distance (M − F means)
Colobus guereza	15.53 − 14.88 = 0.65 $P = 0.031$	0.030 $P = 0.41$
Piliocolobus badius	14.81 − 13.93 = 0.88 $P = 0.022$	0.041 $P = 0.27$
Procolobus verus	12.06 − 12.04 = 0.02 $P = 0.932$	0.037 $P = 0.59$

allometry was investigated by examining the correlations between size and PC scores.

The remaining analyses examine the ontogenetic basis of intra- and inter-taxon adult scaling. Postnatal ontogenetic allometry is examined in each taxon and then compared between taxa. In all taxa PC1 is the only PC that shows any evidence of a significant or pronounced relationship with size. Comparisons of ontogenetic shape trajectories between these taxa were therefore carried out by comparing the angles between the first principal components of shape variation extracted from each following joint GPA. The significance of these angles was assessed by a permutation test. Finally a PCA of the entire ontogenetic series of all taxa combined was carried out to allow further examination of the ontogeny of inter-taxon differences.

Results

The centroid size differences and Procrustes distances between the sexes of each species are tabulated in Table 3.2. In no species is there a significant adult sexual shape difference in the landmark configurations. However, both *Colobus guereza* and *Piliocolobus badius* manifest significant ($P = 0.031$ and $P = 0.022$ respectively) sexual size differences whilst *Procolobus verus* does not.

Size differences and Procrustes distances between pooled-sex adult samples of each taxon together with their significance are given in Table 3.3. Figure 3.1 presents a plot of PC1 vs. PC2 together with smooth rendered visualizations of the warped mean indicating variability represented by these PCs. From Table 3.3 it is apparent that the adult faces of *Procolobus verus* are highly significantly smaller than those of *Colobus guereza* and *Piliocolobus badius*, which in turn differ in size from each other to a smaller yet significant degree. With regard to

Table 3.3. *Differences in centroid size (above the diagonal) and shape (below the diagonal) between the adult facial skeletons of African colobines. The mean adult centroid size for each taxon is given in the first row*

	Colobus guereza	*Piliocolobus badius*	*Procolobus verus*
	15.16 cm	14.44 cm	12.29 cm
Colobus guereza		0.72 cm: $P = 0.023$	2.87 cm: $P < 0.001$
Piliocolobus badius	0.0873: $P < 0.001$		2.15 cm: $P < 0.001$
Procolobus verus	0.1178: $P < 0.001$	0.1056: $P < 0.001$	

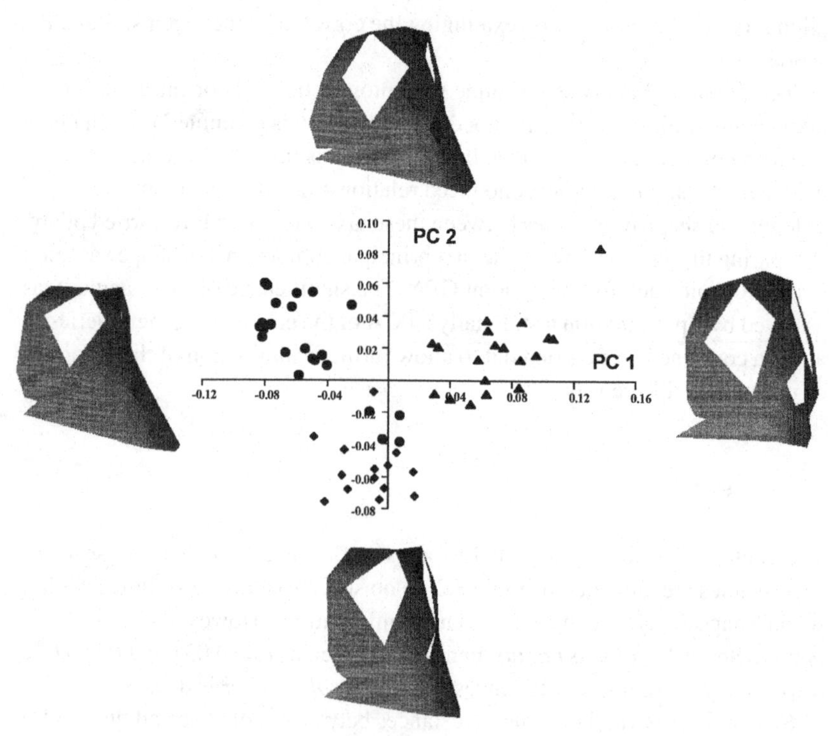

Figure 3.1. Plot of PC1 (37% total variance) vs. PC2 (17% total variance) from GPA/PCA of all adult representatives of *Colobus guereza* (circles), *Piliocolobus badius* (diamonds), and *Procolobus verus* (triangles). Insets indicate the shape variability represented by each PC and are computed by warping the grand mean shape to the extremes of each axis.

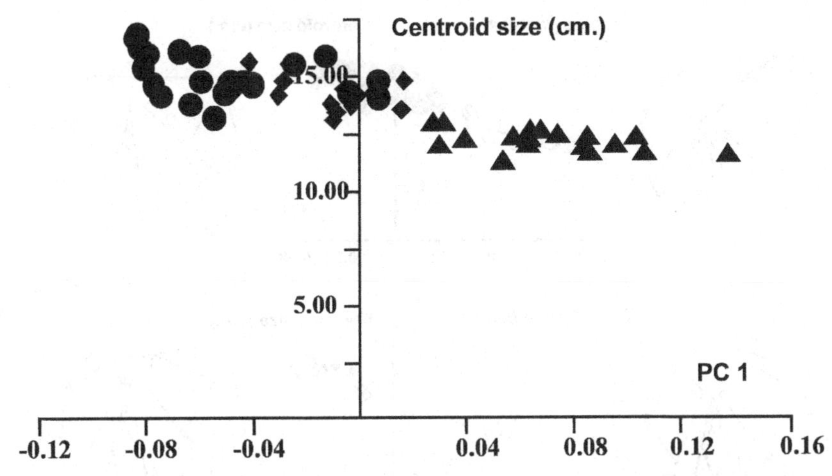

Figure 3.2. Plot of PC1 vs. centroid size ($r = 0.87$, $P < 0.001$) from the analysis of all adult representatives of *Colobus guereza* (circles), *Piliocolobus badius* (diamonds), and *Procolobus verus* (triangles).

shape all are significantly different ($P < 0.001$) in mean shape as assessed by a permutation test. *Colobus guereza* and *Piliocolobus badius* adult means are somewhat more similar than is either to *Procolobus verus*. Consistent with this is the ordination in Figure 3.1 of PC1 vs. PC2 (54% of total variance) extracted from the adult data. From this it is apparent that on the first two PCs the three taxa are almost completely separated although there is some overlap in the distributions of *Colobus guereza* and *Piliocolobus badius*. The inset smooth-rendered warped means, computed and plotted at the extremes of each PC in Figure 3.1, indicate the aspects of shape variation accounted for by these PCs. Thus, PC1 accounts for the differences in facial prognathism that exist between the taxa, with *Procolobus verus* being the least, *Piliocolobus badius* intermediate, and *Colobus guereza* the most prognathic. PC2 differentiates *Piliocolobus badius* on the basis of smaller differences in orbital and maxillary shape.

Static and evolutionary allometry between adults was investigated by examining the correlation between size and PC scores, and Figure 3.2 presents a plot of PC1 vs. centroid size from this analysis. This is the only PC on which specimen scores show a strong and significant correlation ($r = 0.87$, $P < 0.001$) with centroid size. From this plot it appears that the shape differences between and within taxa observed on PC1 (as visualized by the left and right insets of Figure 3.1) are strongly related to facial centroid size.

The remaining analyses examine the ontogenetic basis of the differences between the adults of each species. Figure 3.3 presents results from separate GPA/PCAs of the cross-sectional ontogenetic series of each taxon; these consist

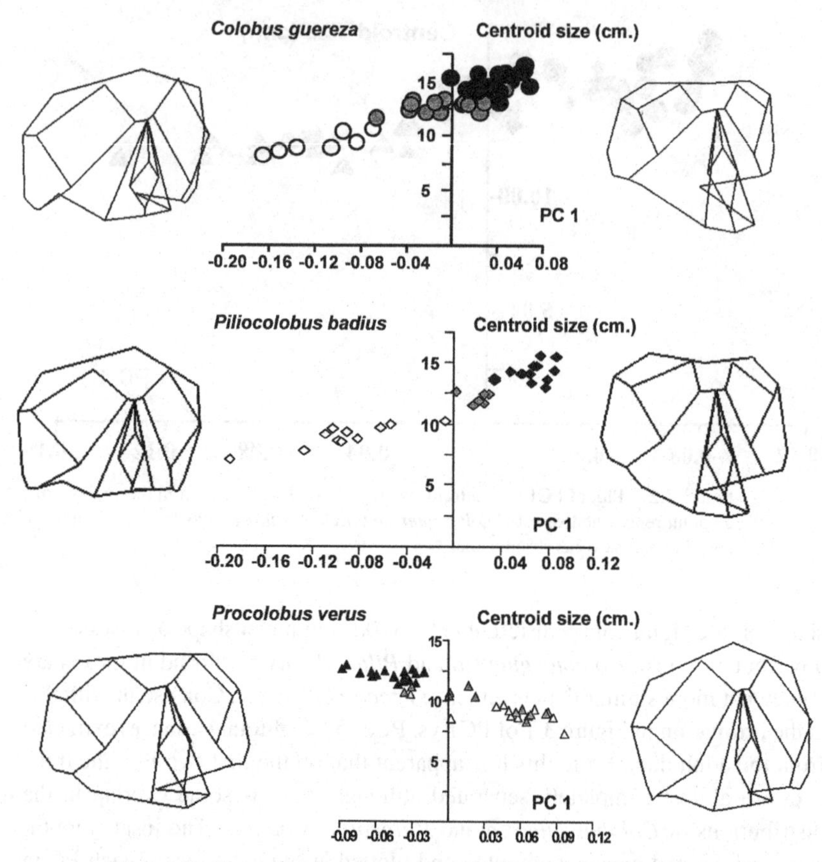

Figure 3.3. Plots of PC1 vs. centroid size from intraspecific GPA/PCA analyses of ontogenetic series in each of *Colobus guereza, Piliocolobus badius*, and *Procolobus verus*. Filled shapes = adults, grey = subadults, and open = infants and juveniles. The aspects of shape variability represented by PC1 in each analysis are indicated by the inset wireframe models drawn to left and right of each plot. Note that PC1 in the analysis of *P. verus* is arbitrarily reflected such that the representation of the shape of the smallest mean is to the right of this figure rather than the left as in *C. guereza* and *P. badius*. PC1 is approximately scaled in length between analyses to facilitate comparison of the magnitude of ontogenetic shape change between taxa.

of plots of PC1 scores vs. centroid size. In each intraspecific analysis PC1 was the only one to show any significant correlation with centroid size (*C. guereza*, $r = 0.92$, $P < 0.001$; *P. badius*, $r = 0.96$, $P < 0.001$; *P. verus*, $r = -0.93$, $P < 0.001$) and so it can interpreted as an axis of ontogenetic shape change. These plots therefore describe ontogenetic allometry, the nature of which is indicated by the inset wireframe models (computed by warping the mean shape of each series to the plotted extremes of PC1). In all taxa this ontogenetic shape

Table 3.4. *Comparison of growth vectors between species: angle between PC1 (ontogenetic allometric vector), below the diagonal; significance of this angle, above the diagonal*

	Colobus guereza	*Piliocolobus badius*	*Procolobus verus*
Colobus guereza		$P = 0.4$	$P < 0.001$
Piliocolobus badius	21.5°		$P < 0.001$
Procolobus verus	43.8°	37.7°	

change consists of a relative reduction in orbital size and increase in maxillary size and prognathism, but it is not possible visually to assess if this is identical in nature between taxa. Certainly, *Procolobus verus* appears to differ from *Colobus guereza* and *Piliocolobus badius* in manifesting a smaller degree of shape and size change between the youngest and adult specimens.

The final analysis addresses the extent to which the PC1s from each taxon after joint GPA represent similar ontogenetic shape variability. The results are presented in Table 3.4 and indicate that *Colobus guereza* and *Piliocolobus badius* show a small and insignificant (as assessed by a permutation test) angle between their facial ontogenetic trajectories whilst that of *Procolobus verus* manifests a significantly large angle with the trajectory of both these species. Joint GPA/PCA of the ontogenetic series generates the plots and visualizations of Figure 3.4. The first three PCs of this analysis account for 42%, 12.5%, and 8.5% (63% in combination) of the total variance and account for most of the inter-taxon differences. The upper frame of this figure shows a plot of PC1 vs. PC2 together with inset visualizations of the aspects of shape variability represented by PC1. In the main, the ontogenetic series are distributed from young to adult along PC1 and the visualizations reflect this in indicating that PC1 relates to shape variability rather similar in nature to that observed along the intra-taxon PC1s of Figure 3.3. Note that the ontogenetic series of *Procolobus verus* does not extend to the positive extremes of PC1 and so adults do not achieve the degree of prognathism found in *Colobus guereza* and *Piliocolobus badius*. On PC2 the ontogenetic series of *Colobus guereza* is separated from those of *Piliocolobus badius* and *Procolobus verus* and on the combination of PC1 and PC2 the three taxa show fairly parallel distributions. The scatter of *Colobus guereza* shows some evidence of curvilinearity in the plot of PC1 vs. PC2 but larger sample sizes are needed to confirm or disprove this.

The scatters of the ontogenetic series of *Colobus guereza* and *Piliocolobus badius* are separated to some degree on PC3 but in neither taxon does this PC differentiate adult from immature specimens. In contrast PC3 separates the

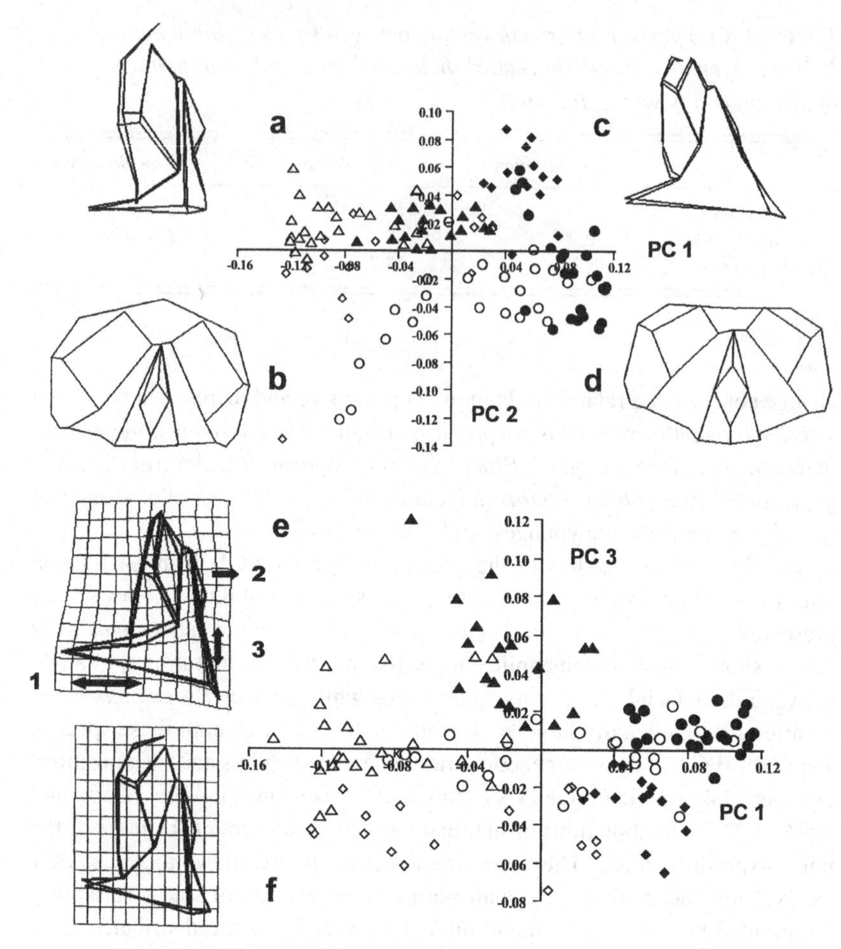

Figure 3.4. Plots of (a) PC1 and PC2 and (b) PC1 and PC3 from GPA/PCA of the combined ontogenetic series of African colobines *Colobus guereza* (circles), *Piliocolobus badius* (diamonds), and *Procolobus verus* (triangles). Filled shapes = adult, open = immature. The wireframe insets adjacent to the top plot are different views of the overall mean shape warped between the negative (a and b) and positive (c and d) extremes of PC1. The wireframes with midline reference (f) and transformation (e) grids adjacent to the lower plot indicate the shape variability represented by PC3 and are computed by warping the mean shape along this PC.

adults but not the infants of *Procolobus verus* from the ontogenetic series of *Colobus guereza* and *Piliocolobus badius*. The correlation between centroid size and PC3 scores for *Procolobus verus* is 0.59 ($P < 0.001$), indicating significant ontogenetic allometry. The nature of this unique aspect of ontogenetic shape change in *Procolobus verus* is visualized in Figure 3.4 in the insets e and f.

Inset f presents a wireframe model of the overall mean face warped to the negative extreme of PC3 with the midline section of a regular 3-D reference grid superimposed. Inset e shows the mean warped to the positive extreme of PC3 with the grid deformed using a 3-D thin-plate spline. This midline grid is shown in preference to the whole grid or any other section of the grid because it neatly captures the salient features of the deformation. The grid is most deformed in the region of the posterior maxilla (Fig. 3.4e, arrow 1) indicating a relative posterior stretching of this region. Other marked deformations include a relative forward siting of the upper nasal region (Fig. 3.4e, arrow 2) and a relative vertical stretching of the lower nasal aperture (Fig. 3.4e, arrow 3). Note that these deformations are computed over the plotted length of PC3 and that the *Procolobus verus* immature-to-adult divergence on this PC spans approximately one-third of it, so the deformation shown in Figure 3.4e exaggerates these aspects of age-related shape change by a factor of approximately 3. These findings indicate that *Colobus guereza* and *Piliocolobus badius* appear to follow nearly parallel ontogenetic trajectories over PCs 1–3, but are differentiated on PC2 and PC3, whilst that of *Procolobus verus* is divergent. This finding echoes that from the computation of angles between trajectories (Table 3.4).

Discussion

The analyses presented here were carried out with the aim of providing data on the extent to which differences in facial form within (dimorphism) and between these colobines might be accounted for by patterning and growth differences occurring pre- and postnatally. In particular they have focused on the influence of allometry in generating observed differences between adults, and allow these differences to be placed in the context of the generally accepted view that red (*Piliocolobus badius*) and olive (*Procolobus verus*) colobus are more closely related to each other than either is to the black and white colobus (*Colobus guereza*).

Thus, with regard to sexual dimorphism no colobine species studied here shows significant dimorphism in the shape of the landmark configuration. However the two species in this study with the largest adult faces (*Colobus guereza* and *Piliocolobus badius*) manifest small but significant size dimorphism. In similar studies of papionins (O'Higgins and Jones, 1998; O'Higgins and Collard, 2002) highly significant size and shape dimorphism was always observed. The degree of shape dimorphism was, however, variably related to ontogenetic scaling because in all taxa sexual dimorphism was shown to arise through a combination of ontogenetic scaling and late growth divergence between males and females. The contrast in result for the African colobines indicates that, relative to other causes of inter-adult variation, any sexual dimorphism in shape arising through ontogenetic scaling is too small to be detected. Further, there is no evidence of late divergence of male and female growth trajectories in these primates. This contrasts with the situation in papionins where up to 50% of sexual dimorphism can be attributed to such a phenomenon. It was hypothesized in an earlier study of papionins (O'Higgins and Collard, 2002) that the divergence in ontogenetic trajectory between subadult males and females could be attributable to sexual differences in dental morphology or to masticatory muscle morphology and force generation. The finding of a lack of such divergence in the African colobines provides a potentially interesting natural experiment for testing of these alternatives. This could be achieved by comparing and contrasting masticatory muscle and dental biomechanics and morphology between African colobines and papionins.

Our analyses have indicated that the inter-taxon differences between adult African colobine faces are such that the two taxa with similarly large faces (*Colobus guereza* and *Piliocolobus badius*) are more similar to each other than either is to the small-faced *Procolobus verus* (Table 3.3, Fig. 3.1). This might imply that the inter-taxon adult differences arise through ontogenetic scaling, such that (most simply) each represents a different end point on the same ontogenetic trajectory (i.e., identical "start shape" and slope). This possibility is also raised by the plot of Figure 3.2. In this, the score of each adult African colobine on PC1 is closely related to its centroid size and the intraspecific slopes appear to grade into the interspecific. However, in Figure 3.1, there are differences between the adult colobine means on PC2 that are not correlated with size and this implies that there are other sources, besides ontogenetic scaling, of inter-taxon variability.

These are investigated through the analyses of ontogenetic series presented in Figures 3.3 and 3.4 and Table 3.4. First, with regard to allometric growth within each colobine species, Figure 3.3 indicates that they all undergo ontogenetic shape changes that share some gross similarities (e.g., increasing maxillary prognathism, increase in maxillary size relative to orbital). However, the magnitude of the ontogenetic size and shape changes observed in *Procolobus verus* is somewhat less than that in the other two colobines. Further, analysis of the angles between the ontogenetic allometric vectors (PC1, Table 3.4) indicates that there are significant differences in growth vector between *Procolobus verus* on the one hand and *Colobus guereza* and *Piliocolobus badius* on the other. This is further substantiated by the findings presented in Figure 3.4, where *Procolobus verus* shows divergent growth with respect to the other two colobines. The nature of this growth divergence is illustrated in Figure 3.4e and begs explanation in future studies through detailed consideration of its potential adaptive role in relation to dietary, behavioral, and ecological factors. Thus, further studies to examine the relationship between facial morphology and functionally or ecologically interesting variables such as body size would be useful to discover if *Procolobus verus* shows different relationships to those of *Colobus guereza* and *Piliocolobus badius*.

Our findings indicate that ontogenetic scaling does not fully explain the differences in facial morphology between African colobines. Rather, *Colobus guereza* and *Piliocolobus badius* share a common ontogenetic allometric trajectory over the age range represented by our samples. Thus, the principal differences between these two species arise earlier in ontogeny and are carried through to adulthood. In contrast, the early established differences between *Procolobus verus* and the other colobines are further accentuated postnatally by its divergent ontogenetic trajectory.

The results of this study are also interesting with regard to phylogeny in that, on several counts, they point to greater similarities between *Colobus guereza* and *Piliocolobus badius* than between either of these and *Procolobus verus*. First, as mentioned above, none of the colobine species appears to show significant sexual dimorphism in shape although both *Colobus guereza* and *Piliocolobus badius* manifest a significant size dimorphism and *Procolobus verus* does not. Second, adult differences in mean shape again indicate that *Colobus guereza* and *Piliocolobus badius* are more similar to each other than either is to *Procolobus verus*. Third, postnatal ontogenetic changes in face shape are more similar in nature and degree between *Colobus guereza* and *Piliocolobus badius* than between either of these and *Procolobus verus*. Thus *Colobus guereza* and *Piliocolobus badius* are more similar to each other than either is to *Procolobus verus* in facial shape, size dimorphism, and postnatal ontogeny. Does this mean that, contrary to the common view, it is more likely that the black and white (*Colobus guereza*) and red colobus (*Piliocolobus badius*) monkeys are sister taxa to the exclusion of the olive colobus (*Procolobus verus*)?

There are several reasons why this possibility should be viewed with caution. First, and most importantly, it must be borne in mind that the facial skeleton is highly adaptable, being formed in membrane and growing under the influence of diverse functional matrices (Moss and Young, 1960; Moss and Salentijn, 1969; Enlow, 1975). Thus it is possible that sister taxa could diverge rapidly in many morphological aspects of the face, leaving one in possession of plesiomorphous aspects of morphology whilst the other rapidly accumulates autapomorphies. Second, there is mounting evidence that, at least within some groups of primates and at lower taxonomic levels, facial morphology, sexual dimorphism, and ontogeny are not useful indicators of phylogeny (Collard and Wood, 2000, 2001). Thus molecular and morphological studies of papionin phylogeny have reached inconsistent conclusions, with the consensus molecular phylogeny indicating that the genus *Cercocebus* is most closely related to *Mandrillus*, whereas

Lophocebus is most closely related to *Papio* and *Theropithecus* (Disotell, 1994, 1996, 2000; van der Kuyl *et al.*, 1994; Harris and Disotell, 1998; Harris, 2000). In contrast, the majority of morphological phylogenies have supported a sister-group relationship between *Papio* and *Mandrillus* and a sister-group relationship between *Cercocebus* and *Lophocebus* (Jolly, 1966, 1967, 1970; Delson, 1975, 1993; Szalay and Delson, 1979; Strasser and Delson, 1987; Delson and Dean, 1993; but see Fleagle and McGraw, 1999; Groves, 2000). Recent studies of papionin facial morphology, sexual dimorphism and postnatal growth using geometric morphometric techniques (Collard and O'Higgins, 2001; O'Higgins and Collard, 2002) lend weight to the view that facial morphology is unreliable as an indicator of phylogeny at least at these taxonomic levels. Thus adult and infant facial morphology, postnatal facial growth vectors, and sexual dimorphism tend to support mangabey monophyly rather than the molecular cladogram.

At higher taxonomic levels the face and its growth may be more reliable as an indicator of taxonomy. Thus, a geometric morphometric study of postnatal facial growth, sexual dimorphism, and adult differences within and between *Cebus apella* and *Cercocebus torquatus* (O'Higgins *et al.*, 2001) indicated similarities between these taxonomically distant primates in the mechanism of ontogeny of sexual dimorphism, but this is superimposed on widely divergent allometric ontogenetic trajectories acting on different early infant morphologies and giving rise to very different patterns of sexual dimorphism and adult facial morphologies. More studies at different taxonomic levels are needed in future to establish reasonable limits of phylogenetic inference from facial morphology and morphogenesis. With regard to the colobines it will be interesting to extend this work to include Asian varieties, examining in more detail craniofacial form, ontogeny, and evolution in light of emerging biochemical and genetic data.

Conclusions

Oxnard's early studies (Ashton *et al.*, 1965; Oxnard, 1972, 1973) indicated the value of quantification and precise statistical analysis in generating novel interpretations of skeletal variations amongst primates. The study presented in this chapter continues the Oxnardian theme of the "mathematical dissection of anatomies" by providing a quantitative account of variations amongst African colobine crania. The results have provided some new insights into the pattern of variations in facial form amongst these monkeys and have indicated how postnatal growth can influence them. In so doing, these studies raise some important questions in relation to sexual dimorphism and the correlation between facial skeletal morphology and phylogeny. The investigation of these is a task for the future but, for now, we can only be certain that the morphometric tradition so elegantly expounded by Oxnard will play a role in their resolution.

Acknowledgments

We wish to thank Paula Jenkins of the Natural History Museum, London for her considerable help in granting us access to the collections in her care. Additionally this work has depended on the continued support of Nicholas Jones, University College London, in providing programming support. R. L. Pan was supported during this work by the Australian Research Council (Large Grants and Research Fellow) and by the Australian Academy of Sciences (Visiting Fellowship). We should also like to thank Rebecca German and Nina Jablonski for their efforts in organizing the symposium at the Annual Meeting of the American Association of Physical Anthropologists from which this volume has arisen. We also thank Fred Anapol for his efforts in the onerous task of editing this volume. The comments of three referees, including Chris Klingenberg, have been very helpful in guiding revision of the manuscript. Our acknowledgments would not be complete without expressing our heartfelt thanks to Charles Oxnard, without whose personal efforts neither of us would be in our chosen field.

References

Ashton, E. H., Healy, M. J. R., Oxnard, C. E., and Spence, T. F. (1965). The combination of locomotor features of the primate shoulder girdle by canonical analyis. *J. Zool. Lond.*, **163**, 319–350.

Bookstein, F. L. (1978). *The Measurement of Biological Shape and Shape Change. Lecture Notes in Biomathematics*. New York, NY: Springer.

(1989). Principal warps: thin-plate splines and the decomposition of deformations. *IEEE Transactions in Pattern Analysis and Machine Intelligence*, **11**, 567–585.

Brandon-Jones, D. (1984). Colobus and leaf monkeys. In: *Encyclopaedia of Mammals*, ed. I. D. Macdonald. London: George Allen and Unwin. pp. 398–408.

Cobb, S. and O'Higgins, P. (2000). Cranial growth in extant large bodied African Hominidae. *Amer. J. Phys. Anthropol.*, **30**, Suppl., 126–127.

Collard, M. and O'Higgins, P. (2001). Ontogeny and homoplasy in the papionin monkey face, *Devel. Evol.* **3**, 322–331.

Collard, M. and Wood, B. A. 2000. How reliable are human phylogenetic hypotheses? *PNAS*, **97**, 5003–5006.

2001. How reliable are current estimates of fossil catarrhine phylogeny? An assessment using extant great apes and Old World monkeys. In: *Hominoid Evolution and Climate Change in Europe, vol. 2: Phylogeny of the Neogene Hominoid Primates of Eurasia*, ed. L. de Bonis, G. D. Koufous, and P. Andrews. Cambridge: Cambridge University Press. pp. 118–150.

Collura, R. V., Auerbach, M. R., and Stewart, C. B. (1996). A quick direct method that can differentiate expressed mitochondrial genes from their unclear pseudogenes. *Current Biol.*, **6**, 1337–1339.

Corruccini, R. S. (1988). Morphometric replicability using chords and Cartesian coordinates of the same landmarks. *J. Zool. Lond.*, **215**, 389–394.

Creel, N. and Preuschoft, H. (1971). Hominoid taxonomy: a canonical analysis of cranial dimensions. *Proceedings of the 3rd International Congress of Primatology, Zurich 1970*, **1**, 36–43.
Delson, E. (1975). Evolutionary history of the Cercopithecidae. *Contrib. Primat.*, **5**, 167–217.
(1993). *Theropithecus* fossils from Africa and India and the taxonomy of the genus. In: *Theropithecus: Rise and Fall of a Primate Genus*, ed. N. G. Jablonski. Cambridge: Cambridge University Press. pp. 157–189.
(1994). Evolutionary history of the colobine monkeys in paleoenvironmental perspective. In: *Colobine Monkeys: Their Ecology, Behavior and Evolution*, ed A. G. Davies and J. F. Oates. Cambridge: Cambridge University Press. pp. 11–43.
Delson, E. and Dean, D. (1993). Are *P. baringensis* R. Leakey, 1969, and *P. quadratirostris* Iwamoto, 1982, species of *Papio* or *Theropithecus*? In: *Theropithecus: Rise and Fall of a Primate Genus*, ed. N. G. Jablonski. Cambridge: Cambridge University Press. pp. 125–156.
Disotell, T. R. (1994). Generic level relationships of the Papionini (Cercopithecoidea). *Amer. J. Phys. Anthropol.*, **94**, 47–57.
(1996). The phylogeny of Old World monkeys. *Evol. Anthropol.* **5**, 18–24.
(2000). Molecular systematics of the Cercopithecidae. In: *Old World Monkeys*, ed. P. F. Whitehead and C. J. Jolly. Cambridge: Cambridge University Press. pp. 29–56.
Dryden, I. L. and Mardia, K. V. (1993). Multivariate shape analysis. *Sankhya*, **55**(A), 460–480.
(1998). *Statistical Shape Analysis*. London: Wiley.
Enlow, D. H. (1975). *Handbook of Facial Growth*. Toronto: Saunders.
Fleagle, J. G. (1988). *Primate Adaptation and Evolution*. New York, NY: Academic Press.
Fleagle, J. G. and McGraw, W. S. (1999). Skeletal and dental morphology supports diphyletic origin of baboons and mandrills. *PNAS*, **96**, 1157–1161.
Good, P. (1993). *Permutation Tests: A Practical Guide to Resampling Methods for Testing Hypotheses*. New York, NY: Springer.
Goodall, C. R. (1991). Procrustes methods and the statistical analysis of shape (with discussion). *J. Roy. Stat. Soc. B*, **53**, 285–340.
Groves, C. P. (1989). *A Theory of Human and Primate Evolution*. Oxford: Clarendon Press.
(2000). The phylogeny of the Cercopithecoidea. In: *Old World Monkeys*, ed. P. F. Whitehead and C. J. Jolly. Cambridge: Cambridge University Press. pp. 77–98.
(2001). *Primate Taxonomy*. Washington, DC: Smithsonian Institution Press.
Harris, E. E. (2000). Molecular systematics of the Old World monkey tribe Papionini: analysis of the total available genetic sequences. *J. Hum. Evol.* **38**, 235–256.
Harris, E. E. and Disotell, T. R. (1998). Nuclear gene trees and the phylogenetic relationships of the mangabeys (primates: Papionini). *Mol. Biol. Evol.*, **15**, 892–900.
Hewett-Emmett, D., Cook, C. N., and Barnicot, N. A. (1976). Old World monkey hemoglobins: deciphering phylogeny from complex patterns of molecular evolution. In: *Molecular Anthropology: Genes and Proteins in the Evolutionary Ascent of the Primates*, ed. M. Goodman, R. E. Tashian and J. H. Tashian. New York, NY: Plenum Press. pp. 257–275.

Jolly, C. J. (1966). Introduction to the Cercopithecoidea, with notes on their use as laboratory animals. *Symp. Zool. Soc. Lond.*, **17**, 427–445.
(1967). The evolution of baboons. In: *The Baboon in Medical Research, vol. II*, ed. H. Vagtborg. Austin, TX: University of Texas Press. pp. 23–50.
(1970). The large African monkeys as an adaptive array. In: *Old World Monkeys: Evolution, Systematics, and Behaviour*, ed. J. R. Napier and P. H. Napier. New York, NY: Academic Press. pp. 139–174.
Kendall, D. G. (1984). Shape manifolds, Procrustean metrics and complex projective spaces. *Bull. Lond. Math. Soc.*, **16**, 81–121.
Kent, J. T. (1994). The complex Bingham distribution and shape analysis. *J. Roy. Stat. Soc*, Series B, **56**, 285–299.
Lele, S. (1993). Euclidean distance matrix analysis: estimation of mean form and form difference. *Math. Geol.*, **25**, 573–602.
Marcus, L. F., Corti, M., Loy, A., Naylor, G. J. P., and Slice, D. (1996). *Advances in Morphometrics*. New York, NY: Plenum Press.
Moss, M. L. and Salentijn, L. (1969). The primary role of functional matrices in facial growth. *Amer. J. of Ortho.*, **55**, 566–577
Moss, M. L. and Young, R. W. (1960). A functional approach to craniology. *Amer. J. Phys. Anthropol.*, **18**, 281–292.
Napier, J. R. and Napier, P. H. (ed) (1967). *A Handbook of Living Primates*. London: Academic Press.
Nowak, R. M. (1991). *Walkers Mammals of the World*. Baltimore, MD: Johns Hopkins University Press.
Oates, J. F., Davies, A. G., and Delson, E. (1994). The diversity of living colobines. In: *Colobine Monkeys: Their Ecology, Behavior and Evolution*, ed. A. G. Davies and J. F. Oates. Cambridge: Cambridge University Press. pp. 45–73.
O'Higgins, P. (2000). Quantitative approaches to the study of craniofacial growth and evolution: advances in morphometric techniques. In: *Vertebrate Ontogeny and Phylogeny: Implications for the Study of Hominid Skeletal Evolution*, ed. P. O'Higgins and P. Cohn. London: Academic Press. pp. 163–185.
O'Higgins, P. and Collard, M. (2002). Sexual Dimorphism and Facial Growth in Papionin Monkeys. *J. of Zool.* **257**: 255–272.
O'Higgins, P. and Dryden, I. (1992). Studies of craniofacial growth and development. *Persp. Hum. Biol.*, *2* / Arch. Oceania, **27**, 95–104.
(1993). Sexual dimorphism in hominoids: further studies of cranial "shape change" in *Pan, Gorilla* and *Pongo*. *J. Hum. Evol.*, **24**, 183–205.
O'Higgins, P. and Jones, N. (1998). Facial growth in *Cercocebus torquatus*: an application of three dimensional geometric morphometric techniques to the study of morphological variation. *J. Anat.*, **193**, 251–272.
O'Higgins, P. and Strand Vidarsdottir, U. (1999). New approaches to the quantitative analysis of craniofacial growth and variation. In: *Human Growth in the Past*, ed. R. Hoppa and C. Fitzgerald. Cambridge: Cambridge University Press. pp. 128–160.
O'Higgins, P., Chadfield, P., and Jones, N. (2001). Facial growth and the ontogeny of morphological variation within and between *Cebus apella* and *Cercocebus torquatus*. *J. Zool.*, **254**, 337–357.

Oxnard, C. E. (1972). Some African fossil foot bones: a note on the interpolation of fossils into a matrix of extant species. *Amer. J. Phys. Anthropol.*, **37**, 3–12.

(1973). *Form and Pattern in Human Evolution: Some Mathematical, Physical and Engineering Approaches*. Chicago, IL: University of Chicago Press.

(1984). *The Order of Man*. New Haven, CT: Yale University Press.

Rohlf, F. J. (1999). Shape statistics: Procrustes superimpositions and tangent spaces. *J. Classification*, **16**, 197–223.

(2000a). On the use of shape spaces to compare morphometric methods. *Hystrix, Italian J. Mamm.*, n.s., **11**, 9–25.

(2000b). Statistical power comparisons among alternative morphometric methods. *Amer. J. Phys. Anthropol.*, **111**, 463–478.

Stewart, C. B. and Disotell, T. R. (1998). Primate evolution in and out of Africa. *Current Biology*, **8**, R582–588.

Strasser, E. and Delson, E. (1987). Cladistic analysis of cercopithecid relationship. *J. Hum. Evol.*, **16**, 81–99.

Szalay, F. S. and Delson, E. (eds.) (1979). *Evolutionary History of the Primates*. New York, NY: Academic Press.

Thompson, D'A. W. (1917). *On Growth and Form*. Cambridge: Cambridge University Press.

van der Kuyl, A. C., Kuiken, C. L., Dekker, J. T., and Goudsmit, J. (1994). Phylogeny of African monkeys based upon mitochondrial 12S rRNA sequences. *J. Mol. Evol.*, **40**, 173–180.

Yang, Z. (1998). Likelihood ratio test for detecting positive selection and application to primate lysozyme evolution. *Mol. Biol. Evol.* **15**, 568–573.

4 *Cranial variation among the Asian colobines*

RULIANG L. PAN
University of Western Australia

COLIN P. GROVES
Australian National University

Introduction

Variation in cranial measurements between species has been commonly used to shed light on controversies in the classification, evolution, phylogeny, and functional adaptation of primates. Variation within species has also been used frequently to reveal differences between populations or sexes in morphology, social activities, behavior, ancestral heritage, size, and sexual selection (e.g., Leutenegger and Kelly, 1977; Oxnard, 1983a; Cheverud *et al.*, 1985; Albrecht and Miller, 1993). Studies involving both inter- and intraspecific variation simultaneously are, however, rare (Pan, 1998; Pan and Oxnard, 2000, 2001a). This approach has proven useful in revealing patterns of variation in different functional units or anatomic regions in the same organ (e.g., skull) and in the analysis of the relationship between species or species groups in terms of shape and functional adaptation of various structures. Such studies, although clearly phenetic, have the potential to shed useful light on controversies relating to phylogeny and classification, especially of closely related primate taxa. Previous studies examining inter- and intraspecific variation were carried out on macaques (Pan, 1998; Pan and Oxnard, 2000, 2001a). In this study, the same approach is applied to the Asian colobines because they appear to have evolved in the same places and roughly at the same time as the macaques, and their evolution was probably influenced by many of the same environmental changes during the Pliocene and Pleistocene (Pan and Jablonski, 1987; Jablonski, 1993).

Asian colobines have been a highly successful radiation, as judged by their diversity and wide distribution. Their biology has received considerable attention in recent years (Strasser and Delson, 1987; Strasser, 1988; Groves, 1989;

Shaping Primate Evolution, ed. F. Anapol, R. Z. German, and N. G. Jablonski. Published by Cambridge University Press.

Jablonski and Peng, 1993; Delson, 1994; Peng and Pan, 1994; Jablonski, 1998) but is still poorly known compared with that of many other primates. One of the main reasons for this is that information on the snub-nosed monkey (*Rhinopithecus*) has been difficult to obtain because of a scarcity of specimens for study. Studies of the anatomy, classification, phylogeny, and evolution of the genus were mostly generated only in the last decade (Ye *et al.*, 1987; Jablonski and Peng, 1993; Pan *et al.*, 1993a, 1993b; Pan and Jablonksi, 1993; Jablonski, 1998).

The first craniometric study of the Asian colobines to include the snub-nosed monkey was carried out by Peng and Pan (1994). In that study all species were allocated to five generic or subgeneric groups (*Rhinopithecus*, *Presbytiscus*, *Presbytis*, *Pygathrix*, and *Nasalis*). Each genus or subgenus was regarded as a unit from which information on the variation between defined groups could be derived, but intraspecific variation (including sexual dimorphism) within each species was ignored. The nature of this variation, however, was found to vary greatly between colobine species (Jablonski and Pan, 1995). Therefore, it was important that variation between and within species be exhibited simultaneously in order to compare the way in which each of the patterns of variation was expressed.

Considerable controversy has surrounded the study of the phyletic relationships and taxonomy of the Asian colobines. This is most clearly reflected in the diversity of current taxonomic schemes for the group: three genera, *Pygathrix*, *Presbytis*, and *Nasalis* (Groves, 1970; Honacki *et al.*, 1982); five genera, *Nasalis*, *Simias*, *Rhinopithecus*, *Pygathrix*, and *Presbytis* (Simpson, 1945; Ellerman and Morrison-Scott, 1951; Buettner-Janusch, 1963; Napier and Napier, 1967; Grzimek, 1975; Walker, 1975); or even as many as nine genera, *Presbytis*, *Semonpithecus*, *Kasi*, *Trachypithecus*, *Pygathrix*, *Simias*, *Nasalis*, *Rhinopithecus*, and *Presbytiscus* (Hill, 1972). Most recently, the group was classified into seven genera by Groves (2001): *Semnopithecus*, *Trachypithecus*, *Presbytis*, *Pygathrix*, *Rhinopithecus*, *Nasalis*, and *Simias*. Studies which have examined gross anatomical and molecular characteristics in a cladistic framework have regarded the douc langurs and snub-nosed monkeys as representing two different genera, *Pygathrix* and *Rhinopithecus* (Ye *et al.*, 1987; Jablonski and Peng, 1993; Jablonski, 1998; Zhang and Ryder, 1998). Other, more traditional comparative anatomical studies have combined these animals into one genus, but different subgenera (Groves, 1993; Brandon-Jones, 1996). There is perhaps even less agreement about the relationships among and between the leaf monkeys, generally relegated to the genera *Semnopithecus*, *Presbytis*, and *Trachypithecus*.

The main purpose of this study was to analyze inter- and intraspecific cranial variation among Asian colobines with morphometric methods, including as

Table 4.1. *Species and specimens used in this study*

Species	Common names	Female	Male	Total
Rhinopithecus roxellana	Golden snub-nosed monkey	8	10	18
R. bieti	Black snub-nosed monkey	9	8	17
R. brelichi	Gray snub-nosed monkey	1	3	4
R. avunculus	Tonkin snub-nosed monkey	2	2	4
Pygathrix nemaeus	Red-shanked douc langur	8	9	17
Nasalis larvatus	Proboscis monkey	3	5	8
Presbytis rubicunda	Maroon leaf monkey	7	5	12
P. comata	Javan leaf monkey	3	4	7
P. melalophos	Banded leaf monkey	5	5	10
Semnopithecus entellus	Gray langur	5	5	10
Trachypithecus vetulus	Purple-faced leaf monkey	5	3	8
T. phayrei	Phayre's leaf monkey	9	8	17
T. francoisi	Francois' leaf monkey	7	8	15
T. obscurus	Dusky leaf monkey	10	10	20
T. cristatus	Silvery leaf monkey	9	8	17

broad a sample of taxa (including snub-nosed monkeys) as possible. Our hopes were to reveal the differences revealed in patterns of intra- and interspecific variation and to see what, if any, bearing these differences might have on the evolutionary relationships between the groups under investigation.

Methods

The specimens and species used in this study are listed in Table 4.1. The allocation of species into genera is based on Groves (2001). Skulls were measured in several institutions (see Acknowledgments). Eight cranial variables (Table 4.2) were chosen, based on their utilitiy in studies of inter- and intraspecific cranial variation in macaques (Pan, 1998; Pan and Oxnard, 2001a).

Multivariate analysis, including discriminant function analysis (DFA) and principal components analysis (PCA), were used. DFA reveals variation between predefined groups (such as species). In order to select variables that could be regarded as good discriminators, a stepwise selection method was used. In this way each cranial variable was evaluated based on its Wilks' lambda score. Only those showing an effective discriminant function were retained for further analysis. In contrast to DFA, PCA examines variation between species, with all species considered as one universal group. For DFA and PCA separately, factor (PCA) and discriminant (DFA) scores for each specimen in the first two axes were recorded. Each sex of a species was regarded as a group

Table 4.2. *Variables used in this study*

CRANL	Cranial length, from the tip of the occipital protuberance to the alveolare.
CALVL	Calvarial length, from glabella to the tip of the occipital protuberance.
CRANW	Cranial width, distance between the points on the left and right suprameatal crests above the external acoustic meatus.
BPORW	Biporionic width, measured as the distance between the most lateral points on the external auditory canals (porion).
BIZYGW	Bizygomatic width, maximum width of the zygomatic arches at the midpoint of the zygomatico-temporal suture.
MIDPARW	Midparietal width, measured as the distance between the points on the left and right parietal bones that lie above the external acoustic meatus, at the maximum expansion of the parietals.
POSTORB	Postorbital constriction, measured as the minimum width of the postorbital constriction.
OCCH	Occipital height, vertical distance from the anterior border of the foramen magnum to the tip of the external occipital protuberance. Distance is measured vertically from the superior edge of a straight edge which is held at the level of the anterior border of the foramen magnum.

in DFA and its mean score was calculated and illustrated in a scatter diagram. The two sexes were analyzed together using PCA, but the mean of the factor scores for each sex of a species was calculated and illustrated in a scatter diagram. The method of analyzing the two sexes simultaneously in DFA and PCA, instead of separately, has been confirmed as a good way of analyzing the variation both within and between species, and for making a comparison between species when these two components are considered together (Pan, 1998; Pan and Oxnard, 2000). The origin and pattern of sexual dimorphism in individual nonhuman primate species has been considered to be associated with different factors. These include the intensity of sexual selection, type of mating systems, terrestriality vs. arboreality, anti-predator defense, and possible passive (allometric) selective forces, such as variance dimorphism. Phylogenetically species evolved from the same ancestor and undergoing similar evolutionary changes tend to show very similar degrees and patterns of sexual dimorphism (Leutenegger and Kelly, 1977; Oxnard, 1983a, 1987; Harvey and Bennett, 1985; Pickford, 1986).

In addition to raw data, size-adjusted data (residuals) derived from allometric analysis were also used. This is because Asian colobines exhibit a great variation in body size (Napier and Napier, 1967; Jablonski and Pan, 1995). It is important to look at the outcomes when relative size (shape-related) data are analyzed because in some cases information closely related to absolute size may conceal variation relevant to special functional adaptations between taxa. This situation is obvious when there exists a great differentiation in size among the

Table 4.3. *Eigenvalues and standardized canonical coefficients in the first two axes of DFA based on raw data*

	DF1	DF2
Eigenvalues	17.88	3.92
Percentage	66.9	17.6
Cumulative %	66.9	84.6
Canonical coefficients		
CRANL	0.368	−1.658
CALVL	0.502	1.493
BIPORW	0.378	−0.154
POSTORB	0.136	0.364

taxa studied (Kay, 1975; Jungers, 1984). Thus, as opposed to an absolute size variable a relative size variable such as a ratio produced by a specific formula or residuals created by allometric study can reveal variation in geometrical (shape), biomechanical, and physiological properties, and in special functional adaptations (Corruccini, 1972; Oxnard, 1983b; de Winter, 1997).

Even though body weight records were available for some specimens in this study, the sample size was too small for an acceptable allometric study to be carried out. Skull length was therefore used as a surrogate for body size because this dimension has been confirmed to be correlated highly with body size (Wood, 1979; Eaglen, 1984; Pan and Oxnard, 2001a). This length, from the tip of the occipital protuberance to the alveolare, was used as a substitute for body size as an independent variable in allometric analysis. The other variables were regarded as dependent.

Results

The eigenvalues and standardized canonical coefficients for the first two axes of DFA based on raw data are presented in Table 4.3. The former shows what proportion of the total variance is accounted for by a specific axis. The latter indicates how much each variable contributes to discrimination. According to the eigenvalues, the first two axes explained 84.6% of the total variance. Thus, only the results dealing with these first axes are described and illustrated (Fig. 4.1). Use of the stepwise selection method resulted in four of the eight cranial variables being chosen as discriminators; the other four were removed from the analysis. The first axis accounted for 66.9% of the total variance.

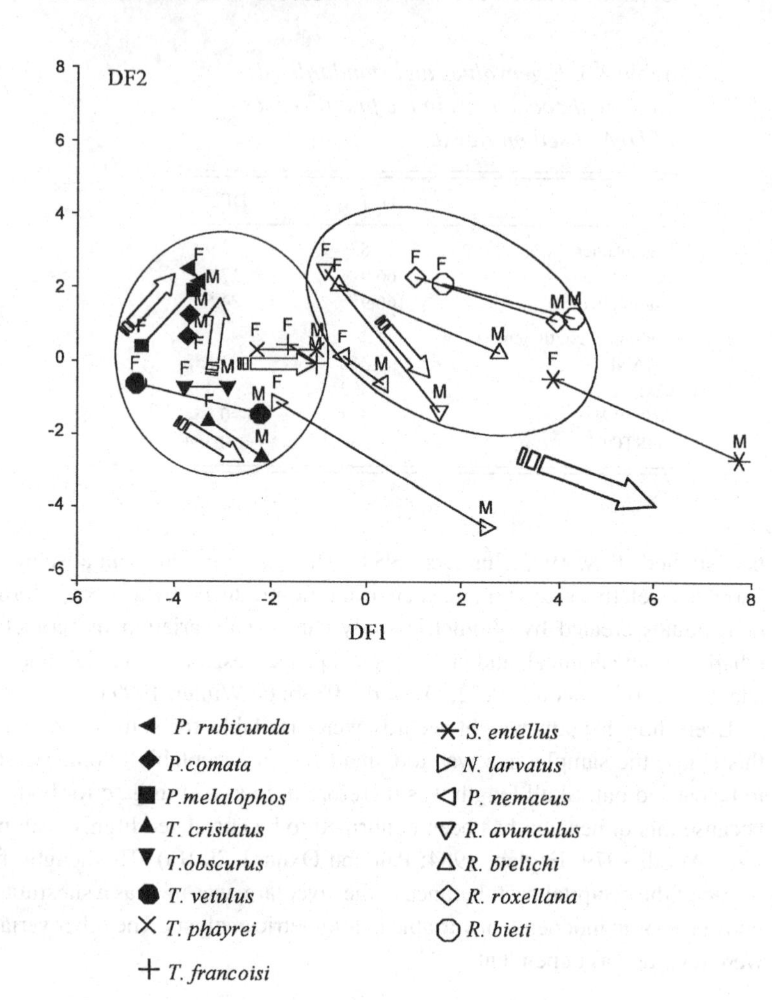

Figure 4.1. Species dispersions on the first two axes of DFA based on raw data. M = male, F = female.

All four chosen variables in this axis had positive coefficients. This meant that they created a dispersion of species along the positive (right) axis in the scatter diagram (Fig. 4.1). The variable calvarial length (CALVL) showed the largest positive canonical coefficient. In the second axis, with 17.6% of the total variance, the variables cranial length (CRANL) and calvarial length (CALVL) showed the largest negative and positive contributions to discrimination of species along this axis.

Figure 4.1 illustrates the dispersion of species based on DFA results. Although there is great variation between species, four clusters can be easily

Table 4.4. *Eigenvalues and eigenvectors of the first two axes of PCA based on raw data*

	PC1	PC2
Eigenvalues	5.57	0.86
Percentage	69.6	10.8
Cumulative %	69.6	80.3
Eigenvectors		
CRANL	0.928	−0.068
CALVL	0.950	−0.098
CRANW	0.960	−0.103
BIPORW	0.824	−0.267
BIZYGW	0.925	−0.058
MIDPARW	0.760	−0.057
OCCH	0.482	0.837
POSTORB	0.730	0.233

identified. The first one includes the species of snub-nosed monkey and douc langur (*Rhinopithecus* and *Pygathrix*); the second contains only the proboscis monkey (*Nasalis larvatus*); the third comprises the gray langur (*Semnopithecus entellus*), and the fourth consists of the other species of leaf monkey (*Presbytis* and *Trachypithecus*). Note, however, that because DF1 is a size factor, the differentiation of the fourth cluster is largely a function of their smaller skull size.

With regard to inter- and intraspecific variations, except to some extent for the Tonkin snub-nosed langur (*R. avunculus*), all species in the first three clusters showed a similar orthogonal direction in the two axes, in which males were larger on DF1 but smaller on DF2 than females. There is some variation between species in the degree of sexual dimorphism in the first cluster. The douc langur exhibited the smallest degree; *R. roxellana* and *R. bieti* displayed the greatest. Of special interest were the axes of sexual dimorphism within the fourth (leaf monkey) cluster. The species of *Trachypithecus* are all dispersed along the first axis, with varying degrees of downward slope, thus paralleling the species in the first three clusters. The species of *Presbytis* were entirely different, however: all three varied little if at all along the first axis, but they did along the second.

Eigenvalues and eigenvectors of the PCA based on raw data are shown in Table 4.4. The first two axes accounted for 80.3% of the total variation. Variable contributions were clearly separated by positive and negative symbols in these two axes. All variables made positive contributions to PC1.

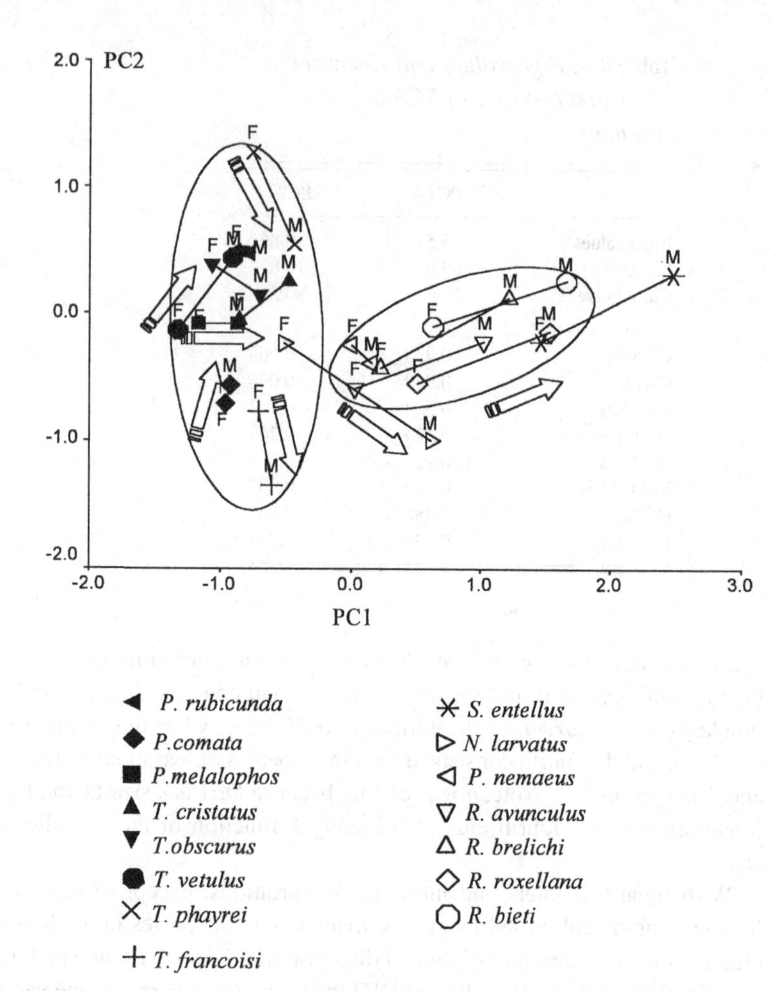

Figure 4.2. Species dispersions on the first two axes of PCA based on raw data. M = male, F = female.

Occipital height (OCCH) showed the lowest value in PC1, with 69.6% of the total variation. This variable, however, made the largest positive contribution to PC2 (with 10.8% of the total variation). The other variables in this axis made negative contributions except for the width of the postorbital constriction (POSTORB).

Figure 4.2 shows the species dispersions in the first two axes of PCA. *Rhinopithecus* and *Semnopithecus* displayed a type that is orthogonal to the two axes. Males were larger than females on both axes. *Pygathrix* and *Nasalis* displayed a similar pattern, perpendicular to the previous two genera.

Table 4.5. *Eigenvalues and standardized canonical coefficients of the first two axes of DFA based on residuals*

	DF1	DF2
Eigenvalues	5.45	2.03
Percentage	56.8	21.1
Cumulative %	56.8	77.8
Canonical coefficients		
BIPORW	−0.133	−0.008
POSTORB	0.531	0.629
OCCH	0.558	−0.924
CALVL	0.321	0.605

Once again there was a division between *Trachypithecus*, which resembled *Nasalius/Pygathrix* but tended to be more perpendicular to PC1, and *Presbytis*, which showed little sexual dimorphism of any kind but varied greatly in what they did exhibit (cf. the different axes of *P. melalophos* and *P. comata*, and the complete absence of any sexual difference in *P. rubicunda*).

Table 4.5 lists eigenvalues and standardized coefficients of the first two axes of DFA based on residuals. 77.8% of the total variance was explained by the first two axes, the first one accounting for 56.8%. Four of the seven variables were selected as discriminators. Except for biporionic width (BIPORW) with a low negative coefficient, all other variables made positive contributions to DF1; POSTORB and OCCH had the greatest influence because of their bigger values. The second axis accounted for 21.1% of the total variance. Two variables, OCCH and BIPORW, showed negative coefficients, the others positive; OCCH dominated in the discrimination in this axis.

Figure 4.3 illustrates species dispersion and variation patterns based on DFA after size was adjusted. Four clusters were defined, as in the previous analyses, but they were less obvious compared with those based on the raw data.

Variation was orthogonal in the first cluster even though *Pygathrix* deviated somewhat from *Rhinopithecus*, and had less sexual dimorphism. *Nasalis* and *Semnopithecus* were similar to each other in a pattern that was almost parallel to DF1, and not greatly different from *Pygathrix/Rhinopithecus* although more dimorphic. The leaf monkeys varied enormously. *Trachypithecus vetulus* and *P. rubicunda*, like *Nasalis* and *Semnopithecus*, had a pattern parallel to DF1; *T. cristatus* deviated only slightly from this. *Trachypithecus phayrei*, *T. obscurus*, and *T. francoisi* showed another type, orthogonal to the axes, in which males were larger than females on DF1, but smaller on DF2. Finally, the pattern in *P. melalophos* was almost perpendicular to DF1, and

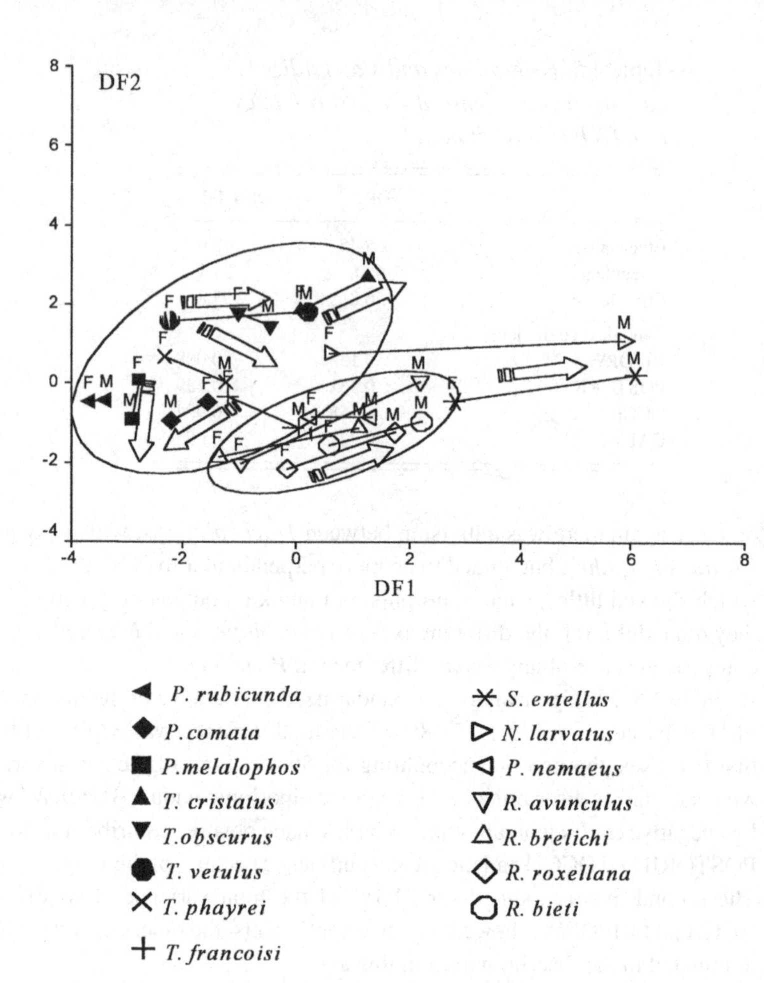

Figure 4.3. Species dispersions on the first two axes of DFA based on residuals. M = male, F = female.

the female was larger than the male on DF2. *Trachypithecus vetulus* exhibited the largest grade of sexual dimorphism; *P. rubicunda* showed the smallest.

Table 4.6 provides eigenvalues and eigenvectors of the first two axes of PCA based on residuals. 75.2% of the total variation was explained by the first two axes, 61.4% of it being accounted for by the first axis, in which all variables made positive contributions. In the second axis, with 13.8% of the total variation, four variables, BIPORW, POSTORB, OCCH, and midparietal width (MIDPARW), made negative, while the others showed positive, contributions

Table 4.6. *Eigenvalues and eigenvectors in the first two axes of PCA based on residuals*

	PC1	PC2
Eigenvalues	4.30	0.97
Percentage	61.4	13.8
Cumulative %	61.4	75.2
Eigenvectors		
CRANW	0.740	0.507
BIPORW	0.805	−0.086
BIZYGW	0.647	0.528
POSTORB	0.871	−0.301
OCCH	0.792	−0.439
CALVL	0.679	0.280
MIDPARW	0.913	−0.250

to the axis. CRANW and bizygomatic width (BIZYGW) showed the greatest influence on this axis.

Figure 4.4 demonstrates the species dispersion and the variation patterns revealed by PCA based on residuals. Species were mainly separated along the second axis. Even though there was some overlap and a considerable variety in term of the variation patterns, four clusters found previously were detected in the two axes.

Four types of variation were evident in the first cluster. Two *Rhinopithecus* species, *R. roxellana* and *R. brelichi*, showed a pattern which is almost parallel to PC1. *Rhinopithecus avunculus* exhibited an orthogonal pattern in which the male was larger than the female on PC1 but smaller on PC2. *Rhinopithecus bieti* displayed another pattern in which the two sexes almost overlapped along PC1, with the females larger than males along PC2. *Pygathrix* also showed an orthogonal type in which the male was larger on both axes. Even though they were remarkably separated along the second axis, *Nasalis* and *Semnopithecus* showed a similar pattern to the first two *Rhinopithecus* species.

Regarding the leaf monkeys, a number of different types of sexual dimorphism were demonstrated: almost perpendicular to PC1 except for a little declination to PC2 (*P. rubicunda*, *T. phayrei*, *T. obscurus*), orthogonal, the male being larger on both axes (*T. vetulus*), and sloping downward on PC2 (*T. francoisi*, *T. cristatus*, *P. comata*). It should be noted here that all species of *Presbytis* exhibited relatively little sexual dimorphism.

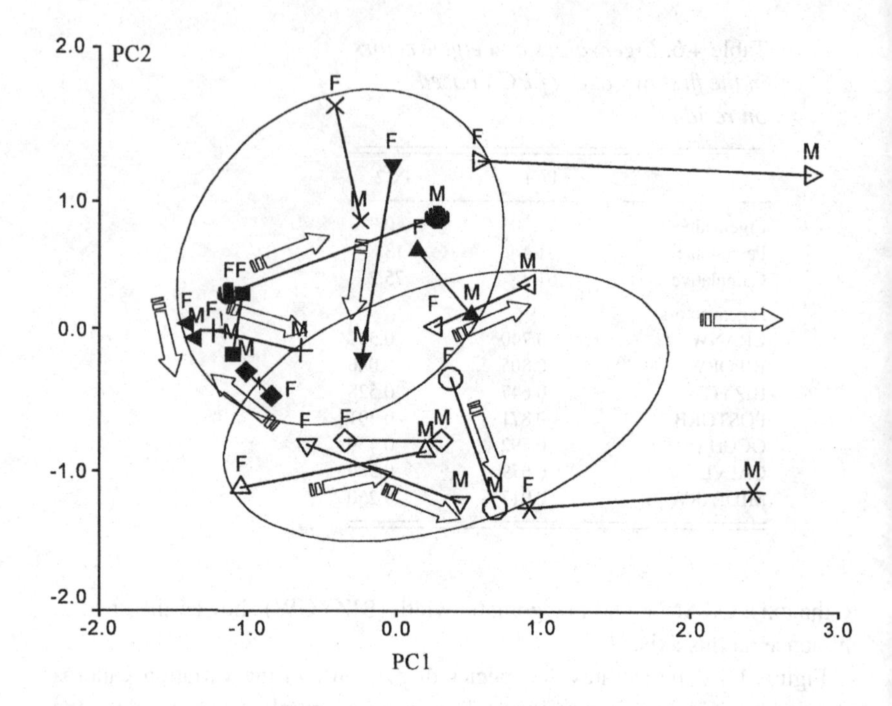

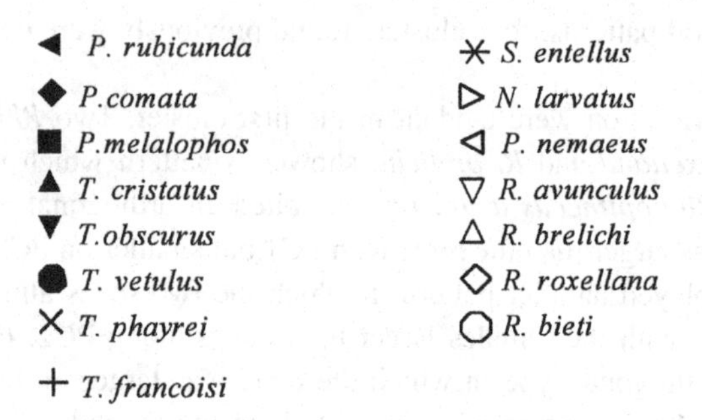

Figure 4.4. Species dispersions on the first two axes of PCA based on residuals. M = male, F = female.

Discussion

The results of this study may shed some light on controversies over the evolutionary relationship among Asian colobines, and may provide a test of some hypotheses proposed by other researchers.

When the raw data were considered, the species were mainly dispersed along the first axes in DFA and PCA. They were, however, mainly allocated along the second axes when the residuals were analyzed. The purpose of the residual is to reveal the variation in relative size (shape) mainly expressed in the second axis. Some differentiation was revealed when different datasets were used. For instance, species were rotated and moved toward the second axis but in different ways in DFA when the residuals were considered (Figs. 4.1 and 4.3). The *Rhinopithecus/Pygathrix* cluster was moved to the lower part of DF2 (below the leaf monkey cluster), and the pattern of sexual dimorphism was rotated anticlockwise. *Nasalis* and *Semnopithecus* were rotated from orthogonal to a position almost parallel to DF1. There were also changes in the leaf monkey cluster, such as the whole ellipse being rotated clockwise and species becoming more dispersed. Its long axis moved from a position perpendicular to DF1 to a location orthogonal to DF1 and DF2. Rotation and movement also occurred in PCA analyses when data were transformed. The whole structure of the *Rhinopithecus/Pygathrix* ellipse changed, with more extension. The elements inside were rotated clockwise in some cases, and the whole cluster was moved to the lower part of the second axis, as was the case in DFA. *Nasalis* was rotated anticlockwise and moved upward (referring to PC2), and forward (referring to PC1). The change in the smaller leaf monkeys was more complicated: as with the DFA, the whole ellipse was rotated from a position with a long diameter perpendicular to PC1 to a location somewhat orthogonal to the two axes, and species were more widely dispersed (Figs. 4.2 and 4.4).

It is hard to explain exactly how rotation occurred after data translation in both DFA and PCA. However, if the total variation/variance explained by the first two axes is compared between the raw data and the residuals, the proportions accounted for by the first and second axes were decreased and increased respectively, as a result of transition. This is because, as expected, the element for size that is normally expressed by the first axis was minimized; that for shape that is normally revealed by the second axis was, however, amplified after data conversion. This change resulted obviously in decreasing variation among species along the first axis. In other words, when residuals were analyzed the main difference among species revealed by DFA and PCA became mainly shape-related, so that the whole structure had to rotate and move toward the second axis.

Different functional units and anatomic regions have different inter- and intraspecific variation patterns in macaques (Pan, 1998; Pan and Oxnard, 1999; Pan and Oxnard, 2000). This may also be the case in *Rhinopithecus*. When PCA results based on the raw data were compared, the patterns in *Rhinopithecus* were very similar to those seen in macaques in the same area (cranium and calvaria): that is, the variations within and between species were orthogonal to each other, and both were proportionally shared by the first two axes.

The species of *Trachypithecus* analyzed in this study showed patterns broadly similar to those of the larger colobines: their axes of sexual dimorphism were shifted lower on the first discriminant function or principal component, but were otherwise parallel to those of the larger taxa. In the PCA and in the DFA residuals, the closely related species *Trachypithecus obscurus* and *T. phayrei* differed from others in that their axes sloped downward on DF2; in the PCA residuals, the variability was too great among all taxa studied to make any firm generalizations.

Presbytis species were characterized throughout by their extremely reduced degree of sexual dimorphism, and the axes of such dimorphism as exist were often quite different from those of any other genus.

The phenetic results showing inter- and intraspecific variation in this study may clarify some controversies over the classification and taxonomy of colobines, which is unsettled and has been subjected to continuing revision (Napier and Napier, 1967, 1985; Groves, 1970, 1993, 2001; Oates *et al.*, 1994). There is consensus neither on the number of genera, nor on the relationship of the genus-level groups to one another.

Rhinopithecus and *Pygathrix* were closely approximated and displayed very similar patterns; there was good separation between them and other species. These results might support the conclusion that they are congeneric (Groves, 1970, 1993). On the other hand, there exist some differences between them, most conspicuously the smaller degree of sexual dimorphism (with the exception being the PCA residuals diagram, Fig. 4.4). We will maintain a generic separation between them, as advocated by several studies over the past 20 years (Ye *et al.*, 1985, 1987; Peng *et al.*, 1988; Jablonski and Peng, 1993; Jablonski, 1995, 1998; Groves, 2001). A close relationship between *Rhinopithecus* and *Pygathrix* may confirm some proposed evolutionary scenarios for the groups. They have been postulated to have evolved from the same ancestor(s). The separation between the two genera might have occurred in the Pliocene (Delson, 1994) in southern and southwestern China (Jablonski and Pan, 1988; Pan and Oxnard, 2001b). In this group, according to Jablonski and Peng (1993), the douc langur maintains a larger number of primitive features in the *Pygathrix–Rhinopithecus* clade, while the snub-nosed monkeys, especially the Chinese species, show more derived features. Further evidence from other parts of the skeleton, and from soft-tissue anatomy, is still needed to test this proposed relationship.

In a study involving *R. roxellana*, *R. bieti*, *P. nemaeus*, *T. phayrei*, and *T. francoisi* using rDNA data, the snub-nosed monkeys were found to be more closely related to the leaf monkeys than to *Pygathrix* (Wang *et al.*, 1995). This finding was quite different from that reported by Zhang and Ryder (1998), in which *Rhinopithecus* showed a closer genetic similarity to *Nasalis* than to *Pygathrix*. This result was similar to the results found in this study.

The relationship between *Nasalis* and *Pygathrix* revealed in this study is different from that revealed in the cladistic analysis of Jablonski (1998), in which *Nasalis* was shown to be closer to *Pygathrix* than to *Rhinopithecus*. In the present study, *Nasalis* and *Pygathrix* showed quite different patterns of intraspecific variation regardless of research methods (Figs. 4.1–4.4). The former actually showed a pattern which was similar to that of *Semnopithecus* except for the profile of PCA based on raw data. *Nasalis* has long been recognized as unique in many respects. For example, it is the only colobine genus in which the interorbital pillar is narrow and cercopithecine-like; it also exhibits a long, narrow nasal structure and a long muzzle, and a back-tilted ascending ramus (Delson, 1994). It has been considered the sister group to other colobines, and was placed in a subfamily, Nasalinae, apart from the other genera by Groves (1989). Kuhn (1967) also suggested that the genus *Nasalis* might be the sister taxon to other Asian colobines.

The two genera of leaf monkeys (*Presbytis* and *Trachypithecus*) are closely approximated in all four diagrams regardless of dataset. The separation between them and other colobines was remarkable when the raw data were considered (Figs. 4.1 and 4.2). This can be considered to be associated with their smaller body size as compared with the others so that they were dispersed on the left side of the DFA and PCA. This close approximation may be related to the fact that their evolutionary relationships are still heatedly debated by taxonomists, as mentioned above. These two genera, however, do show a great variety in the forms of inter- and intraspecific variation although, broadly speaking, *Presbytis* differed from both *Semnopithecus* and *Trachypithecus* (Figs. 4.1, 4.3, 4.4). *Trachypithecus* is seen as essentially a reduced version of *Semnopithecus* and both adhered much more closely to the general Asian colobine pattern than did *Presbytis*. Within both, there was interspecific variability, but this was insignificant in *Presbytis* considering that sexual dimorphism was extremely low, where it existed at all. In *Trachypithecus*, on the other hand, there were some significant differences between species in the pattern of sexual dimorphism, greater than between some of them (*T.cristatus*, *T. vetulus*) and *Semnopithecus entellus*, which may argue for their being combined into one genus, as long advocated by Brandon-Jones (1984, 1996).

Delson (1994) and Groves (2001) regarded the Colobinae as consisting of two subtribes or groups: Colobina for the African colobus species, and Presbytina for the Asian extant colobine species and the fossils linking the African and Asian groups. According to Delson (1994) the Asian colobines evolved from an ancestor that split into three stocks in the Middle Pliocene. One resulted in the snub-nosed monkey and the douc langur, one gave rise to the proboscis monkey and pig-tailed langur (*Simias concolor*), and the third was ancestral to the leaf monkeys (including the gray langur).

The evidence from the fossil record is still equivocal. A fossil species related to *Semnopithecus entellus* might be *Mesopithecus pentelicus* in terms of the similarities on cranial and postcranial skeleton morphology. It has been found from the sediments from 2–10 million years ago in Eurasia (Delson, 1994). *Rhinopithecus* fossils are known from China (Jablonski and Gu, 1991; Jablonski, 1993, 1998; Jablonski and Peng, 1993), including *R. lantianensis* and *R. tingianus*. The latter has been considered a subspecies of extant *R. roxellana* (Colbert and Hooijer, 1953), but the type is a juvenile specimen lacking definitive criteria of species or subspecies. A possible douc langur has been recorded in a Late Pleistocene cave at Shan Bei Yan, the same district in which a possible fossil *R. avunculus* was found (Fooden and Feiler, 1988; Jablonski, 1996, 1998). *Nasalis* is similar to *Dolichopithecus ruscinensis* from the Miocene to Middle Pliocene of Europe: they share the long muzzle and narrow interorbital pillar (Szalay and Delson, 1979; Delson, 1994). Some fossils closely related to extant leaf monkey species were found in the deposits of Middle and Late Pleistocene and Holocene caves in Java, Sumatra, Borneo, and Vietnam (Hooijer, 1962; Kahlke, 1973).

Even though the fossils mentioned above are supposed to be related to the extant species, the exact association between fossils and living species and the relationship between fossil species is far from clear. It is still problematic to find the linkage between most extinct colobine taxa and extant living forms (Delson, 1994). The species clusters found in this study may provide some clues to support Delson's hypothesis, but the unique phenetic position of *Presbytis* is a new factor that has not been recognized hitherto.

The variables used in this study showed a range of discriminating abilities in the cranial variation analysis. When raw data were analyzed in DFA four variables, namely CRANW (cranial width), BIZYGW (bizygomatic width), MIDPARW (midparietal width), and OCCH (occipital height), were filtered out from analysis because they did not meet the criterion required by Wilks' lambda (Table 4.3). The other four selected variables, CRANL (cranial length), CALVL (calvarial length), BPORW (biporionic width), and POSTORB (postorbital constriction) also showed great contributions to the first axis in PCA (Table 4.4). This meant that these selected four variables were very good characters for the study of variation in Asian colobines if raw data were considered. When residuals were considered, four variables, CALVL, BIPORW, POSTORB, and OCCH, were chosen in the DFA analysis. These variables also made great contributions to the first axis of PCA. Three variables, CALVL, BIPORW, and POSTORB were the features selected by Wilks' lambda and made great contributions to the first axes of DFA and PCA regardless of the dataset. Thus, it seems that the main differences (both size and shape) among colobines were expressed both in cranial length and width.

Conclusions

Based on the morphometric results found in this study and evidence from other studies, some conclusions related to the relationship among Asian colobines can be drawn:

(1) Four clusters were found – *Rhinopithecus/Pygathrix*, *Nasalis*, *Semnopithecus*, and *Trachypithecus/Presbytis* – based on the profiles of species dispersion in the first axes of PCA and DFA.

(2) *Rhinopithecus* and *Pygathrix* showed a close relationship and similar intraspecific patters of variation, but were not identical. The degree of sexual dimorphism of the douc langur was markedly less than that seen in the snub-nosed monkeys.

(3) *Nasalis* exhibited a sexual dimorphism pattern that was quite different from the other odd-nosed colobines, though broadly similar to *Semnopithecus*. This may support its basal position in the other Asian colobine radiation.

(4) The differences between the Southeast Asian species regarded as *Presbytis* and *Semnopithecus* by Brandon-Jones (1996) or *Presbytis*, *Semnopithecus*, and *Trachypithecus* by Groves (1989, 1993) are considerable: they are similar in size, but they show some differentiation in sexual dimorphism. Those in *Presbytis* are smaller.

(5) This study may support the separation of the gray langur as a monotypic genus (*Semnopithecus*) as proposed by Groves (1989, 1993), Oates *et al.* (1994) and Rowe (1996). Differences among species referred to *Trachypithecus* are very considerable, and it remains possible that some are more similar to *Semnopithecus* than others.

(6) This study highlights some issues relevant to the variation among Asian colobine species. Further studies with more species and specimens of leaf monkey and including the pig-tailed langur (*Simias concolor*) are, however, necessary to confirm and extend the results.

Acknowledgments

We are especially grateful to Dr. N. G. Jablonski who provided some data collected from the American Museum of Natural History, New York, the Natural History Museum, London, the Field Museum of Natural History, Chicago, the Museum National d'Histoire Naturelle in Paris, the National Museum of Natural History, Washington, DC, and the Royal Ontario Museum, Toronto. Our thanks also to the Shanghai Natural History Museum, the Nanchong Teachers College (Sichuan), the East China Normal University (Wuhan), and the Institute of Zoology (Beijing) for their help in data collection. We also sincerely thank Charles Oxnard for his enthusiastic support of this project. This work was supported by the Australian Research Council (Large Grant and Research Fellowship).

References

Albrecht, G. H. and Miller, J. M. A. (1993). Geographic variation in primates: a review with implication for interpreting fossils. In: *Species, Species Concepts, and Primate Evolution*, ed. W. H. Kimbel and L. B. Martin. New York, NY: Plenum Press. pp. 123–161.

Brandon-Jones, D. (1984). *Colobus* and leaf monkeys. In: *Encyclopaedia of Mammals*, ed. I. D. Macdonald. London: George Allen and Unwin. pp. 398–408.

Brandon-Jones, D. (1996). The Asian colobinae (Mammalia: Cercopithecidae) as indicators of Quaternary climate change. *Biol. J. Linn. Soc.*, **59**, 327–350.

Buettner-Janusch, J. (ed.) (1963). *Evolutionary and Genetic Biology of Primates*. New York, NY: Academic Press.

Cheverud, J. M., Malcolm, M. D., and Leutenegger, W. (1985). The quantitative assessment of phylogenetic constraints in comparative analysis: sexual dimorphism in body weight among primates. *Evol.* **38**, 1335–1351.

Colbert E. H. and Hooijer, D. A., (1953). Pleistocene mammals from the limestone fissures of Szechwan, China. *Bul. Amer. Mus. Nat. Hist.*, **102**, 1–134.

Corruccini, R. S. (1972). Allometry correlation in taximetrics. *Syst. Zool.*, **21**, 375–383.

de Winter, W. (1997). *Perspectives on Mammalian Brain Evolution: Theoretical and Morphometrical Aspects of a Controversial Issue in Current Evolutionary Thought.* Ph.D. thesis, University of Western Australia.

Delson, E. (1994). Evolutionary history of the colobine monkeys in paleoenvironmental perspective. In: *Colobine Monkeys: Their Ecology, Behavior and Evolution*, ed. A. G. Davies and J. F. Oates. Cambridge: Cambridge University Press. pp. 11–43.

Eaglen, R. H. (1984). Incisor size and diet revisited: the view from a platyrrhine perspective. *Am. J. Phys. Anthropol.*, **64**, 263–275.

Ellerman, J. R. and Morrison-Scott, T. C. S. (1951). *Checklist of Palaearctic and Indian Mammals.* London: British Museum (Natural History).

Fooden, J. and Feiler, A. (1988). *Pygathrix nemaeus* in Hainan? New evidence, no resolution. *Int. J. Primatol.*, **9**, 275–279.

Groves, C. P. (1970). The forgotten leaf-eaters, and the phylogeny of the colobinae. In: *Old World Monkeys: Evolution, Systematics and Behavior*, ed. J. R. Napier and P. H. Napier. New York: Academic Press. pp. 555–587.

(1989). *A Theory of Human and Primate Evolution*. Oxford: Clarendon Press.

(1993). Order primates. In *Mammal Species of the World: a Taxonomic and Geographic Reference*, ed. D. E. Wilson and D. M. Reeder, pp. 243–277, Washington and London: Smithsonian Institute Press.

(2001). *Primate Taxonomy*. Washington, DC: Smithsonian Institution Press.

Grzimek, B. (ed.) (1975). *Grzimek's Animal Life Encyclopedia, vol. 10, Mammals, I.* New York, NY: Van Nostrand Reinhold.

Harvey, P. and Bennett, P. (1985). Sexual dimorphism and reproductive strategies. In: *Human Sexual Dimorphism*, ed. J. Ghesquiere, R. Martin, and F. Newcombe. London: Taylor and Francis. pp. 43–59.

Hill, W. C. O. (ed.) (1972). *Evolution Biology of the Primates*. London: Academic Press.

Honacki, J. H., Kinman, K. E., and Koeppl, J. W. (ed.) (1982). *Mammal Species of the World, a Taxonomic and Geographic Reference*. Lawrence, Kan: Allen Press and the Association of Systematics Collections.

Hooijer, D. A. (1962). Quaternary langurs and macaques from the Malay archipelago. *Zool. Verhandelingen*, **55**, 1–64.

Jablonski, N. G. (1993). Quaternary environments and the evolution of primates in East Asia, with notes on two new specimens of fossil Cercopithecidae from China. *Folia Primatol.*, **60**, 118–132.

(1995). The phyletic position and systematic of the douc langurs of Southeast Asia. *Amer. J. Primatol.*, **35**, 185–205.

(1996). A diverse anthropoid fauna of probable Late Pleistocene age from Luoding, Guangdong, P. R. China. *Amer. J. Phys. Anthropol.*, **22**, (suppl.), 130.

(1998). The evolution of the doucs and snub-nosed monkeys and the question of the phyletic unity of the odd-nosed colobines. In: *The Natural History of Doucs and Snub-Nosed Monkeys*, ed. N. G. Jablonski. Singapore: World Scientific. pp. 13–52.

Jablonski, N. G. and Gu, Y. M. (1991). A reassessment of *Megamacaca Lantianensis*, a large monkey from the Pleistocene of north-central China. *J. Hum. Evol.*, **20**, 51–66.

Jablonski, N. G. and Pan, R. L. (1995). Sexual dimorphism in the snub-nosed langurs (Colobinae: *Rhinopithecus*). *Amer. J. Phys. Anthropol.*, **96**, 251–272.

Jablonski, N. G. and Pan, Y. R. (1988). The evolution and palaebiogeography of monkeys in China. In: *The Paleoenvironment of East Asia from the Mid-Tertiary*, ed. J. S. Aigner, N. G. Jablonski, G. Taylor, D. Walker, and W. Pinxian. Hong Kong: University of Hong Kong. pp. 849–867.

Jablonski, N. G. and Peng, Y. Z. (1993). The phylogenetic relationships and classification of the doucs and snub-nosed langurs of China and Vietnam. *Folia Primatol.* **60**, 36–55.

Jungers, W. L. (1984). Aspects of size and scaling in primate biology with special reference to the locomotor skeleton. *Yrbk. Phys. Anthropol.*, **27**, 73–97.

Kahlke, H. D. (1973). A review of the Pleistocene history of the Orang-Utan (*Pongo* Lacépède, 1799). *Asian Perspectives*, **15**, 5–14.

Kay, R. F. (1975). Allometry and early hominids. *Science*, **189**, 63.

Kuhn, H. J. (1967). Zur systematik der Cercopithecidae. In: *Neue Ergebnisse der Primatologie*, ed. D. Starck, R. Schneider, and H. J. Kuhn. Stuttgart: Gustav Fischer. pp. 25–46.

Leutenegger, W. and Kelly, J. T. (1977). Relationship of sexual dimorphism in canine size and body size to social behaviour, and ecological correlates in anthropoid primates. *Primates*, **18**, 117–136.

Napier, J. R. and Napier, P. H. (eds.) (1967). *A Handbook of Living Primates*. London: Academic Press.

(eds.) (1985). *The Natural History of the Primates*. London: British Museum.

Oates, J. F., Davies, A. G., and Delson E. (1994). The diversity of living colobines. In: *Colobine Monkeys: Their Ecology, Behavior, and Evolution*, ed. A. G. Davies and J. F. Oates. Cambridge: Cambridge University Press. pp. 45–73.

Oxnard, C. E. (1983a). Sexual dimorphism in the overall proportions of primates. *Amer. J. Primatol.*, **4**, 1–22.

(1983b). *The Order of Man: A Biomathematical Anatomy of the Primates*. Hong Kong: University of Hong Kong Press.

(ed.) (1987). *Fossils, Teeth, and Sex: New Perspectives on Human Evolution*. Seattle, WA: University of Washington Press.

Pan, R. L. (1998). *A Craniofacial Study of the Genus* Macaca, *with Special Reference to the Stump-Tailed Macaques*, M. arctoides *and* M. thibetana*: a Functional Approach*. Ph.D. thesis, University of Western Australia.

Pan, R. L. and Jablonski, N. G. (1993). Scaling of limb proportions and limb bone diameters in three species of Chinese snub-nosed langurs (Genus *Rhinopithecus*). *Folia Primatol.*, **60**, 56–62.

Pan, R. L. and Oxnard, C. E. (1999). A strategy for the morphometric analysis of the skull: implication for macaques. *Amer. J. Phys. Anthropol.* **28**, (suppl.), 217.

(2000). Craniodental variation of macaques (*Macaca*): size, function and phylogeny. *Zool. Res.*, **21**, 14–28.

(2001a). Radiation and evolution of three macaque species, *Macaca fascicularis*, *M. radiata* and *M. sinica*, as related to geographic changes in the Pleistocene of Southeast Asia. In: *Faunal and Floral Migrations and Evolution in SE Asia–Australalasia*, ed. I. Metcalfe, J. M. B. Smith, M. Morwood, and K. Hewison. Lisse: Swets and Zeitlinger. pp. 337–355.

(2001b). Cranial morphology of the golden monkey (*Rhinopithecus*) and douc langur (*Pygathrix nemaeus*). *J. Hum. Evol.*, **16**, 199–223.

Pan, R. L., Peng, Y. Z., Ye, Z. Z., and Wang, H. (1993a). The relationship between cranial and tooth size in the golden monkey (*Rhinopithecus*). *Prim. Report*, **35**, 49–62.

Pan, R. L., Peng, Y. Z., and Ye, Z. Z. (1993b). Sexual dimorphism of the shoulder girdle and upper limb in golden monkey (*Rhinopithecus*). *Prim. Report*, **35**, 31–47.

Pan Y. R. and Jablonski, N. G. (1987). The age and geographical distribution of fossil cercopithecids in China. *Hum. Evol.*, **2**, 59–69.

Peng, Y. Z. and Pan, R. L. (1994). Systematic classification of Asian colobines. *Hum. Evol.*, **9**, 25–33.

Peng, Y. Z., Ye, Z. Z., Zhang, Y. P., and Pan, R. L. (1988). The classification and phylogeny of snub-nosed monkey (*Rhinopithecus* spp.) based on gross morphological characters. *Zool. Res.*, **9**, 39–248.

Pickford, M. (1986). Sexual differences in higher primates: a summary statement. In: *Sexual Dimorphism in Living and Fossil Primates*, ed. M. Pickford and B. Chiarelli. Firenze: Sedicesimo. pp. 191–199.

Rowe. N. (ed.) (1996). *The Pictorial Guide to the Living Primates*. New York, NY: Pogonias Press.

Simpson, G. G. (1945). The principles of classification and a classification of mammals. *Bull. Amer. Mus. Nat. Hist.*, **85**, 1–350.
Strasser, E. (1988). Pedal evidence for the origin and diversification of cercopithecid clades. *J. Hum. Evol.*, **17**, 25–245.
Strasser, E. and Delson, E. (1987). Cladistic analysis of cercopithecid relationship. *J. Hum. Evol.*, **16**, 81–99.
Szalay, F. S. and Delson, E. (eds.) (1979). *Evolutionary History of the Primates*. New York, NY: Academic Pres.
Walker, E. P. (ed.) (1975). *Mammals of the World. Vol 2*. Baltimore, MD: Johns Hopkins University Press.
Wang, W., Su. B., Lan, H., *et al.* (1995). rDNA differences, and phylogenetic relationship among two species of golden monkeys and three species of leaf monkeys. In: *Primate Research and Conservation*, ed. W. P. Xia and Y. P. Zhang. Beijing: China Forestry Publishing House. pp. 77–80.
Wood, B. A. (1979). Analysis of tooth and body size relationship in five primate taxa. *Folia Primatol.*, **31**, 187–211.
Ye, Z. Z., Peng, Y. Z., Zhang, Y. P., and Liu, R. L. (1985). Some morphological characteristics of *Rhinopithecus*. *Acta Anthropol. Sinica*, **4**, 346–351.
(eds.) (1987). *The Anatomy of the Golden Monkey* (Rhinopithecus). Kunming, China: Yunnan Science and Technology Press.
Zhang, Y. P. and Ryder, O. A. (1998). Mitochondrial cytochrome b gene sequences of Old World monkeys: with special reference on evolution of Asian colobines. *Primates* 3939–49.

5 *Craniometric variation in early* Homo *compared to modern gorillas: a population-thinking approach*

JOSEPH M. A. MILLER
University of California, Los Angeles

GENE H. ALBRECHT
University of Southern California

BRUCE R. GELVIN
California State University

Introduction

Controversy has surrounded *Homo habilis* since its inception. Leakey *et al.* (1964) first described the taxon based on fossils discovered at Olduvai Gorge in Tanzania, East Africa (e.g., OH 7, OH 13, and OH 16). The first major debate concerned whether the fossils were representative of previously known taxa such as *Australopithecus africanus* and *Homo erectus* (e.g., Robinson, 1966; Brace *et al.*, 1973). With the discovery of additional specimens in the 1970s from Tanzania (e.g., OH 24), Kenya (e.g., ER 1470 and ER 1813), and South Africa (e.g., Stw 53), *H. habilis* gained general acceptance as a valid taxon (Tobias, 1991).

In the 1980s and 1990s, however, a new controversy arose. Some workers concluded that the craniometric variation among early *Homo* crania is too great in degree or too different in pattern for a single species (Wood, 1985; Stringer, 1986; Lieberman *et al.*, 1988; Wood, 1991, 1993; Kramer *et al.*, 1995; Grine *et al.*, 1996). Others (Tobias, 1991; Miller, 1991, 2000) concluded that the available data provide no basis for rejecting the single-species hypothesis.

The present study uses a population-thinking approach to compare craniometric variation among the most complete early *Homo* crania (KNM-ER 1470, KNM-ER 1813, OH 24, and Stw 53) to intraspecific variation in gorillas (see Fig. 5.1 for geographic distribution of fossils and gorillas). Our approach, first

Shaping Primate Evolution, ed. F. Anapol, R. Z. German, and N. G. Jablonski. Published by Cambridge University Press.

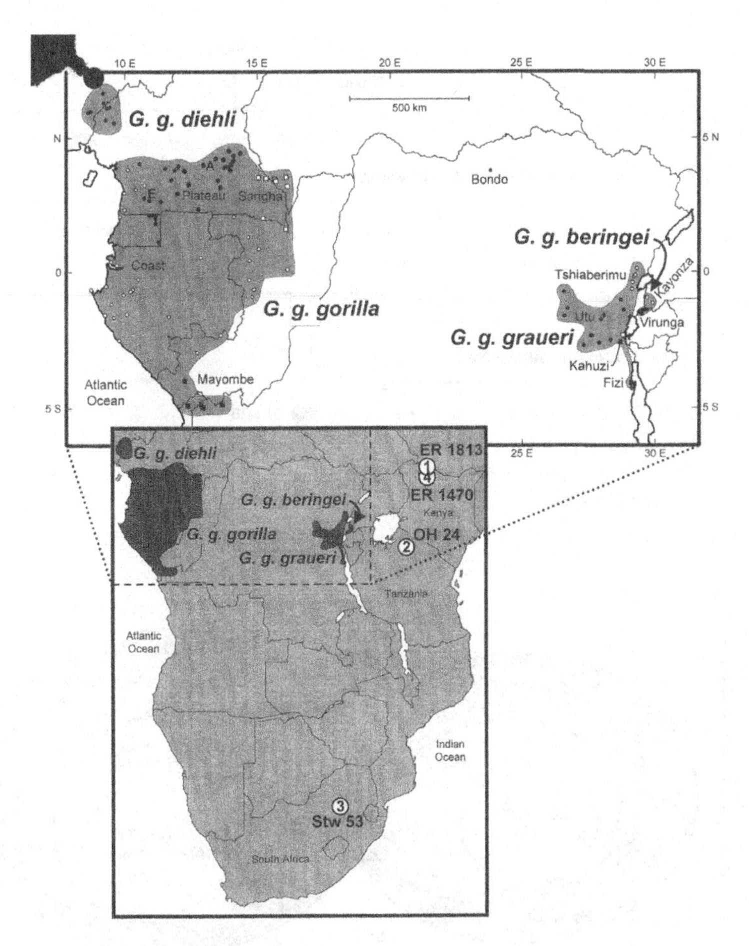

Figure 5.1. Central and southern Africa, showing geographic distributions of the early *Homo* fossil specimens and the localities, demes, and subspecies of *Gorilla gorilla*. Numbers for the fossil specimens are the same as those used in Figures 5.4–5.9. Numbers of localities and sample sizes for each deme and subspecies of gorilla are given in Table 5.3. Localities with known latitude and longitude are mapped as either symbols or single capital letters:
G. g. gorilla demes: Bondo (separated from main subspecies range), Coast (open circles), Mayombe (solid squares), Plateau (solid circles; A = Abong-Mbang, E = Ebolowa), and Sangha (open squares)
G. g. diehli (solid circles)
G. g. graueri demes: Fizi (solid square), Kahuzi (open square), Tshiaberimu (open circles), and Utu (solid circles)
G. g. beringei demes: Kayonza (open circles) and Virunga (solid circles).

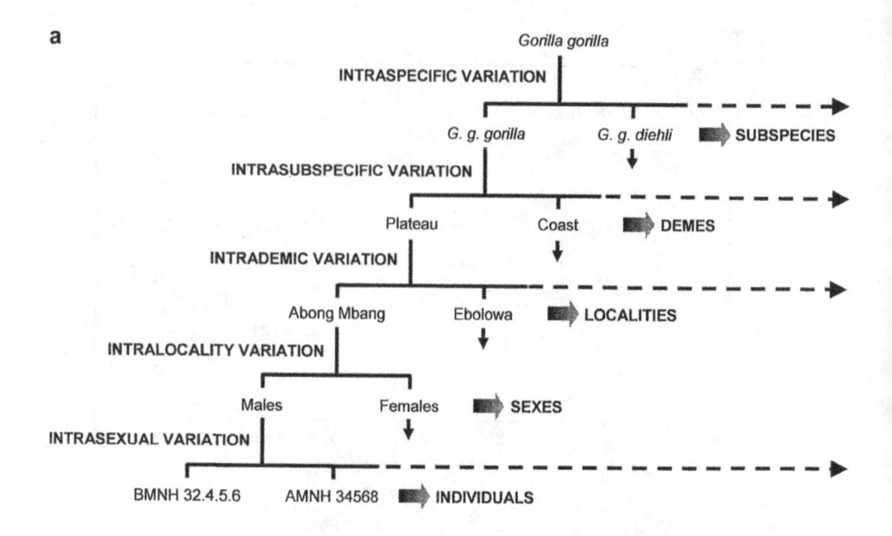

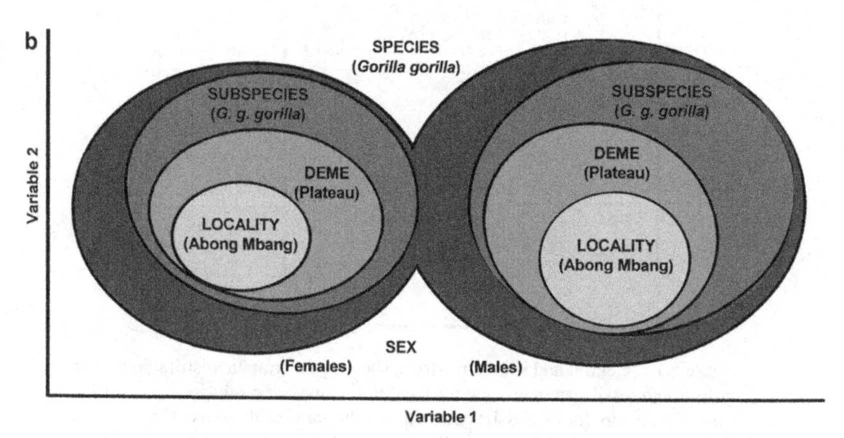

Figure 5.2. (a) The hierarchy of morphological variation in a polytypic species (adapted from Figure 1 in Albrecht and Miller, 1993, and Figure 14.1 in Albrecht *et al.*, 2003). (b) Nested hierarchy of intraspecific variation in a sexually dimorphic, polytypic species portrayed by shaded ellipses representing the hypothetical distributions of specimens in a bivariate morphometric data space.

introduced in Albrecht *et al.* (2003), differs from previous, traditional studies in two fundamental ways. First, population thinking provides characterizations of intraspecific variation that reflect the population taxonomy and population structure of a species. The result is a composite, comprehensive picture of the analog species, including subspecific, demic, locality, sexual, and individual levels of variation (Fig. 5.2). Second, a population-thinking approach allows variation

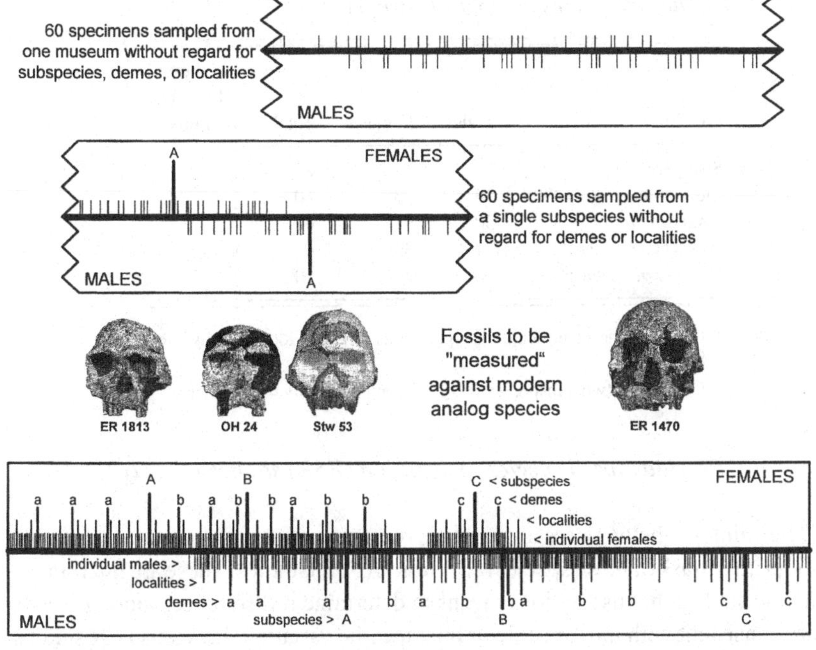

Figure 5.3. Yardsticks of intraspecific morphological variation for a hypothetical analog species used to assess species number in a sample of four early *Homo* crania (adapted from Figure 4.10 in Albrecht *et al.*, 2003). Measurement values of individual males and females for a hypothetical cranial variable are shown as the shortest thin lines on either side of the ruler's central axis. Successively longer and thicker lines indicate mean values for localities, demes (lowercase letters), and subspecies (uppercase letters). The upper two yardsticks are typical of the traditional approach of using a poorly calibrated yardstick based on only 30 specimens of each sex. The poorly calibrated yardsticks illustrate two common strategies for assembling analog samples: (1) measure an assortment of readily available specimens at some convenient museum without attention to locality, deme, or subspecies attributions; (2) invoke some level of control by limiting the comparative sample to a single subspecies. The lower yardstick is typical of the population-thinking approach of using a well calibrated yardstick based on 500 specimens that represent a reasonable sampling of the localities and demes of the three subspecies of the analog species.

among fossils to be compared to and measured against well-characterized infraspecific levels of intraspecific variation in the modern analog species (Fig. 5.3). Applying population thinking to fossil species recognition studies provides a more comprehensive and biologically meaningful interpretation of fossil variation.

Table 5.1. *Samples used to characterize intraspecific variation in modern gorillas for five influential FSR studies of species number in early* Homo

FSR Study	Males	Females	Total	Number of subspecies identified[a]
Stringer (1986)	15	14	29	?
Lieberman *et al.* (1988)	20	20	40	?
Kramer *et al.* (1995)	30	31	61	?
Wood (1991, 1993)	34	30	64	1
Grine *et al.* (1996)[b]	26	24	50	2

[a] Question marks indicate that specimens were not identified by subspecies.
[b] One of the two subspecies was represented by only 4 specimens.

Previous fossil species recognition (FSR) studies of early Homo

Determining whether a fossil sample is representative of one or more species is the primary task of fossil species recognition (FSR) studies. These are microtaxonomic studies that use various means to determine if differences among fossils are consistent with intraspecific or interspecific variation. Previous FSR studies of early *Homo* focused on the application of different methods of data analysis. These methods included probability approaches (Lieberman *et al.*, 1988), canonical variates analysis (Wood, 1991, 1993), variability profiles (Kramer *et al.*, 1995), exact randomization (Kramer *et al.*, 1995; Grine *et al.*, 1996), and principal components analysis (Miller, 2000). These FSR studies are a testimony to the legacy of Professor Oxnard (1973, 1975, 1983, 1987, Oxnard *et al.*, 1990), who has long championed the power and promise of sophisticated morphometric techniques in physical anthropology.

The quantitative sophistication of FSR studies is laudable but has obscured fundamental deficiencies in applying the evolutionary principles of population thinking to understanding and characterizing intraspecific variation in modern analog species. The core problem with traditional FSR studies is reliance on relatively small, uncontrolled samples to characterize the entirety of variation across wide-ranging, polytypic analog species. This practice is evident in five influential studies listed in Table 5.1 that rejected the single-species hypothesis for *H. habilis*. The sample sizes for these studies ranged from 29 to 64 gorilla specimens. Three studies failed to specify how many subspecies the gorilla sample represented, one limited analysis to a single subspecies, and another used two subspecies of gorillas (this last study had only four specimens for the second subspecies). None included all four gorilla subspecies (see Fig. 5.1), none took into account the known range of variation in gorilla skull morphology (see photographs of gorilla skulls in Coolidge, 1929, and the multivariate craniometric results of Groves, 1967, 1970), and none made any attempt to understand intraspecific population structure of gorillas as an analog species (see Albrecht *et al.*, 2003).

Population thinking and yardsticks of morphological variation

The use of small, uncontrolled samples to characterize intraspecific variation in modern analog species is symptomatic of the general lack of population thinking in FSR studies. Population thinking is a primary Darwinian concept expounded upon by the architects of the modern synthesis (e.g., Mayr, 1976, pp. 26–29). Albrecht *et al.* (2003, p. 62) argued that "the concept of 'population thinking' should have a central role in fossil species recognition studies." We define population thinking as "the theory and practice of population systematics in which biological species are thought of as aggregates of interbreeding natural populations comprising individuals that vary genetically and phenetically" (p. 62).

The biological organization of the natural populations of a polytypic species is reflected in its population taxonomy and population structure. Population *taxonomy* refers to "the study of the biological organization of species that recognizes species as geographically variable aggregates of populations" (Albrecht *et al.*, 2003, p. 65). Polytypic species are naturally divided up into subspecies, which are aggregates of local populations (demes), which are aggregates of individuals at different localities. Population *structure* refers to "the geographic arrangement of local populations across the species' range" (p. 65). Population structure can be described in terms of the population continuum (where there is continuous contact among local populations), geographic isolates, and zones of secondary intergradation (hybrid zones) (Mayr and Ashlock, 1991). As we emphasized earlier (Albrecht *et al.*, 2003), population thinking provides the appropriate theoretical basis for understanding the nature of morphological variation comprising intraspecific variation in a polytypic species. Albrecht and Miller's (1993) hierarchy of morphological variation is a useful paradigm for understanding the biological organization of a species that is reflective of its population taxonomy and population structure (Fig. 5.2a).

Population thinking provides the theoretical basis that has been missing from FSR studies. The prerequisite of applying a population-thinking approach to FSR studies is an understanding of the population taxonomy and population structure of each analog species. Albrecht *et al.* (2003) provide the requisite information about the degrees and patterns of intraspecific variation for gorillas. We characterized the hierarchy of morphological variation (Fig. 5.2a) using 657 gorillas from 196 collecting localities representing 12 demes of all four subspecies. As conceptually summarized in Figure 5.2b, overall intraspecific gorilla variation can be partitioned into nested distributions at various hierarchical infraspecific levels that reflect the population taxonomy and population structure of gorillas as a species.

Our earlier analysis of intraspecific variation in gorillas makes clear the artifact and fallacy of the traditional approach to FSR studies. The small, uncontrolled samples of traditional FSR studies "cannot possibly represent the complexities of variation among the sexes, localities, demes, and subspecies that comprise the totality of intraspecific variation in a sexually dimorphic, geographically variable, polytypic species" (Albrecht et al., 2003, p. 92). This means that the "yardsticks" that have been traditionally promoted as reliable estimates of intraspecific variation in an analog species are woefully inadequate for assessing the significance of variation among fossils. Since the population-thinking approach rests on a sound theoretical basis and is comprehensive in its characterization of intraspecific variation, it results in the well-calibrated yardstick depicted in Figure 5.3. The traditional approach of characterizing intraspecific variation using small, uncontrolled samples yields the poorly calibrated yardsticks of Figure 5.3 in which the limits of variation in the analog species are underestimated or unknown (i.e., how long is the yardstick?). Furthermore, poorly calibrated yardsticks provide little if any information about what the sample of the analog species actually represents in terms of the infraspecific levels of variation (i.e., what are the measurement units?). The well-calibrated yardstick not only reflects the limits of variation for the analog species more accurately, but also shows the degrees and patterns of variation at each hierarchical level of intraspecific species structure.

Fossil species recognition studies are suspect if not based on a strong foundation of knowledge about the population taxonomy and population structure of the analog species to which the fossils are being compared. Clearly, as depicted in Figure 5.3, incomplete knowledge about the degree and pattern of variation in the analog species results in poorly calibrated yardsticks for evaluating morphological differences among fossils. For example, since no one museum usually has a comprehensive sampling across a species' range, selecting gorillas from one or two convenient collections is insufficient to characterize overall species variation. Likewise, using only a single subspecies provides an equally deficient estimate of variation in a polytypic, geographically wide-ranging species like gorillas. The inevitable consequence of underestimating analog species variation is overestimating species diversity in the fossil sample. Reliance on poorly calibrated yardsticks probably explains why traditional FSR studies concluded that early *Homo* crania are too variable to represent a single species. In contrast, as we will show, a well-calibrated yardstick provides different results.

This paper represents the second of a two-part FSR study using a population-thinking approach to evaluate the question of species diversity in early *Homo* crania compared to modern gorillas. The first step was to understand the population taxonomy and population structure of gorillas using reasonably comprehensive, representative samples (Albrecht *et al.*, 2003). That work resulted in a well-calibrated yardstick reflecting the degree and pattern of craniometric differences among gorillas at the various hierarchical levels of species structure (individual, sexes, localities, demes, and subspecies). We now apply this well-calibrated gorilla yardstick to the problem of determining species number among the most complete early *Homo* skulls.

Gorillas as a hominid analog

Gorillas are a useful analog for interpreting morphological variation in fossil hominids because they are: (1) a polytypic species whose taxonomy is relatively well understood (Groves, 1967, 1970, 1986; papers in Taylor and Goldsmith, 2003); (2) genetically closely related to early hominids (Marks, 1993, 1994; Rogers, 1993; Deinard *et al.*, 1998; Deinard and Kidd, 1999); (3) sexually dimorphic as some early hominids are known to be (e.g., *Australopithecus afarensis* and robust australopithecines; McHenry, 1991; Leakey and Walker, 1988; Walker and Leakey, 1988; Lockwood *et al.*, 1996); (4) distributed across a wide geographic range as were early hominids (Fig. 5.1); (5) increasingly well-studied in terms of their morphology, ecology, behavior, and genetics (Taylor and Goldsmith, 2003); (6) known from large samples of sexed, wild-caught specimens representing numerous localities for different demes and subspecies (Groves, 1967, 1970); and (7) already well-characterized using a population-thinking approach to construct a well-calibrated yardstick of intraspecific craniometric variation (Albrecht *et al.*, 2003). Gorillas, however, are by no means the only appropriate hominid analog (see Miller, 2000, for additional discussion). We have reported preliminary results using modern humans (Miller *et al.*, 1998) and modern chimpanzees (Miller *et al.*, 2002). An examination of orangutans, though more distantly related to early hominids than are the African apes, would be instructive since they are also a large-bodied, sexually dimorphic, hominoid species. Other primate species also might provide additional perspectives.

Table 5.2. *Sixteen craniofacial measurements*

Greatest length of skull (prosthion to opisthocranium)
Cranial length (glabella to opisthocranium)
Skull breadth (biporionic breadth)
Nuchal breadth (breadth of nuchal surface in mastoid region)
Biorbital breadth
Palate breadth (outside first molars)
Upper nasal breadth (greatest breadth of nasal bones)
Nasal height (top of piriform aperture to prosthion)
Bicanine breadth (across canine alveoli)
Facial height (top of supraorbital torus to prosthion)
Orbital height (inside)
Orbital breadth (inside one orbit)
Postorbital breadth (at postorbital constriction)
Lower nasal breadth (greatest breadth of piriform aperture)
Interorbital breadth
Supraorbital thickness (normal thickness of supraorbital torus)

Materials and methods

Measurements

We use a 16-measurement subset of the 30 craniofacial dimensions measured by Colin Groves (Table 5.2). The 16 variables characterize the overall dimensions of the cranium and are replicable on the fossil casts and gorilla skulls we measured. When possible, we checked our measurements taken on fossil casts against published dimensions of the actual fossils. We did not use Groves' mandibular data because no lower jaws are associated with the fossil crania.

Fossils

The fossils used in this study are the four most complete crania at the epicenter of the *H. habilis* controversy (Fig. 5.1): (1) KNM-ER 1470 and KNM-ER 1813 from Koobi-Fora in Kenya, (2) OH 24 from Olduvai Gorge in Tanzania, and (3) Stw 53 from the Sterkfontein Caves in South Africa. The fossil localities stretch about 3700 km along the eastern side of the continent from Kenya to South Africa and date between 1.6 and 1.9 million years ago (Fiebel *et al.*, 1989; Grine *et al.*, 1996; Tobias, 1991). In contrast, gorillas range about 2400 km discontinuously across equatorial Africa (Fig. 5.1) with museum specimens collected over the last 150 years. Thus, the fossils exceed museum collections of gorillas in geographic range and time depth.

Gorillas

We recognize four subspecies of gorilla (Fig. 5.1; Albrecht *et al.*, 2003): *Gorilla gorilla gorilla* (the western lowland gorilla), *G. g. diehli* (the Cross River gorilla), *G. g. graueri* (the eastern lowland gorilla), and *G. g. beringei* (the mountain gorilla). The two males from the enigmatic Bondo locality are treated as a deme of *G. g. gorilla*. We do not believe that the current morphological and genetic evidence warrants differentiation of the western and eastern gorillas as separate species, although they may represent semispecies that have acquired some but not all of the characteristics of species rank (see Tuttle, 2003; Groves, 2003).

Albrecht *et al.* (2003) describe fully the gorilla samples consisting primarily of 605 adult gorillas from 33 museums measured by Groves (1967, 1970). We obtained Groves' data from the archives of the Natural History Museum, London. We measured an additional 52 adult gorilla skulls at the US National Museum. The total sample of 657 individuals comes from 196 collecting localities divisible into 12 demes of 4 gorilla subspecies (Table 5.3).

All 657 gorillas are wild-caught specimens and all but 22 are from known collecting localities (Albrecht *et al.*, 2003; Fig. 5.1 and Table 5.3). The localities are subdivided into the same geographic groups recognized by Groves (1967, 1970), who based his partitioning of the gorilla subspecies on geographic and ecological criteria thought to be relevant for gorilla systematics. We call these geographic subgroups "demes" in an attempt to represent an intermediate level of species structure between the collecting locality and the subspecies that approximates the deme of population biology (i.e., individuals in a given area that form a single interbreeding community).

Large sample sizes make *G. g. gorilla* the primary focus of our FSR study below the subspecies level (Table 5.3). The widespread Coast deme has 36 females and 84 males and the more geographically restricted Plateau deme has 112 females and 135 males. The two localities with the largest samples are Abong Mbang (28 females, 37 males) and Ebolowa (15 females, 24 males) in the Plateau deme. Each of these localities has about the same number of specimens as used in traditional FSR studies to represent the entire diversity of all gorillas.

Methods

Albrecht *et al.* (2003) discuss fully the methods for investigating the degree and pattern of intraspecific variation in gorillas. We used covariance-based principle components analysis (PCA) as a simple, straightforward, well-known

Table 5.3. *Numbers of gorilla localities and gorilla specimens by sex, deme, and subspecies*

Subspecies	Deme[a]	Localities	Females	Males	Total
G. g. gorilla	Bondo	1	0	2	2
	Coast	53	36	84	120
	Mayombe	5	4	14	18
	Plateau	52	112	135	247
	Sangha	22	15	46	61
	unassigned[b]	4	3	20	23
	Subtotal	137	170	301	471
G. g. diehli	Cross River	10	25	26	51
G. g. graueri	Fizi	4	10	12	22
	Kahuzi	3	1	2	3
	Tshiaberimu	6	15	12	27
	Utu	14	10	21	31
	unassigned[b]	0	1	1	2
	Subtotal	27	37	48	85
G. g. beringei	Kayonza	2	1	2	3
	Virunga	15	16	24	40
	unassigned[b]	0	1	0	1
	Subtotal	17	18	26	44
G. g. graueri / beringei	unassigned	5	1	5	6
	Grand Total	196	251	406	657

[a] See Figure 5.1 for mapping of demes.
[b] Specimens without precise locality information or whose locality cannot be attributed to a particular deme, but which can be reasonably assigned to a subspecies based on the available information about geographic provenance.

ordination technique to display graphically the metric relationships among the gorilla specimens. The PCA using 16 craniofacial dimensions for all 657 gorilla specimens comes from Albrecht *et al.* (2003). This one PCA is the basis for the different graphs of PC1 versus PC2 shown in Figures 5.4–5.9, which plot the four fossils against a background of gorilla variation at different infraspecific hierarchical levels. We proceed by making sequential comparisons of the fossils to overall species-level variation (Figs. 5.4–5.5), subspecies-level variation (Figs. 5.6–5.7), demic-level variation (Fig. 5.8), and locality-level variation (Fig. 5.9). At each level, we examine both sexual dimorphism (i.e., how do the fossils compare to variation between female and male gorillas?) and intrasexual variation (i.e., how do the fossils compare to variation within each sex of gorillas?).

The fossils are interpolated into the PC plots for gorillas by using the PC eigenvectors to calculate each fossil's PC scores on the PC axes. Given the

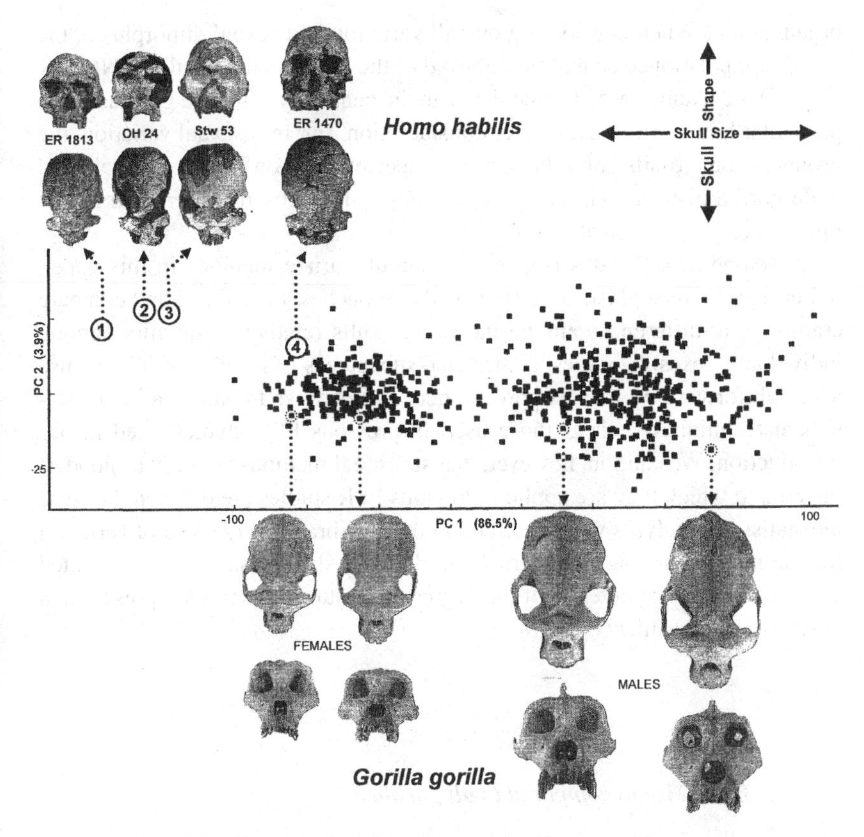

Figure 5.4. Craniometric variation in a sample of four early *Homo* crania compared to overall craniometric variation in *Gorilla gorilla* at the species level. The plot shows the first two principal-component axes, which account for 90.4% of the variation from the covariance-based PCA of 16 dimensions for 657 adult gorillas. Gorillas are plotted as closed squares and the fossil hominids are plotted as circled numbers. The first PC axis reflects increasing skull size from the smallest on the left to the largest on the right. The second and subsequent PC axes reflect size-independent shape variation. The dorsal and frontal views of the four gorilla and four fossil skulls correspond to the individual points identified in the plot.

small number of fossils compared to the large number of gorillas, it makes little difference if the fossils are included directly or interpolated into the PCA. We first show the fossils in their actual interpolated positions on the first two PC axes (Fig. 5.4). In subsequent plots (Figs. 5.5–5.9), without changing the metric distances among the fossils themselves, we move the four fossils as a group to superimpose them over various subgroups of gorillas. This allows graphic comparison of the craniometric differences among the fossils to the degree and pattern of variation among gorillas at the various hierarchical levels of species organization. When considering overall variation and sexual dimorphism, the fossils are positioned so that the centroid of the three smaller skulls (KNM-ER 1813, OH 24, and Stw 53) coincides with the centroid for female gorillas at the particular hierarchical level under consideration. For intrasexual variation, the group of four fossils is duplicated and superimposed on both the female and male gorilla distributions such that the range of the fossils is centered on the range of each sex of gorillas for both PC axes.

We emphasize the descriptive use of multivariate methods in this paper, rather than more sophisticated statistical approaches, to ordinate and compare craniometric differences among the fossil skulls relative to variation among individuals, sexes, localities, demes, and subspecies of gorillas. While we use a straightforward graphical approach here, we endorse the application of statistical techniques such as those used in previous FSR studies cited in our introduction. We caution, however, that statistical methods are only as good as the data to which they are applied. Previous FSR studies are deficient because the statistical analyses were based on poorly calibrated yardsticks of variation for the modern analogs. Our graphical displays show what a well-calibrated yardstick looks like in terms of the degree and pattern of variation present in a polytypic species like *G. gorilla*.

Results

Early Homo *compared to all gorillas*

The PC plot of the four early *Homo* specimens and all 657 gorillas is shown in Figure 5.4. The first two PC variates account for 90.4% of the overall variation. The pictures of the four gorilla skulls mapped onto the gorilla distribution emphasize how metric variation shown in the plots represents actual morphological differences among the skulls themselves. The first PC axis reflects increasing size from small skulls on the left to large skulls on the right. The second and subsequent PC axes reflect non-size-related shape differences in gorilla crania. The only apparent structure is two clusters of gorillas separated by a less populated intermediate region.

Figure 5.4 also shows the result of interpolating the four *early* Homo specimens into the PCA. The fossils are offset to the left of the gorilla distribution along PC1 consistent with the smaller size of their skulls compared to gorillas. The obvious shape difference between the fossil and gorilla skulls is reflected in the fossils being displaced upward on PC2. Subsequent PC axes, which are shape variables not shown here, combine to create a distinct morphological separation of the fossils from the gorillas.

The focus of FSR studies, however, is to compare the within-group structure of fossils and analog species as opposed to confirming the obvious morphological differences between them. The three smaller fossils (KNM-ER 1813, OH 24, Stw 53), which are often thought of as females, cluster close together with considerable similarity in their skull morphology. These three fossils are far from the larger KNM-ER 1470 specimen, which is often thought of as a male or a separate species ("*Homo rudolfensis*"; e.g., Wood, 1993). However, the size differences between the smallest and largest gorillas greatly exceed the distance between the smallest (KNM-ER 1813) and largest (KNM-ER 1470) of the fossils. The greater metric differences among gorillas only confirms what is obvious from the skulls themselves as can be seen in the dorsal and frontal photographs of the fossils and gorillas in Figure 5.4.

Figure 5.5a compares the fossils to overall species-level variation and sexual dimorphism in gorillas. The graph is the same as Figure 5.4 except that the gorillas are identified by sex. The fossils, with their morphometric relationships intact, are superimposed on the gorilla distribution in order to compare them directly with the analog species. With regard to overall species level variation, the differences among the early *Homo* specimens are far exceeded by overall variation in gorillas. With regard to sexual dimorphism, the separation of KNM-ER 1470 from the three smaller fossils (KNM-ER 1813, OH 24, and Stw 53) mirrors the pattern of differences between female and male gorillas but is far less extreme in magnitude. Indeed, if the three smaller fossils are considered average gorilla females, then KNM-ER 1470 would be a very small gorilla male.

Figure 5.5b compares the fossils to intrasexual variation at the species level among gorillas. The plot of gorillas is identical to Figure 5.5a except a "discriminant function" (vertical line) separates female from male gorillas. The fossils are duplicated and superimposed on both the female and male clouds of gorilla specimens. The spread among the four fossils is just within the margins of variation for female gorillas (left side of Figure 5.5b) but is much less than the variation among male gorillas (right side of Figure 5.5b).

Early **Homo** *compared to* **Gorilla** *subspecies*

Figure 5.6 compares the fossils to overall variation and sexual dimorphism in each gorilla subspecies. In Figure 5.6a, early *Homo* is compared to the large sample of *G. g. gorilla* with results similar to those at the species level. Craniometric differences among the fossils are much less than overall variation in *G. g. gorilla*. The separation of KNM-ER 1470 from the three smaller fossils is consistent with the pattern of sexual dimorphism in *G. g. gorilla*, but the degree of

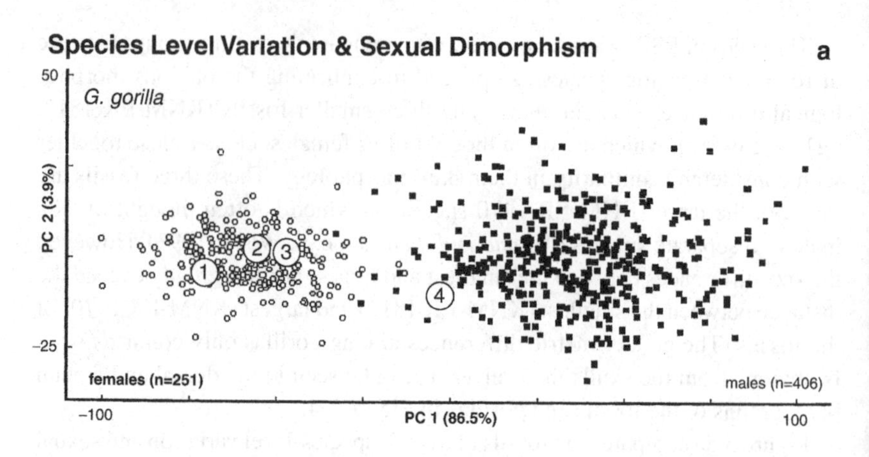

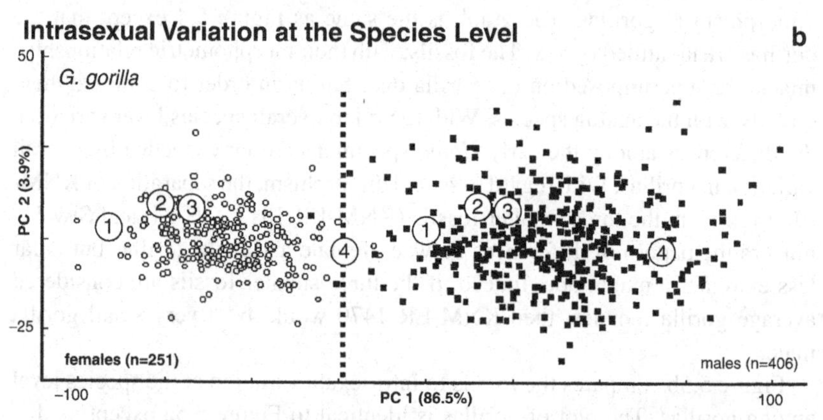

Figure 5.5. (a) Early *Homo* sample compared to species-level variation and sexual dimorphism in *Gorilla gorilla*. The gorilla points are the same as shown in Figure 5.4 except specimens have been identified by sex (females = circles, males = squares). The four fossils have been moved as a group from their position in Figure 5.4 and superimposed on the gorillas such that the centroid of the three smaller specimens (KNM-ER 1813, OH 24, and Stw 53) is centered on the centroid of the *G. gorilla* females. (b) Early *Homo* sample compared to intrasexual variation at the species level in *G. gorilla*. The plot is the same as Figure 5.5a except a discriminant function is added in the form of a vertical dashed line that best separates females from males. The group of four fossils has been duplicated and one set of fossils superimposed on *G. gorilla* females and the other on *G. gorilla* males such that the range of the fossils is centered on the range of each sex of gorillas on the two PC axes.

dimorphism is much less. Figure 5.6b–d shows similar results for the other three subspecies, whose sample sizes are considerably smaller. In each case, gorillas exceed the fossils in the overall degree of subspecies-level variation in skull morphology. The degree of sexual dimorphism suggested by the separation of the fossils into three smaller "females" (1–3) and a single larger "male" (4)

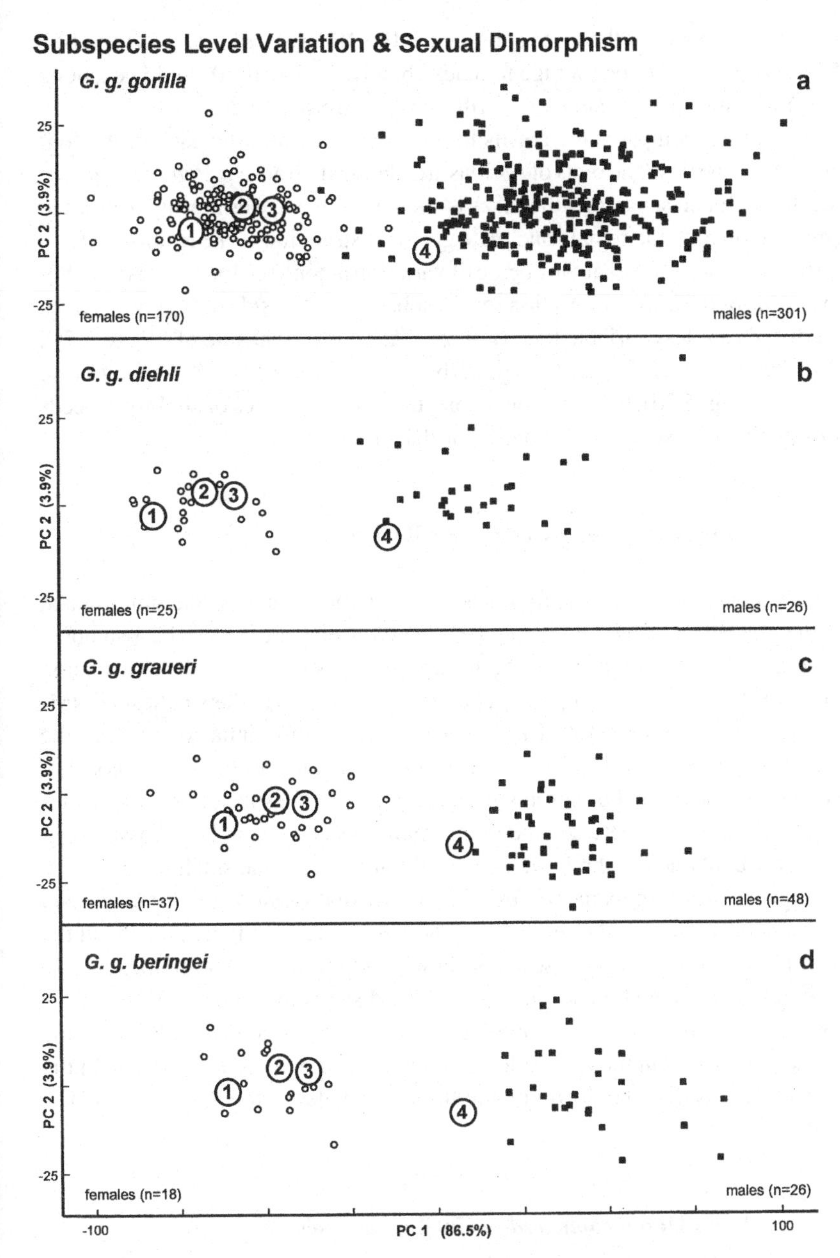

Figure 5.6. Early *Homo* sample compared to subspecies-level variation and sexual dimorphism in *Gorilla gorilla gorilla* (a), *G. g. diehli* (b), *G. g. graueri* (c), and *G. g. beringei* (d). Conventions are the same as in Figure 5.5a except the group of four fossils is superimposed on each subspecies rather than on the species as a whole.

is less than in any of the three subspecies. If KNM-ER 1813, OH 24, and Stw 53 are considered to be average females, then KNM-ER 1470 would be among the smallest of males using any gorilla subspecies as a model.

Figure 5.7 compares the fossils to intrasexual variation in each of the four gorilla subspecies. The plots of gorillas are identical to those in Figure 5.6 except the discriminant separating the sexes is shown and the fossils are duplicated and centered on both the female and male gorilla distributions. The results for *G. g. gorilla* females show that the degree of variation among the four fossils matches the limits of intrasexual variation for the subspecies (left side of Figure 5.7a) but the fossils are less variable than *G. g. gorilla* males (right side of Figure 5.7a). With respect to *G. g. diehli* (Fig. 5.7b), *G. g. graueri* (Fig. 5.7c), and *G. g. beringei* (Fig. 5.7d), the variation among the fossils matches or slightly exceeds subspecific intrasexual variation in gorillas of either sex.

Early **Homo** *compared to* **Gorilla** *demes*

Figure 5.8a–b compares the fossils to overall variation and sexual dimorphism in the Plateau and Coast demes of *G. g. gorilla*, which are the two best-sampled gorilla demes. Superimposing the group of fossils on the gorilla distribution makes it clear that the morphological difference between the smallest (KNM-ER 1813) and largest (KNM-ER 1470) fossils is substantially less than the broad scope of demic-level variation from the smallest to the largest gorillas. With respect to sexual dimorphism, as is the case at the species and subspecies levels, the disparity between the three small fossils and one large fossil could be easily explained as three average-sized females and one small male.

Figure 5.8c–d compares the fossils to intrasexual variation in the Plateau and Coast demes. The gorillas are in the same positions as in Figure 5.8a–b but the discriminant separating the sexes is shown and the fossils are duplicated and centered on both the female and male gorilla distributions. For both demes, the fossils are at the limits of variation among females as shown on the left side of the two plots. On the right side of Figure 5.8c–d, the differences among the fossils are less than the distributional limits for males of the Plateau and Coast demes.

Early **Homo** *compared to* **Gorilla** *localities*

Figure 5.9a–b compares the fossils to overall variation and sexual dimorphism at Abong Mbang and Ebolowa in the Plateau deme of *G. g. gorilla*, which are the two best-sampled gorilla collecting localities. The craniometric variation of

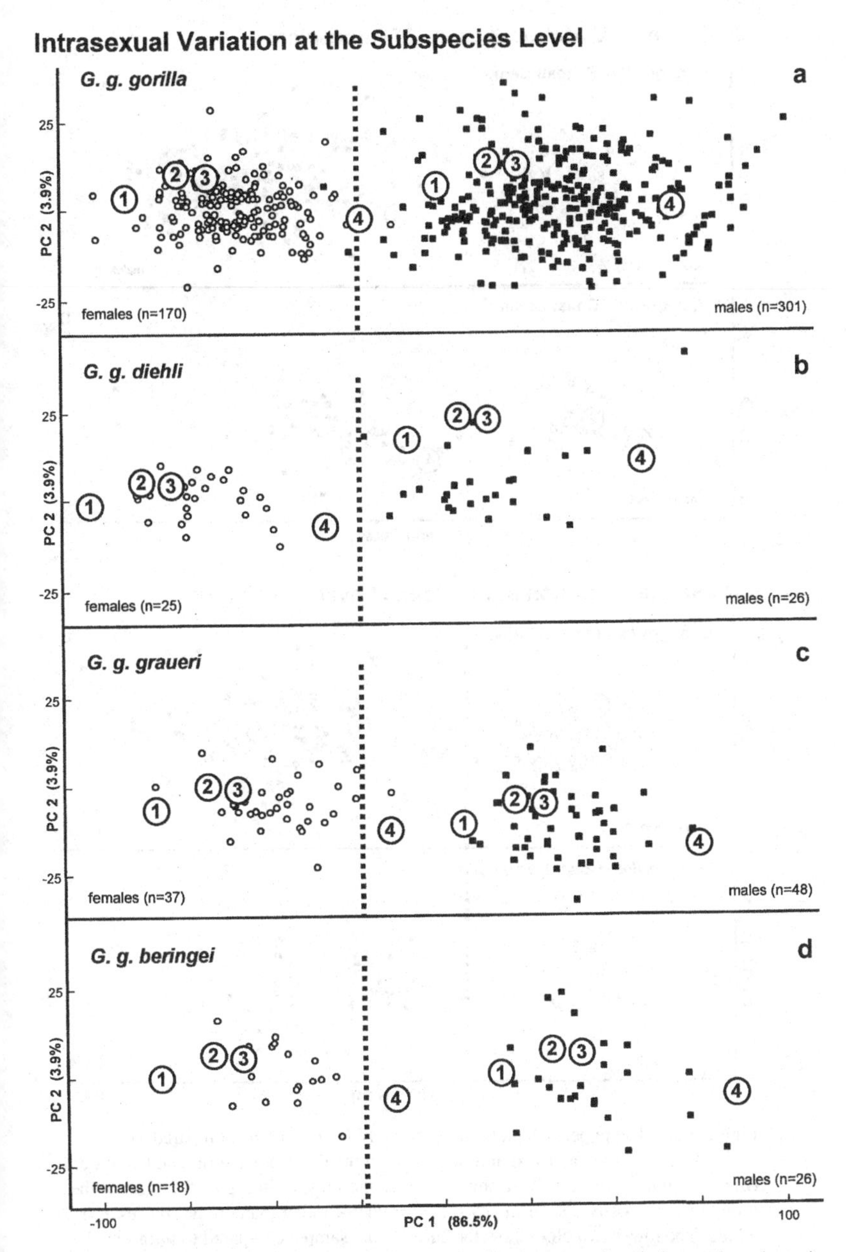

Figure 5.7. Early *Homo* sample compared to intrasexual variation at the subspecies level in *Gorilla gorilla gorilla* (a), *G. g. diehli* (b), *G. g. graueri* (c), and *G. g. beringei* (d). The plots are the same as Figure 5.6 except the sexual discriminant of Figure 5.5b is added and the group of fossils is centered on the range of each sex for each subspecies. Other conventions are the same as in Figure 5.5b.

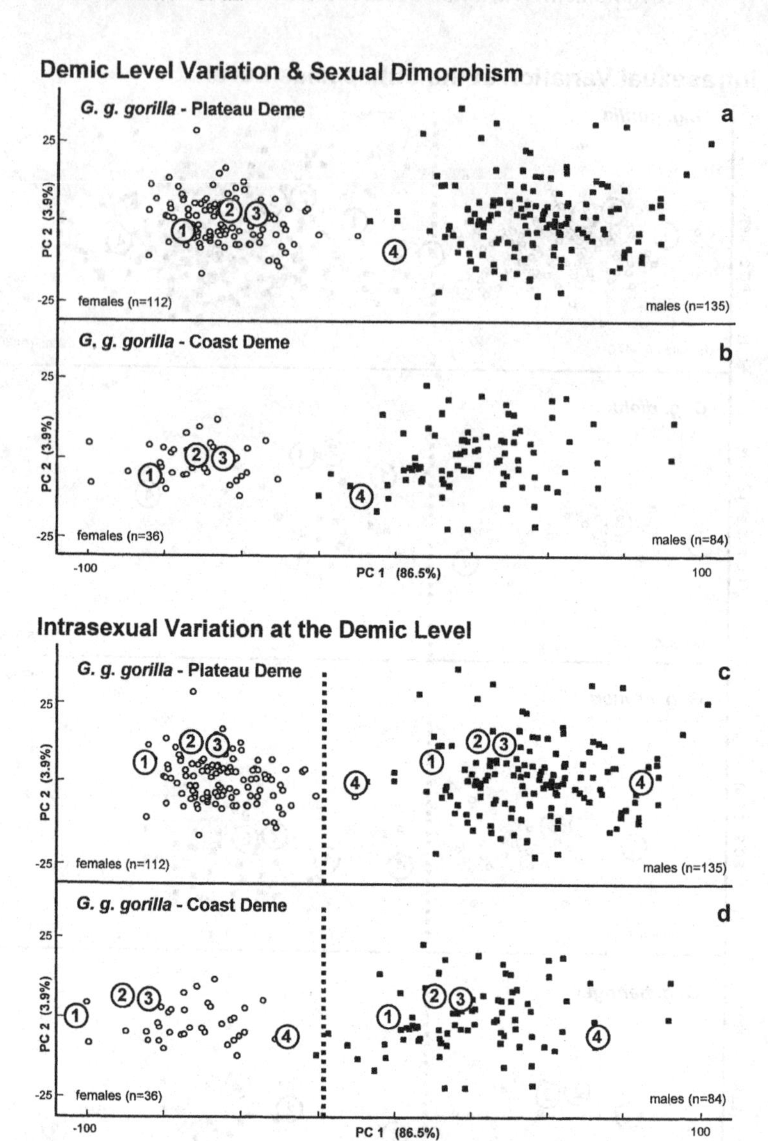

Figure 5.8. The upper two plots show the early *Homo* sample compared to demic-level variation and sexual dimorphism in the Plateau deme (a) and Coast deme (b) of *Gorilla gorilla gorilla*. Conventions are the same as in Figure 5.5a except the group of four fossils is superimposed on each deme rather than on the species as a whole. The lower two plots show the early *Homo* sample compared to intrasexual variation at the demic level in the Plateau deme (c) and Coast deme (d) of *G. g. gorilla*. The plots are the same as Figures 5.8a–b except the sexual discriminant of Figure 5.5b is added and the group of fossils is centered on the range of each sex for each deme. Other conventions are the same as in Figure 5.5b.

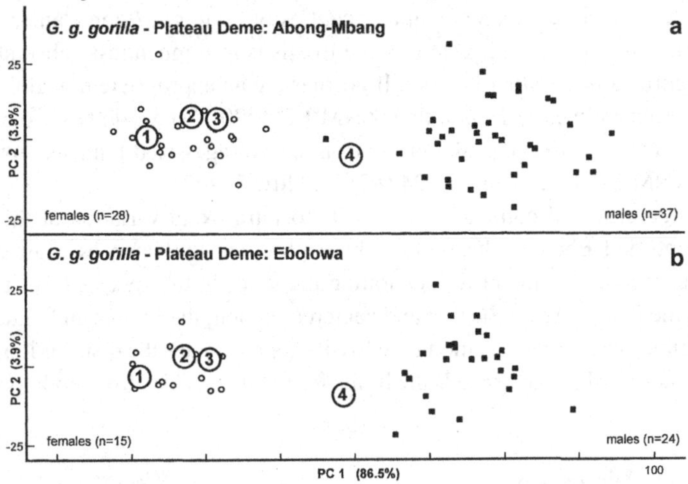

Intrasexual Variation at the Locality Level

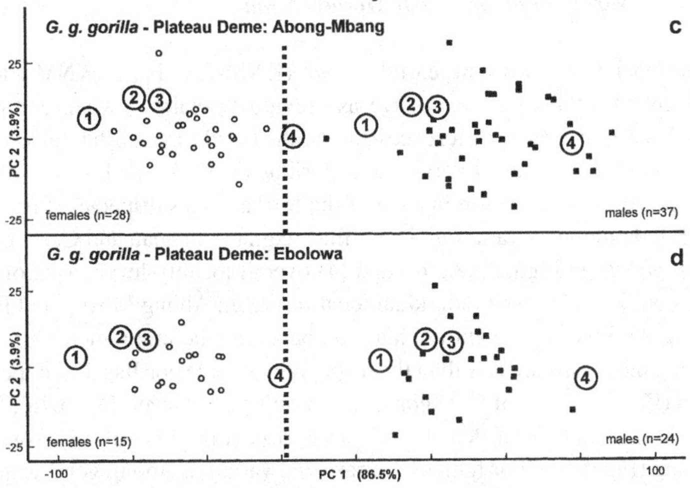

Figure 5.9. The upper two plots show the early *Homo* sample compared to locality-level variation and sexual dimorphism at Abong Mbang (a) and Ebolowa (b), which are collecting sites in the Plateau deme of *Gorilla gorilla gorilla.* Conventions are the same as in Figure 5.5a except the group of four fossils is superimposed on each locality rather than the species as a whole. The lower two plots show the early *Homo* sample compared to intrasexual variation at the locality level for Abong Mbang (c) and Ebolowa (d) from the Plateau deme of *G. g. gorilla.* The plots are the same as Figures 5.9a–b except the sexual discriminant of Figure 5.5b is added and the group of fossils is centered on the range of each sex for each locality. Other conventions are the same as in Figure 5.5b.

the fossil sample, which includes specimens collected thousands of kilometers apart, is much less than at single localities of *G. g. gorilla* in Cameroon. The sexual dimorphism suggested by the fossils is also much less substantial than for gorillas from either locality. If gorillas are an appropriate model for sexual dimorphism in early *Homo*, then KNM-ER 1470 (#4) would have to be a very small male compared to the group of three average-sized females represented by KNM-ER 1813 (#1), OH 24 (#2), and Stw 53 (#3).

Figure 5.9c–d compares the fossils to intrasexual variation at the Abong Mbang and Ebolowa localities. These plots are the same as Figure 5.9a–b except the discriminant for separating the sexes at the species level is shown and the fossils are duplicated and centered on both the female and male gorilla distributions. Variation among the fossils approximates the distributional limits for male gorillas but exceeds the limits for female gorillas from both localities.

Discussion

Variation among early Homo *crania*

Craniofacial variation in early *Homo* (KNM-ER 1470, KNM-ER 1813, OH 24, Stw 53) is not excessive as previously claimed when compared to gorillas. Craniometric differences among the fossils are substantially less than: (1) overall species-level variation in gorillas (Figs. 5.4 and 5.5a); (2) overall subspecies-level variation in each of the four gorilla subspecies (Fig. 5.6); (3) overall demic-level variation for gorillas (e.g., the Plateau and Coast demes of *G. g. gorilla* in Figure 5.8a–b); and (4) overall locality-level variation for gorillas collected at some individual localities (e.g., Abong Mbang and Ebolowa in Figure 5.9a–b). For intrasexual comparisons, the craniometric differences among the fossils are less than the dispersion of male gorillas: (1) at the species level (Fig. 5.5b); (2) of *G. g. gorilla* at the subspecies level (Fig. 5.7a); (3) from the Plateau and Coast demes of *G. g. gorilla* (Fig. 5.8c–d); (4) from Abong Mbang (Fig. 5.9c). For females, intrasexual variation among gorillas generally exceeds the differences among the fossils with the exception of *G. g. gorilla* for which the fossil and gorilla variation is coequal (Fig. 5.7a).

These comparisons with gorillas at different hierarchical levels of intraspecific variation provide new perspectives on variation in early *Homo*. On the one hand, the four fossils taken together are less variable than male gorillas from a single locality, from a single deme, from a single subspecies, and from the species as a whole. Therefore, it is certainly possible that the fossils might all be males of a single hominid species. On the other hand, if KNM-ER 1470 is a male and the other three fossils (KNM-ER 1813, OH 24, and Stw 53) are females, then their pattern of sexual dimorphism resembles the differences between male and female gorillas at each infraspecific level. However, the degree of sexual dimorphism exhibited by the fossils is dramatically less than is seen in gorillas. Therefore, it is also certainly possible that the four fossil crania represent a male and three females of a single early *Homo* species that was much less sexually dimorphic than gorillas. This possibility is supported by OH 16, which is a fragmentary fossil skull from Olduvai Gorge in Tanzania that is considered to be "male" by Tobias (1991). Smaller in size than KNM-ER 1470, OH 16 presumably would be morphometrically intermediate between KNM-ER 1470 and the three smaller fossils in our PCA if it were complete enough to include in our study. Thus, the dimorphism of *H. habilis* may be even less than implied by the use of KNM-ER 1470 alone as the only "male" representative of the fossil species.

Presently, the results are consistent with two possibilities: (1) the fossils represent males and females of an early *Homo* species that is far less sexually dimorphic than gorillas (Fig. 5.5a); or (2) the fossils are members of the same sex of an early *Homo* species that is as intrasexually variable as gorillas (Fig. 5.5b). Both possibilities belie previous claims that early *Homo* crania are more variable than gorillas in degree and/or pattern of variation. Our results expose such ideas as perceptual artifacts resulting from a lack of population thinking and woefully inadequate samples for constructing a meaningful yardstick of intraspecific variation in gorillas.

Our results comparing early *Homo* crania to a gorilla analog are contrary to the conclusions of the FSR studies listed in Table 5.1. All rejected the single-species hypothesis for *H. habilis* fossils based on poorly calibrated yardsticks that inadequately characterize gorilla variation. Instead, we show that the degree and pattern of fossil variation, when compared to large, comprehensive, well-characterized samples of gorillas, is consistent with the hypothesis that the early *Homo* crania represent a single, polytypic species. Our results cast a new light on the debate about species number in early *Homo* by showing how fossil variation actually compares to a biologically accurate representation of intraspecific variation in a sexually dimorphic hominoid species like gorillas.

The population-thinking approach to FSR studies

Our work emphasizes the inadequacies of the traditional FSR approach for characterizing intraspecific variation in analog species. Traditional FSR studies using relatively small, uncontrolled samples from one or two museums yield fragmentary, poorly calibrated yardsticks of morphological variation for which the limits, degree, and pattern of intraspecific variation are unknown

(Fig. 5.3). Since poorly calibrated yardsticks underestimate intraspecific variation in the analog species, then applying such yardsticks to fossils will overestimate the degree of variation among the fossils. Left unchecked, this practice will inevitably lead to the needless recognition of new species. For example, traditional FSR studies claim that early *Homo* crania are more variable than gorillas. This misperception was a factor in erecting "*H. rudolfensis*" as a new taxon to accommodate the "excessive" variation beyond what could be attributed to *H. habilis*. The recognition of superfluous taxa only complicates and confuses phylogenetic reconstruction and biological scenarios about human evolution.

The sophisticated quantitative techniques of traditional FSR studies have created a false impression of statistical "rigor." Such is the case with probability methods like exact randomization when they are based on the relatively small, uncontrolled samples that fail to encompass the breadth and diversity of intraspecific variation in an analog species. Since the traditional FSR approach seriously underestimates intraspecific variation to begin with, then the statistics and probabilities calculated to assess variability among fossils relative to an analog species are essentially meaningless. As in all scientific pursuits, the results are only as good as the data upon which they are based, irrespective of the analytic methods used.

The population-thinking approach avoids three fundamental problems of traditional FSR studies. First, population thinking is "a way of thinking about intraspecific variation in fossil species that is consistent with the biology of polytypic species" (Albrecht *et al.*, 2003, p. 93). As such, it derives from a theoretically sound understanding of biological species as they exist in nature. Species have a certain population taxonomy and population structure that can be represented operationally by the hierarchy of intraspecific morphological variation to provide a biologically accurate picture of a species' biological organization (Fig. 5.2; Albrecht and Miller, 1993; Albrecht *et al.*, 2003). Any attempt to characterize intraspecific variation that does not reflect the population taxonomy and population structure of an analog species will lead to spurious, meaningless results.

Second, the population-thinking approach provides the only sound basis upon which to employ morphometric statistical methods. For example, the gorilla yardstick generated by the population-thinking approach can be used with exact randomization techniques to calculate the probability of sampling gorilla pairs as different as fossil pairs. For the present study, this was unnecessary since visual inspection of the graphed data is sufficient to determine that variation among the fossils is less than in gorillas. Whatever analytic technique is used, the formulation of a sampling strategy based on the population-thinking approach is the prerequisite of any FSR study. No quantitative technique, however sophisticated, can compensate for inadequate, poorly characterized analog samples. Indeed, any FSR study is suspect if not based on comprehensive analog samples that reflect population thinking.

Third, the population-thinking approach provides a rigorous and biologically meaningful strategy of comparison (Albrecht *et al.*, 2003). By using large, well-controlled samples that characterize the breadth and complexity of intraspecific variation, the population-thinking approach allows comparison of fossil variation to each infraspecific level of species structure in the analogs. The result will be new perspectives on the possible biological significance of morphological differences among fossils vis-à-vis analog variability at the species, subspecies, demic, and locality levels.

In the case of early *Homo*, for example, the four fossils are less variable than males of a single locality, of a single deme, of a single subspecies of gorillas. This hierarchical perspective provides fresh insights about the significance of variation in early *Homo* crania compared to gorillas not evident from the traditional FSR approach. The fossils are not from a single place (locality) and they are probably not members of the same deme. However, they could represent males of a single widespread subspecies of a hominid species as variable as gorillas. Alternatively and perhaps more likely, they could represent two sexes of a single hominid subspecies that was far less sexually dimorphic than gorillas. Even more likely still, given the geographic range of the fossils, we suspect they are males and females of different subspecies of a single, polytypic hominid species that was far less sexually dimorphic than gorillas. As with gorillas, there may have been morphological overlap and perhaps not much geographic variation among these different postulated *H. habilis* subspecies despite considerable geographic separation. Such a hypothesis is supported by the great similarity among the "female" *H. habilis* crania despite their geographic dispersion (KNM-ER 1813 from Kenya, OH 13 from Tanzania, and Stw 53 from South Africa). The population-thinking approach provides a powerful, much-needed tool for paleoanthropologists to increase both the resolution and the rigor of FSR studies. The result will be more meaningful biological insights and interpretations about the fossils that paleoanthropologists strive to understand (Albrecht *et al.*, 2003).

Integrating the modern synthesis into paleoanthropology

Given the importance of population thinking and its position as a cornerstone of the modern evolutionary synthesis, we wonder about its absence from paleoanthropological FSR studies. Even though the modern synthesis was formulated and expounded 60–70 years ago, there is little evidence of its principles being integrated into FSR studies. For example, the five influential FSR studies of

Table 5.1 that rejected the single-species hypothesis for *H. habilis* are largely devoid of population thinking in their sampling strategy, collection of data, application of analytic techniques, and interpretation of results. Moreover, FSR studies of other hominid taxa have been equally inattentive to incorporating population thinking in any meaningful, operative way. Why is this the case? We are not really sure. Whatever the reason, we cannot agree with Tattersall's (2000) effort to blame the ills of paleoanthropology on the modern synthesis, which "was doomed to harden, much like a religion, into dogma: a dogma whose heavy hand continues to oppress the science of human origins a half-century later." Rather, Foley (2001) is far closer to the truth that "paleoanthropology, rather than being shackled for the last 50 years by the modern synthesis, has in fact remained blithely innocent of most theoretical issues, and that this, rather than rigid dogma, was the central problem."

Given the absence of population thinking in paleoanthropological FSR studies, what is to be done? We make the following suggestions:

(1) The principles of the modern synthesis, with population thinking as its foundation, must be appreciated by paleoanthropologists and taught to graduate students. These principles are the theoretical base for all fields of organismal biology including paleoanthropology.

(2) Students interested in fossil species recognition must be exposed to the nature and extent of intraspecific variation within and among modern primate species. They will benefit by studying the rigorous, detailed, often overlooked work of primate systematists who apply population thinking to large, comprehensive samples of skins and skeletons to derive their taxonomic conclusions (e.g., see the work of Jack Fooden, Philip Hershkovitz, Richard Thorington, and Colin Groves). The point is not to memorize arcane primate taxonomies but to appreciate the nature and extent of primate variation as the prerequisite to understanding variation among fossils.

(3) The theoretical perspective of population thinking must be translated into a procedural strategy for sampling the intraspecific variation of modern analog species. A useful model for operationalizing population thinking in practical terms is the hierarchy of morphological variation (Fig. 5.2a; Albrecht and Miller, 1993; Albrecht *et al.*, 2003).

(4) The need for large, comprehensive samples in FSR studies has been initially satisfied through the generosity of William Howells and Colin Groves, who have made available the datasets they collected on recent humans (Howells, 1973) and great apes (Groves, 1967, 1970; Shea *et al.*, 1993). One problem, however, is previous FSR studies that based their analyses on small samples of the full datasets rather than constructing well-calibrated yardsticks from the hundreds of available specimens

so as to best model overall intraspecific variation. A second problem is that these datasets were not necessarily compiled for answering questions about earlier hominids. New comprehensive studies of intraspecific variation of hominoid taxa and other extant primates, using a population-thinking approach, must be valued and supported by funding agencies and grant reviewers.

Traditional FSR studies emphasize how little appreciated is the true extent of intraspecific variation in gorillas despite Groves' (1967, 1970) readily available morphometric work (also see the new studies of gorilla variation in Taylor and Goldsmith, 2003). The same is no doubt true of chimpanzees, orangutans, and other primates. A central activity of those involved in FSR studies must be the in-depth study of intraspecific variation in hominoid and other primate species. The ability to interpret fossil variation meaningfully is limited by prior experience in studying the variation of living species and practical knowledge of how to apply the principles of population thinking. We hope others will see the value of our suggestions for integrating the principles of the modern evolutionary synthesis into FSR studies.

Caveats and considerations

The population-thinking approach focuses on how to characterize variation in modern analog species in a manner consistent with their actual biology. It does not address the philosophical question about what are appropriate modern analogs for a hominid species. Although gorillas are traditionally used, some workers argue, unconvincingly in our view, that gorillas are inappropriate analogs for early hominid species (e.g., Grine *et al.*, 1996). We suspect similar arguments will be made against other hominoid species, such as chimpanzees and even recent humans, when the true extent of their intraspecific variation becomes known. If these arguments prevail, then paleoanthropology will devolve to a data-free, theory-free field based on nothing but the word and authority of this or that paleoanthropologist who uses differences in this or that morphological trait between this or that fossil to define each new fossil species. Such typological thinking, which persists in paleoanthropology today, will not advance the field at a time when the discovery of more and more fossils requires increasingly fine-grained analyses to answer questions about their significance.

The population-thinking approach, as applied here, does not deal with anagenetic variation. Consider that the four fossils included in this study represent a time range of about 300,000 years. Modern gorillas, on the other hand, represent a time range of only about 150 years of collecting by museums. The gorilla yardstick we developed in Albrecht *et al.* (2003) and applied to the fossils in this study may be well-calibrated but it does not take into account additional anagenetic variation that probably occurs in hominids across 300,000 years of time. One solution for dealing with anagenetic variation is to use knowledge about the degree and pattern of intraspecific variation in a modern analog species to model the effects of evolutionary processes that may have been at work during the time range of the fossils (e.g., stasis, gradual change, speciation, and shifting geographic ranges in response to ecological change).

The population-thinking approach provides a sound theoretical basis for constructing morphological yardsticks for analog species used in FSR studies. There is, however, no necessary reason to believe that fossil species were limited to the degrees and patterns of variation seen in modern species (Kelley, 1993). Indeed, Oxnard (1987; Oxnard *et al.*, 1985) argued that each hominoid species (including fossil hominids) has a unique pattern of variation. This is not to say that understanding intraspecific variation in modern species is unimportant. On the contrary, it only means that paleoanthropologists must study intraspecific variation for as many extant primate species as possible to best understand the possible complexities of extinct species.

Do not confuse the population-thinking approach with the particular method that we used to ordinate the data (PCA). The population-thinking approach is not a new quantitative method of analysis. Rather, it is a sampling strategy that follows the principles of population thinking to construct an accurate, well-calibrated yardstick of intraspecific morphological variation in a modern species. The population-thinking approach is also a strategy of analysis that compares fossil variation to each infraspecific level of intraspecific variation in the analog species. This powerful tool is compatible with many quantitative methods.

Our results and conclusions apply only to the analysis of craniometric variation among the four most complete early *Homo* crania. The isolated mandibular, dental, and postcranial samples also attributed to *H. habilis* require separate examination and may provide different results. New fossil discoveries may raise new questions about species number in early *Homo*. No matter which specimens are under investigation, the application of population thinking must be a foundational construct for every FSR study.

Conclusion

In this paper, we apply the population-thinking approach to the problem of interpreting craniofacial variation in *H. habilis* using intraspecific variation in gorillas as an analog. We analyze metric differences among the four most complete early *Homo* fossil crania relative to individual and sexual variation in gorillas at the species, subspecific, demic, and locality levels. The results provide a clear picture of the degree and pattern of cranial variation among the fossils as compared to a modern, sexually dimorphic, geographically variable hominoid species whose variation is relatively well known. If the four early *Homo* fossils are thought to be one sex, then they are less variable than males of a single locality, a single deme, or a single subspecies of *G. gorilla*. Alternatively, if KNM-ER 1470 is a male and KNM-ER 1813, OH 13, and Stw 53 are females, then their pattern of sexual dimorphism resembles gorillas except the degree of dimorphism among the fossils is far less substantial. In general, the four fossils are much less variable than gorillas, contrary to the claims made in previous FSR studies of the same fossils that used poorly calibrated yardsticks of morphological variation in gorillas. Have we proven that these fossils represent a single species? No, but our results provide a very good reason why comparative craniometrics cannot be used to reject the single-species hypothesis for *H. habilis* at this time.

Acknowledgments

We express our deepest gratitude to Professor Charles Oxnard, the academic father of one us, the academic grandfather of another, and colleague and friend to each of us. He has been instrumental in our careers by nurturing, guiding, encouraging, inspiring, and otherwise showing us how research as one of life's endeavors has great measures of fun as well as satisfaction. He has modeled for us the value of communicating one's views, regardless of how unorthodox they may seem. Of this, he is a champion who loves to play the maverick. Perhaps his only regret is that his ideas about hominid evolution, regarded as so unorthodox by so many at one time, have been vindicated over the years. So now he is beginning to look rather orthodox. If so, however, it is not because he conformed himself to the orthodoxy of others, but rather because others found themselves conforming, through their own research, to his unorthodoxy! In this, as his academic offspring, his colleagues, and his friends, we find no end of delight.

We express our gratitude to Dr. Colin Groves, whose data on gorillas provided the foundation for our work. His example of data collecting provides a standard for investigations of intraspecific variation in primates. Professor Groves' generosity in sharing his data with others is a model for unselfish collegiality that should be more widespread in our field. Additionally, we thank Dr. Richard Thorington of the US National Museum of Natural History for access to the museum's gorilla collection that allowed us to supplement Colin Groves' dataset. Last, but not least, we commend Dr. Fred Anapol for his unending patience, encouragement, and willingness in leading this project that honors our mentor, Professor Charles Oxnard.

References

Albrecht, G. H. and Miller, J. M. A. (1993). Geographic variation in primates: a review with implications for interpreting fossils. In: *Species, Species Concepts, and*

Primate Evolution, ed. W. H. Kimbel and L. B. Martin. New York: Plenum Press. pp. 123–161.

Albrecht, G. H., Gelvin, B. R., and Miller, J. M. A. (2003). The hierarchy of intraspecific craniometric variation in gorillas: a population-thinking approach with implications for fossil species recognition studies. In: *Gorilla Biology: a Multidisciplinary Perspective*, ed. A. Taylor and M. Goldsmith. Cambridge: Cambridge University Press. pp. 62–103.

Brace, C. L., Mahler, P. E., and Rosen, R. B. (1973). Tooth measurements and the rejection of the taxon "*Homo habilis*". *Yrbk. Phys. Anthropol.*, **17**, 50–68.

Coolidge, H. J. (1929). Revision of the genus *Gorilla*. *Mem. Mus. Comp. Zool. Harvard U.*, **50**, 291–381.

Deinard, A. and Kidd, K. (1999). Evolution of a HOXB6 intergenic region within the great apes and humans. *J. Hum. Evol.*, **36**, 687–703.

Deinard, A. S., Sirugo, G., and Kidd, K. K. (1998). Hominoid phylogeny: inferences from a sub-terminal minisatellite analyzed by repeat expansion detection (RED). *J. Hum. Evol.*, **35**, 313–317.

Fiebel, C. S., Brown, F. H., and McDougall, I. (1989). Stratigraphic context of fossil hominids from the Omo group deposits: Northern Turkana Basin, Kenya and Ethiopia. *Amer. J. Phys. Anthropol.*, **78**, 595–622.

Foley, R. (2001). In the shadow of the modern synthesis? Alternative perspectives on the last fifty years of paleoanthropology. *Evol. Anthropol.*, **10**, 5–14.

Grine, F. E., Jungers, W. L., and Schultz, J. (1996). Phenetic affinities among early *Homo* crania from East and South Africa. *J. Hum. Evol.*, **30**, 189–225.

Groves, C. P. (1967). Ecology and taxonomy of the gorilla. *Nature*, **213**, 890–893.

(1970). Population systematics of the gorilla. *J. Zool.*, **161**, 287–300.

(1986). Systematics of the great apes. In: *Comparative Primate Biology. Vol. 1. Systematics, Evolution, & Anatomy*, ed. D. R. Swindler and J. Erwin. New York: Alan R. Liss. pp. 187–217.

(2003). A history of gorilla taxonomy. In: *Gorilla Biology: a Multidisciplinary Perspective*, ed. A. Taylor and M. Goldsmith. Cambridge: Cambridge University Press. pp. 15–34.

Howells, W. W. (1973) Cranial variation in man. A study of multivariate analysis of patterns of differences among recent human populations. *Papers Peabody Mus. Archaeol. and Ethnol.*, **67**, 1–259.

Kelley, J. (1993). Taxonomic implications of sexual dimorphism in *Lufengpithecus*. In: *Species, Species Concepts, and Primate Evolution*, ed. W. H. Kimbell and L. B. Martin. New York: Plenum Press. pp. 429–458.

Kramer, A., Donnelly, S. M., Kidder, J. H., Ousley, S. D., and Olah, S. M. (1995). Craniometric variation in large-bodied hominoids: testing the single-species hypothesis for *Homo habilis*. *J. Hum. Evol.*, **29**, 443–462.

Leakey, L. S. B., Tobias, P. V., and Napier, J. R. (1964). A new species of the genus Homo from Olduvai Gorge. *Nature*, **202**, 7–9.

Leakey, R. E. F. and Walker, A. C. (1988). New *Australopithecus boisei* specimens from Lake Turkana, Kenya. *Amer. J. Phys. Anthropol.*, **76**, 1–24.

Lieberman, D. E., Pilbeam, D. R., and Wood, B. A. (1988). a probabilistic approach to the problem of sexual dimorphism in *Homo habilis*: a comparison of KNM-ER 1470 and KNM-ER 1813. *J. Hum. Evol.*, **17**, 503–511.

Lockwood, C. A., Richmond, B. G., Jungers, W. L., and Kimbel, W. H. (1996). Randomization procedures and sexual dimorphism in *Australopithecus afarensis*. *J. Hum. Evol.*, **31**, 537–548.
Marks, J. (1993). Hominoid heterochromatin: terminal C-bands as a complex, genetic trait linking chimpanzee and gorilla. *Amer. J. Phys. Anthropol.*, **90**, 237–246.
(1994). Blood will tell (won't it?): a century of molecular discourse in anthropological systematics. *Amer. J. Phys. Anthropol.*, **94**, 59–79.
Mayr, E. (1976). *Evolution and the Diversity of Life*. Cambridge, MA: Belknap Press.
Mayr, E. and Ashlock, P. D. (1991). *Principles of Systematic Zoology*. New York, NY: McGraw-Hill.
McHenry, H. M. (1991). Sexual dimorphism in *Australopithecus afarensis*. *J. Hum. Evol.*, **20**, 21–32.
Miller, J. A. (1991). Does brain size variability provide evidence of multiple species in *Homo habilis*? *Amer. J. Phys. Anthropol.*, **84**, 385–398.
Miller, J. M. A. (2000). Craniofacial variation in *Homo habilis*: an analysis of the evidence for multiple species. *Amer. J. Phys. Anthropol.*, **112**, 103–128.
Miller, J. M. A., Albrecht, G. H., and Gelvin, B. R. (1998). A hierarchical analysis of craniofacial variation in *Homo habilis* using a modern human analog. *Amer. J. Phys. Anthropol.*, **Suppl. 26**, 163.
(2002). Craniofacial variation in *Homo habilis* compared to modern chimpanzees. *Amer. J. Phys. Anthropol.*, **Suppl. 34**, 113.
Oxnard, C. E. (1973). *Form and Pattern in Human Evolution*. Chicago, IL: University of Chicago Press.
(1975). *Uniqueness and Diversity in Human Evolution. Morphometric Studies of Australopithecines*. Chicago, IL: University of Chicago Press.
(1983). *The Order of Man*. New Haven, CT: Yale University Press.
(1987). *Fossils, Teeth, and Sex*. Seattle, WA: University of Washington Press.
Oxnard, C. E., Lieberman, S. S., and Gelvin, B. R. (1985). Sexual dimorphisms in dental dimensions of higher primates. *Amer. J. Primatol.*, **8**, 127–152.
Oxnard, C. E., Crompton, R. H., and Lieberman, S. S. (1990). *Animal Lifestyles and Anatomies*. Seattle, WA: University of Washington Press.
Robinson, J. T. (1966). The distinctiveness of *Homo habilis*. *Nature*, **209**, 957–960.
Rogers, J. (1993). The phylogenetic relationships among *Homo*, *Pan*, and *Gorilla*: a population genetics perspective. *J. Hum. Evol.*, **25**, 201–215.
Shea, B. T., Leigh, S. R., and Groves, C. P. (1993). Multivariate craniometric variation in chimpanzees: implications for species identification in paleoanthropology. In: *Species, Species Concepts, and Primate Evolution*, ed. W. H. Kimbel and L. B. Martin. New York: Plenum Press. pp. 265–296.
Stringer, C. B. (1986). The credibility of *Homo habilis*. In: *Major Topics in Primate and Human Evolution*, ed. B. Wood, L. Martin, and P. Andrews. Cambridge: Cambridge University Press. pp. 266–294.
Tattersall, I. (2000). Paleoanthropology: the last century. *Evol. Anthropol.*, **9**, 2–16.
Taylor, A. and Goldsmith, M. (2003). *Gorilla Biology: A Multidisciplinary Perspective*. Cambridge: Cambridge University Press.
Tobias, P. V. (1991). *Olduvai Gorge. Vol. 4. The Skulls, Endocasts, and Teeth of* Homo habilis. Cambridge: Cambridge University Press.

Tuttle, R. H. (2003). An introductory perspective: gorillas – how important, how many, how long? In: *Gorilla Biology: A Multidisciplinary Perspective*, ed. A. Taylor and M. Goldsmith. Cambridge: Cambridge University Press. pp. 11–14.

Walker A. C. and Leakey R. E. (1988). The evolution of *Australopithecus boisei*. In: *Evolutionary History of the "Robust" Australopithecines*, ed. F. E. Grine. New York, NY: Aldine de Gruyter. pp. 247–258.

Wood, B. (1985). Early *Homo* in Kenya, and its systematic relationships. In: *Ancestors: the Hard Evidence*, ed. E. Delson. New York: Alan R. Liss. pp. 206–214.

(1991). *Koobi Fora Research Project. Vol. 4*. Oxford: Clarendon Press.

(1993). Early *Homo*: how many species? In: *Species, Species Concepts, and Primate Evolution*, ed. W. H. Kimbell and L. B. Martin. New York: Plenum Press. pp. 485–522.

Part II
Organ structure, function, and behavior

My own earliest studies tried to understand functional adaptation in the locomotor system, primarily of bones and joints, through morphometrics. As a preliminary, however, I did attempt to find out what I could of the muscles that moved the bone–joint unit. In those days all we did was dissect muscles and measure them through relative lengths, directions of pull, relative weights, and frequencies of (what were called in those days) muscular anomalies. Limited though such studies were, they were incredibly time-consuming. In all, I dissected 52 shoulders representing 28 primate species, 145 arms and forearms representing 27 primate species, and 167 hips and thighs in 33 primate species. That took many years. Yet even such primitive data provided useful initial information about muscles in relation to the respective bone–joint units.

But it was always clear to me, and I wrote about it without ever doing it, that better muscular studies would be necessary. Such studies would need to understand much more about muscular architecture than simply relative muscle weights, much more about muscle activity and functions through studies of living muscle than just anatomical inferences, much more about movements, postures, and overall behaviors of the living animals than just simple classifications of locomotion.

However, I certainly did not, in those early days, envisage the possibility of studies of the biomechanics of the jaws and teeth and, further, the biomechanics of food being masticated by them. And I knew nothing at all of functional adaptations in the brain.

A few examples of such investigations follow in this section. Again, my students and colleagues have gone so much further than was possible for me. Yet notwithstanding, stimulated by them, I have been able in these latter years to enter some of these areas myself.

Charles Oxnard

6 *Fiber architecture, muscle function, and behavior: gluteal and hamstring muscles of semiterrestrial and arboreal guenons*

FRED ANAPOL, NAZIMA SHAHNOOR, J. PATRICK GRAY
University of Wisconsin–Milwaukee

Introduction

Charles Oxnard championed the use of statistical techniques to reduce a multitude of variables into a comprehendible pattern in order to demonstrate some aspect of primate behavior or evolution. In reality, Professor Oxnard's interests and writings have spanned an enormity and diversity of topics. By analogy, one might conclude that the "first principal component" of his work is reflected by the succinct statement on page 6 of his 1975 book, *Uniqueness and Diversity in Human Evolution*: "There is no doubt that the locomotor behavior of an animal is, on a gross level, controlled by the anatomy of the animal." This point of departure for the research of many contributors to this volume, has influenced our own endeavor to understand how the internal morphology of a muscle itself, i.e., fiber architecture and histochemical fiber type composition, is associated with organ function and animal behavior. In this chapter, we consider the relationship of locomotor behavior, muscle function, and the fiber architecture of the gluteal and hamstring muscle groups.

Many classes of polymorphic variables determine the internal morphology of whole skeletal muscles. These include the relative composition of histo/immunocytochemical fiber types (see Chapter 7), neurological compartmentalization related to fascial partitioning (English, 1984) by the branching pattern of motoneurons (e.g., Herring *et al.*, 1989, 1991), and the spatial arrangement of muscle fibers and tendons, otherwise known as *fiber architecture* (Gans and Bock, 1965; Gans, 1982; Richmond, 1998). Muscle fiber architecture

Shaping Primate Evolution, ed. F. Anapol, R. Z. German, and N. G. Jablonski. Published by Cambridge University Press.

provides an interesting way to contemplate how muscles divide the labor in a functional complex, thereby further substantiating the relationship between morphology, function, and behavior.

Muscle fiber architecture

For this chapter, muscle fiber architecture includes whole-muscle weight, lengths of fasciculi (groups of myofibers), the extent to which fibers are arranged in parallel or pinnately (i.e., at an angle to the distal tendon of attachment), the extent to which pinnate muscles range from uni- to multipinnate (Gans and Bock, 1965; Gans, 1982; Anapol, 1984; Anapol and Jungers, 1986; Anapol and Barry, 1996), and the relative proportions of the length of a tendon and the length of its associated muscle fasciculus. The many permutations of these components account for much of the functional heterogeneity among muscles, thereby effecting the division of labor among the homologous heads of a synergistic muscle group (e.g., Anapol and Jungers, 1986; Anapol and Barry, 1996; Anapol and Gray, 2003). What is useful about these highly variable architectural components is that they comprise a group of quantifiable morphologic features that are interpretable in physiologic terms (e.g., Gans and Bock, 1965; Gans, 1982; Anapol and Jungers, 1986; Gans and de Vree, 1989; Anapol and Herring, 1989). In essence, contractile properties of whole muscles, and their functional implications, can be estimated from measurements taken on cadavers.

Whole muscles, those whose fibers insert at an angle to the distal tendon of attachment, are described as having a pinnate, or feather-like, architecture. By contrast, muscles having all fibers in parallel between proximal and distal attachment sites are referred to as parallel-fibered. The pinnate arrangement serves to increase maximum force output by increasing the number of fibers, thus increasing the number of cross-bridges in parallel (Gans and Bock, 1965) while enabling force to be concentrated into a smaller area of attachment. Besides increasing this *physiologic cross-sectional area*, pinnation also may provide for equivalent placement of sarcomeres during shortening, possibly associated with coordination for both recruiting and maintaining force (Gans and de Vree, 1989). For example, one benefit of not maintaining all sarcomeres positioned in maximum overlap at a single jaw position is that sufficient tension can be generated in all functional positions, e.g., as the mandible travels in three dimensions throughout the masticatory cycle. This is likely the case in pig masseter, where the maximum tension that can be produced by direct stimulation of the muscle nerve falls well below what one would predict from the weight of the muscle alone (Anapol and Herring, 1989), albeit before reduction by fasciculus length and pinnation angle, indicating that not all fasciculi are positioned to deliver maximum force concurrently.

Early studies of muscle architecture endeavored to establish basic mathematical relationships between measurable aspects of whole-muscle morphology and the visual observation of both potential force and excursion (e.g., Weber, 1851; A. Fick, 1860; R. Fick, 1910; Haines, 1932; Pfuhl, 1937; Benninghoff and Rollhäuser, 1952). These were based exclusively on *in situ* measurements of cadaverous material but were neither substantiated, disproved, nor augmented until later physiologic experiments on living tissue (e.g., Ramsey and Street, 1940; Hoyle and Smyth, 1963; Willemse, 1963). Gans and Bock (1965) review this early work at length in their classic treatise on muscle fiber architecture.

Focusing on isolated muscles, more recent studies (e.g., Goslow *et al.*, 1977; Peters and Rick, 1977; Spector *et al.*, 1980; Walmsley and Proske, 1981; Muhl, 1982) have somewhat clarified the extent to which the fiber architecture of a whole muscle influences its contractile properties. In general, parallel-fibered muscles are considered to be best suited for speed of contraction and/or excursion (correlated to the number of sarcomeres in series), while, as noted above, pinnate-fibered muscles are thought to either increase force (correlated to the number of cross-bridges in parallel) or concentrate force into a smaller allocated area of attachment. With respect to muscle, force and velocity generally are incompatible in that a hyperbolic relationship exists between force and velocity, i.e., the higher the load, the slower velocity of contraction (Hill, 1938). Although this model is derived from results of experiments using frog sartorius, a parallel-fibered muscle, the force–velocity curve is basically the same shape for rabbit digastric, a unipinnate muscle (Anapol *et al.*, 1987). Increasing fasciculus length, however, also can increase potential velocity of a muscle otherwise constructed for tremendous force, e.g., vastus lateralis of *Lemur fulvus* (Anapol and Jungers, 1986).

The trade-off between velocity and force within a whole muscle, however, is not quite so reciprocal. For example, the length–active tension curve of the parallel-fibered m. tibialis anterior exhibits high tension over a broader range of muscle lengths than the pinnately arranged m. extensor digitorum longus (Goslow *et al.*, 1977). Muhl (1982), however, demonstrates that the range of excursion of a pinnate muscle (rabbit digastric) is greater than that of its constituent fibers, thereby allowing the sarcomeres to remain within a range of optimal overlap. This can be explained by the tendency of the myofibers of a pinnate muscle to rotate, as they contract, about an axis passing through their proximal end as they draw the tendon of insertion proximally during whole-muscle shortening (Gans, 1982). Based on these findings, Muhl attributes the results of Goslow *et al.* (1977) to the longer muscle fibers of m. tibialis anterior, by contrast to those of m. extensor digitorum longus, rather than to the parallel/pinnate dichotomy as concluded in the original study. This is consistent with Walmsley and Proske (1981) who compared two pinnate muscles in a cat, attributing the flatter length–tension curve in m. soleus to its having more sarcomeres per fiber than m. medial gastrocnemius.

Physiologic implications of muscle fiber architecture

Several physiologic parameters can be estimated by examining fiber architecture. These generally include estimates of potential maximum force output, potential maximum excursion or velocity, expense of contraction, and whether a muscle is more likely employed isotonically (constant force, changing length) or isometrically (constant length, changing force). Muscle architecture appears to influence peak isometric tension and maximum velocity of shortening even more than histochemical fiber types (fast/slow twitch) (Spector *et al.*, 1980). When normalized for pinnation angle, fiber length, and muscle mass, the disparity in tension and velocity between fast (medial gastrocnemius) and slow (soleus) twitch muscles, measured *in vivo* at the tendon of insertion in cat, is considerably reduced (Spector *et al.*, 1980).

What is the maximum force a muscle can generate?

Relative muscle weights (∼ mass) are often used, and often useful, in interpreting gross behavioral comparisons, e.g., braking vs. propulsive function (Haxton, 1944), quadrupedalism vs. leaping (Fleagle, 1977) (also see Appleton, 1921; Ashton and Oxnard, 1963; Grand, 1967, 1968a, 1968b, 1977; Grand and Lorenz, 1968; Stern, 1971; Jenkins and Weijs, 1979; Anapol and Herring, 1989; Swartz and Tuttle, 1990). However, isolated muscle weight indices (individual muscles as a proportion of some arbitrarily determined standard) are of questionable interpretability and tend to be inconsistently applied (Stern, 1971). Furthermore, the potential maximum force of a muscle is dramatically affected by its intrinsic morphology, especially since fiber length will contribute significantly to the weight of a muscle but not to its force. The relative contributions of individual muscle bellies toward the total force output of the group can be greatly altered when reduced physiologic cross-sectional areas, rather than muscle weights alone, are considered, as exemplified by the four synergistic bellies of quadriceps femoris (Anapol, 1984; Anapol and Jungers, 1986; Anapol and Barry, 1996). This will, of course, affect the interpretation of relative muscle function during locomotion, especially if a “developmental constraint” limits the total mass per muscle group of common embryogenic origin (Swartz and Tuttle, 1990).

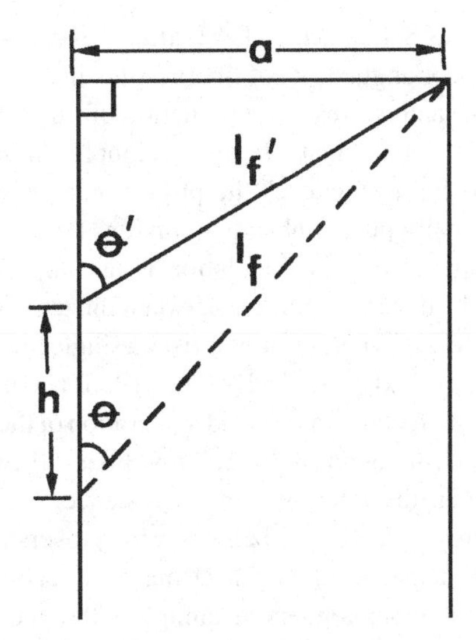

Figure 6.1. Calculation of angle of pinnation. When a pinnate muscle shortens distance h, the fasciculi shorten from l_f to $l_{f'}$, increasing the resting angle of pinnation from θ to θ'. After reconstructing l_f from $l_{f'}$ (see Materials and methods), θ can be calculated by measuring a, the distance normal to the tendon of insertion from the proximal attachment of the fiber: arcsin (a/l_f) (from Anapol and Barry, 1996).

Along the length of a single myofiber, maximum force (a physiologic parameter) is generated at the length at which maximum overlap of cross-bridges between the myosin heads and actin filament (a morphologic parameter) occurs (Gordon *et al.*, 1966). Likewise, maximum potential force output of a whole muscle can be estimated by calculation from morphologic measurements. The maximum force a muscle can generate is equal to the total cross-sectional area of all its constituent fibers, or what is referred to as its *physiologic cross-sectional area*, by contrast to that of a mid-belly plane normal to the long axis. For a parallel-fibered muscle, these would be one and the same, by contrast to a pinnate muscle, as illustrated in Figure 6.1.

An obvious approach to estimating physiologic cross-sectional area would be to measure cross-sectional areas of myofibers from photomicrographs of whole-muscle serial sections (Quest, 1987), or by displacement (Buchner, 1877). An estimate of physiologic cross-sectional area also can be determined indirectly by geometric reduction. Muscle weight divided by the specific gravity of muscle (1.0564 g cm^{-3}: Murphy and Beardsley, 1974) yields a volume. Dividing the volume by the working length (proximal to distal attachments) of the fasciculi yields an area value. The slightly more complex, but decidedly more accurate, variant of this calculation, "reduced" physiologic cross-sectional area (RPCA)

(Haxton, 1944) modifies Schumacher (1961, after Weber, 1851) in its consideration of the pinnation, or angulation, of fibers to the direction of contraction or pull (~ line of action) and the resulting reduction of linear force by the exclusion of a lateral vector. For a pinnate muscle, weight is multiplied by the cosine of the angle of pinnation to "reduce" the physiologic cross-sectional area to estimate the force actually put to substrate (Haxton, 1944).

One overlooked aspect of this calculation is the importance of including fasciculus length as the distance spanned between the tendons (or bone) by the muscle tissue, regardless of whether it comprises a single fiber or a train of fibers with overlapping ends (Huxley, 1957; Fawcett, 1986; Gaunt and Gans, 1992). Tissue preparation often includes nitric acid maceration of the whole muscle and direct microscopic measurement of "freed" myofibers. If two or three shorter overlapping fibers span the inter-attachment distance, the measured lengths of these fibers will result in RPCA being severely overstated when divided into volume (~ weight/specific gravity). Using a direct measurement of the distance between the tendons renders meaningless the question of how many fibers span this distance. Furthermore, unless pinnation angle is measured prior to maceration, these data will be lost and pre-maceration measurements will not be verifiable. Without pinnation angle, physiologic cross-sectional area will be further overestimated. Although precluded by our small sample size, the possibility exists that correlation of morphologically similar muscles can yield, e.g., pinnation angle and fiber length.

How rapidly can the muscle move a limb segment, i.e., put force to substrate?

The velocity of muscle contraction is a function of excursion over time. Because all the sarcomeres in a fiber are approximately the same length and contract more or less the same distance simultaneously, so do lengths of fibers and whole muscles in their entirety. Consequently, relative velocity (distance/time) is proportional to fiber or fasciculus length, i.e., the number of sarcomeres in series. Because of their generally greater length, muscles with fibers arranged in parallel are considered to be adapted for higher speeds (longer excursion) of contraction, i.e., longer fibers mean higher velocity. While antigravity muscles are largely pinnate (Anapol and Jungers, 1986; Anapol and Barry, 1996), limb flexors appear to be more parallel-fibered. This also is associated with the relative proportion of slow- and fast-twitch muscle fibers found in particular functional classes (Reiser *et al.*, 1985). Antigravity muscles have higher proportions of slow-twitch myofibers more concentrated in the deeper regions of the whole muscle, while limb flexors have primarily fast-twitch fibers with the fewer slow fibers dispersed throughout the belly (Burke and Edgerton, 1975; Armstrong *et al.*, 1982; Armstrong and Phelps, 1984).

In addition to fiber or fasciculus length, the calculation to estimate linear velocity in pinnate muscles also must take into account the angle of pinnation (Benninghhoff and Rollhäuser, 1952) plus an adjustment for the discordance between whole-muscle excursion and that of its constituent fibers (see Materials and methods; Muhl, 1982). While this calculation, normalized to belly length, is useful for comparing potential velocity, *per se* between muscles, the ratio of muscle weight to estimated tetanic force, derived from the calculation of reduced physiologic cross-sectional area, provides some insight into whether an individual muscle is more dedicated to force or velocity, with higher values indicating a stronger propensity for velocity (Sacks and Roy, 1982). i.e., how much of a muscle's mass results from fiber length, rather than girth.

What is the energy expense of muscle function?

Because they are flexible and highly resistant to stress, tendons are well suited to transmit muscle forces to bone. Modifying the relative proportions and shape of muscle bellies and tendons can satisfy a variety of functional requirements. For example:

(1) the force of the largest of muscles can be concentrated into a relatively small area of attachment;
(2) forces from several synergistic muscles with different actions can be applied to a relatively small portion of a bone;
(3) muscle mass can be positioned in a kinetically preferable location (e.g., nearer the trunk, as in the swiftly running cheetah) while retaining the ability to transmit force to distal limb segments;
(4) by comparison to muscle girth, the compactness of tendon allows force to be transmitted across joints without sacrificing joint mobility so that the small size of manipulative structures, e.g., manual digits, can be maintained with adequate strength by situating their effecting muscle mass further proximally in the antebrachium;
(5) tendons also allow force to be transmitted around "corners" by passing through ligamentous pulleys (e.g., retinacula) or beneath bony protrusions (e.g., sustentaculum tali) (Hildebrand, 1974).

How muscle and tendon proportionately contribute to work also appears to be associated with the ability to store elastic strain energy, e.g., as in human running (reviewed by Cavagna, 1969; Cavagna and Kaneko, 1977; Cavagna *et al.*, 1980). If a muscle and tendon lengthen as the force on them increases, elastic strain energy is stored. Conversely, energy is returned through elastic recoil by their shortening as the force falls. When both events occur during the contact phase of a step cycle, energy is conserved (Alexander and Bennet-Clark, 1977; Biewener *et al.*, 1981; Biewener and Baudinette, 1998; Biewener, 1998).

The extent to which the elasticity within the sarcomeres participates in energy storage, however, remains somewhat equivocal (see Cavagna *et al.*, 1980).

Morphologically, the question of energy storage or savings can be addressed by comparing the length of a muscle fasciculus with the total length of tendon through which both its proximal and/or distal ends are attached to bone. When shorter contractile fibers transmit tension to the bone through longer non-contractile tendons, as opposed to the length of the muscle being composed entirely of contractile tissue, the active energy cost, the currency of which is the exothermic hydrolysis of adenosine triphosphatase (ATP), is reduced (Hill, 1938; Wilkie, 1968; Paul, 1983). Furthermore, a lower fasciculus : extrinsic-tendon ratio is correlated to the tendency of a muscle to be used relatively isometrically, rather than isotonically. Thus, a more isometrically designed muscle is able to take advantage of the strain energy passively stored by its series and parallel elastic elements for use when the muscle is stretched during loading (Biewener *et al.*, 1981). This proportion varies in a functionally interpretable manner, both among synergistic heads of muscle complexes and interspecifically (Anapol and Jungers, 1986; Anapol and Barry, 1996).

Functional implications of muscle fiber architecture in locomotion

Only a few investigators have touched upon the behavioral consequences of interspecific variation in muscle architecture with respect to locomotion, and these only briefly. Stern (1971) noted that restricted proximal attachments characterized vasti peripherales (v. medialis, v. lateralis) in the habitually leaping cebid monkeys. This implies fewer, but longer, muscle fibers, by contrast to more, but shorter, fibers in the more extensive proximal attachments of the relatively non-leaping prehensile-tailed forms. However, with aponeurotic tendons, larger angles of pinnation can dramatically reduce muscle fiber length without necessarily altering the area of attachment site, such as in the vastus lateralis of the brown lemur (Anapol and Jungers, 1986).

Alexander (1974) predicted that, within a limb, either parallel-fibered or long-fibered pinnate muscles would be located most proximally and would do substantial work in acceleration and jumping. Short-fibered pinnate muscles would be located more distally and would save energy by elastic storage. However, within the quadriceps femoris of brown lemurs (Anapol and Jungers, 1986), vervets, and red-tailed monkeys (Anapol and Barry, 1996), the most proximal attachment is that of the two-joint rectus femoris on the pelvis (the three vasti attach to the femur proximally), which has the shortest fibers of the four. Furthermore, Anapol and Barry (1996) found that the more distally located triceps surae generally had greater maximum potential excursion/velocity than the quadriceps femoris in the semiterrestrial vervets, by contrast to the arboreal red-tailed monkeys in which both muscle groups had more or less the same excursion potential. Thus, fiber length may be an adaptation more to specific locomotor preferences, rather than to limb segment location.

Most studies of primate limb muscles tend to focus on the topography and/or functional morphology of whole agonists and synergists during leaping and arboreal and terrestrial quadrupedalism (e.g., Jungers *et al.*, 1980, 1983; Kimura *et al.*, 1983; Anapol and Jungers, 1987), and forelimb muscles in quadrumanous climbing and armswinging (e.g., Ashton and Oxnard, 1963; Grand, 1968a; Tuttle and Basmajian, 1978a, 1978b; Stern *et al.*, 1977; Fleagle *et al.*, 1981; Larson and Stern, 1986, 1987). Despite a significant literature on the architectural underpinnings of nonmammalian (e.g., Willemse, 1977; Wineski and Gans, 1984) and nonprimate mammalian (e.g., Gonyea and Ericson, 1977; Spector *et al.*, 1980; Sacks and Roy, 1982; McClearn, 1985; Lieber and Blevins, 1989) muscles, the functional significance of the internal architecture of primate locomotory muscles remains largely overlooked. The few exceptions have substantiated the relationship between documented observations of species' locomotion in the wild and the physiologic implications of the internal architecture of the antigravity hind-limb muscles of those species (e.g., Anapol and Jungers, 1986; Anapol and Barry, 1996; Anapol and Gray, 2003).

Behavioral implications of muscle fiber architecture in locomotion

In this study of primate locomotor morphology, fiber architecture of the gluteal (gluteus superficialis, g. medius, g. minimus) and hamstring (biceps femoris [flexor cruris lateralis], semimembranosus accessorius [presemimembranosus], semimembranosus proprius [postsemimembranosus], semitendinosus) muscles are compared between two species of African guenons that exploit somewhat different habitats in the wild: the semiterrestrial vervet (*Chlorocebus* [formerly *Cercopithecus*] *aethiops*) (Gray, 1870; Groves, 1989, 2000) and the arboreal red-tailed monkey (*Cercopithecus ascanius*). Comparative studies of vervets and red-tailed monkeys find that the muscle force potentially generated by antigravity muscles is more medially directed toward the substrate, which is less than body width, in the arboreal form; the homologous muscles of the semiterrestrial vervets are constructed more for velocity and isotonic contraction (Anapol and Barry, 1996; Anapol and Gray, 2003).

The morphological results presented here can be interpreted within the context of previously published observations and electromyography studies of primate locomotor function. The emphasis is on propulsion and braking in the negotiation of terrestrial and more complex arboreal environments, ascent and descent on diagonally oriented branches, and in the transposition between ground and canopy. The working hypothesis of this study is that muscle fiber architecture will differ between arboreal and semiterrestrial guenons and will complement the biomechanical configuration of the musculoskeletal system in both cases. Comparisons are hierarchical: (1) among muscles within synergistic groups, (2) between functional muscle groups in the same animal, and (3) between homologous muscles between species. Several questions are addressed in this study:

(1) Do interspecific differences in the physiologic implications of muscle fiber architecture reflect differences in documented modes of locomotion in the wild for these two species?
(2) Can the "fine-tuning" of morphological studies, by examining intramuscular morphology, provide more comprehensive interpretations of muscle function?
(3) What role does the inherent flexibility of muscle tissue, vis-à-vis skeletal morphology, play in a species' transition from one locomotor niche to another?
(4) Is "semiterrestrial" in some cases a uniquely legitimate locomotor category, or merely a descriptor for terrestrial and/or arboreal species that occasionally inhabit the habitat of the other?

Materials and methods

Six embalmed (10% buffered formalin) cadavers[1] each of *Chlorocebus aethiops* and *C. ascanius*, used in two previous studies (Anapol and Barry, 1996; Anapol and Gray, 2003), were dissected. The following muscles were identified (following Howell and Straus, 1933) and removed from the left hind limb: gluteus superficialis, g. medius, g. minimus, biceps femoris, semimembranosus accessorius, semimembranosus proprius, and semitendinosus. Prior to excision, proximal and distal extents of the distal attachment of each muscle on the femur and tibia were measured, as were the lengths of these bones (following Stern, 1971).

Upon excision, muscles were measured lengthwise ($l_{b'}$) and sampled following Anapol and Barry (1996). Because these cadavers had been embalmed without regard to limb position, most joints were either flexed or extended in deviation from the postural position observed in living animals anticipating

[1] Nine cadavers were from the Neil C. Tappen collection in the Department of Anthropology at the University of Wisconsin–Milwaukee (Tappen, 1991), two were from the Barbados Monkey Crop Control Damage Program, and one was from the Primate Research Center of Tulane University.

locomotor progression. Consequently, whole-muscle and fiber lengths were either shortened or stretched from resting length, i.e., the muscle length in postural position (Gordon *et al.*, 1966; Fawcett, 1986), at which maximum overlap occurs between the actin thin filaments and myosin heads (heavy meromyosin, subfragment 1), and where force is of maximum potential. Furthermore, as diagrammed in Figure 6.1, when a pinnate muscle, for example, shortens (distance h), not only do the fibers (or fasciculi) shorten from l_f to $l_{f'}$, they also swivel proximally, changing the angle of pinnation from its value at resting length, θ, to θ'. Therefore, *in situ* muscle fasciculus lengths and pinnation angles require adjustment to re-establish resting values.

In brief, this was accomplished (following Anapol and Barry, 1996) by paraffin-embedding small ($\sim 2\ mm^3$) tissue samples removed from proximal, middle, and distal regions of the muscle belly, and staining 6-μ serial sections with phosphotungstic acid hematoxylin (PTAH), following Prophet *et al.* (1992). PTAH enhances visibility of the striation pattern, thus improving visibility and measurability of sarcomeres in series. The length of ten consecutive sarcomeres was measured (in micrometers, μ) and divided by 25 μ (the presumed length of 10 resting sarcomeres: see Anapol and Barry, 1996) to derive a percentage estimate of how much the measured fasciculus length deviated from expected resting length. A numerical restoration of resting length (l_f) in each sampled region of the muscle belly then was computed. The angle of pinnation, θ, also was restored to its resting value by calculation: $\theta = \arcsin(a/l_f)$, where l_f is the length of a fasciculus *after* restoration to resting length (Fig. 6.1).

For each fasciculus measured directly in a muscle, the following data were recorded (Fig. 6.2): (1) length of muscle fasciculus ($l_{f'}$) between proximal and distal myotendinous junctions, and/or fleshy attachment to bone; (2) perpendicular distance (a) from the proximal myotendinous junction to the tendon of distal attachment; (3) length of tendon (t_p) from the proximal attachment of the whole muscle to the proximal myotendinous junction; (4) length of tendon (t_d) from the distal myotendinous junction to the distal attachment of the whole muscle. Lastly, muscles were blotted dry, trimmed of excess tendon and fascia, and weighed twice (Schumacher, 1961; Scapino, 1968; Anapol, 1988).

The following four variables were calculated for each sample in each muscle:

(1) Reduced physiologic cross-sectional area (RPCA) (Schumacher, 1961, after Weber, 1851, and adjusted by Haxton, 1944): RPCA (cm^2) = [wet weight (g) × cos θ] / [l_f (cm) × specific density]. The specific density of muscle is 1.0564 g cm^{-3} (Murphy and Beardsley, 1974). For parallel-fibered muscles, cos $\theta = 1.0$. RPCA of each muscle was divided by the total RPCA for each muscle group (shoulder or arm) to yield percentage

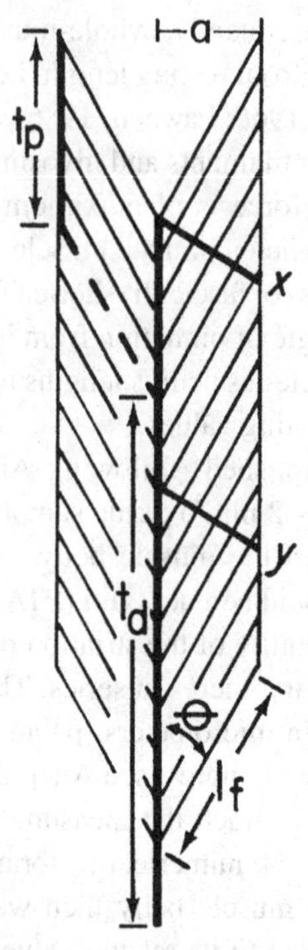

Figure 6.2. The measured adjusted variables are illustrated on this diagram of a bipinnate muscle. l_f, length of muscle fasciculus; *a*, distance normal to tendon of distal muscle attachment from proximal myotendinous junction; θ, angle of pinnation; t_p, length of tendon attaching to the most proximal end of fasciculus; t_d, length of tendon attaching to the most distal end of fasciculus; *x* and *y* (normal to l_f), representative portions of the true physiologic cross-sectional area (from Anapol and Barry, 1996).

reduced physiologic cross-sectional area (%RPCA), which allows interspecific comparisons regardless of size differences.

(2) Potential excursion of a whole muscle (h) (Benninghoff and Rollhäuser, 1952): $h(\text{mm}) = l_f(\cos\theta - \sqrt{\cos^2\theta + n^2 - 1})$ where n is the coefficient of contraction,[2] length of muscle fiber *after* contraction/resting fiber length, or 0.769. To account for inter-specimen size differences, h for any muscle was normalized by dividing it by the resting length of the belly[3]

[2] The coefficient of contraction is generally estimated to be 70% for whole muscle (Gans and Bock, 1965). Muhl (1982) determined *in vivo* that the muscle fibers of the pinnate rabbit digastric lengthened an average of 0.77 mm per 1 mm lengthening of the muscle belly. Therefore, the coefficient of contraction is recomputed for fibers of pinnate muscles as .769 (calculated as 1 − ((.77)(.30)).

[3] Anapol and Barry (1996) expressed h as a proportion of muscle weight, thus departing from Anapol (1984), Anapol and Jungers (1986), and Anapol and Gray (2003), in which muscle belly length was the standard used. The present paper follows the latter protocol.

(l_b) reconstructed from its length at excision ($l_{b'}$), following Muhl (1982): $l_b \approx l_{b'} + [(1.0/0.77)(l_f - l_{f'})]$.

(3) The extent to which a muscle is dedicated to force or velocity (W/P_0), where W is weight and P_0, the approximate tension a whole muscle is capable of generating, is estimated as RPCA × 2.3 N cm^{-2}.

(4) Total tendon length per muscle fasciculus ($l_t/l_f + l_t$) where $l_t = t_p + t_d$.

Means, standard deviations, and standard errors were computed for six specimens of each species. A two-tailed student's *t*-test was used to determine statistically significant interspecific differences ($P < 0.05$).[4] For %RPCA and $l_t/l_f + l_t$, means tested were of arcsin-transformed values of individual muscles (Sokal and Rohlf, 1981). For each calculated variable, the coefficient of variation adjusted for small-sample bias (V*) (Sokal and Rohlf, 1981) was calculated for each muscle group and species.[5]

Randomization tests (see Edgington, 1995; Good, 1999; Manly, 1997) were conducted on the species differences for the four calculated muscle variables as a check on the robustness of the conclusions derived from the parametric statistic tests. For W/P_0, h/l_b, and $l_t/l_f + l_t$ the procedure was as follows:

(1) A value for the muscle of each animal was calculated by taking the mean of all measurements for that muscle.

(2) A species mean was calculated for each species using the means calculated in step 1.

(3) The difference between means of the species was taken as the observed value for the data.

(4) A replicate dataset was constructed by randomly selecting a single measurement from each animal, randomly assigning the measurements to one of the two species, calculating species means, and taking the difference between those means. The difference was recorded.

(5) Step 4 was repeated 9999 times, creating a distribution of 10 000 mean differences (including the observed value).

(6) The null hypothesis that the observed value was significantly greater (or smaller) than expected, given the distribution of differences obtained by random placement of data into two species, was tested by calculating the number of differences that were larger (or smaller) than the observed

[4] When the probability of statistically significant difference falls between 0.05 and 0.10, its calculated value is provided.

[5] For V*, significance of differences between gluteal and hamstring muscles was not tested. Ordinarily this test is accomplished by calculating an *F*-statistic that compares the logarithm (base 10) of the variances (Zar, 1984, after Lewontin, 1966). However, this test depends on the normal distribution of the logged data as tested, e.g., by the *W* statistic (Shapiro and Wilk, 1965), for which the sample size here ($n = 4$) is too small for relevant testing.

value. The probability of obtaining a difference greater than or equal to the observed value was obtained by counting the number of times the observed value was equaled or surpassed and dividing by 10 000.

The randomization procedure for %RPCA was conducted as above with the following modification. For each muscle of interest a single random selection of a measurement of that muscle was made for each animal. For each animal a random selection was made of a single measurement for each of the remaining muscles needed to calculate the %RPCA of the muscle of interest. These %RPCA values were then randomly assigned to the two species and the species differences calculated. The original observed value was located in the distribution of differences as described above.

Logistic regressions were computed for the four calculated muscle variables, RPCA (not %RPCA, in the absence of intragroup independence for this variable), W/P_0, h/l_b, and $l_t/l_f + l_t$, and separately for previously published measurements of the proximal hind-limb skeleton from the same species: distal index and dorsal index of pelvis (following Fleagle and Anapol, 1992), crural index, femoral robusticity (femoral shaft diameter/femur length), height of greater trochanter (Anapol and Bischoff, 1992). The regression equations were used to predict class membership for each animal. All computations and statistical analyses were accomplished using Statistical Analysis Software (SAS, SAS Institute, Cary, NC) on the University of Wisconsin–Milwaukee IBM mainframe UWM-3270 computer.

Results

Gross morphology

Male (4260 g) and female (2980 g) vervets are slightly larger than male (3700 g) and female (2920 g) red-tailed monkeys (Fleagle, 1999), with greater sexual dimorphism occurring in the more terrestrial vervets. The gross morphology of the gluteal group and posterior flexors of the thigh (hamstrings) is similar enough to that of the rhesus macaque to allow easy recognition and dissection using Howell and Straus (1933). Interspecific differences with respect to muscle location and orientation are not remarkably different between vervets and red-tailed monkeys.

For this study, the gluteal group includes gluteus superficialis separated into anterior and posterior portions, g. medius, and g. minimus with the exclusion of tensor fasciae latae and piriformis. Although not as “fan-like” as in humans, the convergence of their fasciculi towards insertion allows a wide variety of actions.

As a group, they strongly retract and abduct (Howell and Straus, 1933), and rotate and flex (Stern, 1971; Vangor, 1979) the thigh.

Gluteus superficialis attaches proximally to a broad aponeurosis dorsal to the sacrum. The muscle is situated in parallel with the femoral shaft and is separable into anterior and posterior portions, which are treated separately for the architectural analysis in this study. Gluteus superficialis anterior arises from an extensive attachment, which includes the iliac crest, the sacrotuberous ligament, and the overlying dorsal aponeurosis. While some anterior fibers of the anterior portion fuse with the tensor fascia lata, most fibers fuse directly on fascia lata itself. The caudal fibers of g. superficialis converge beneath the cranial margin of biceps femoris, attaching distal to lateral lip of the linea aspera on the femoral shaft. The distal attachment of g. superficialis is positioned for retraction and lateral rotation of the femur. No conspicuous interspecific differences occur although the distal attachment of g. superficialis posterior extends further distally ($P < 0.05$) on the femur (see below) in vervets than in red-tailed monkeys (Table 6.1).

Gluteus medius is the most robust muscle of the group; its fibers of various lengths occupy most of the gluteal fossa. Proximally, the most superficial fibers attach to the iliac crest beneath the tensor fascia lata, while the deeper fibers attach to the dorsal and acetabular borders of the ilium. The fibers converge distally to attach to the greater trochanter via a strong tendon. Gluteus medius is positioned for extension, medial rotation, and, to a lesser degree, abduction of the thigh. In red-tailed monkeys, the fleshy fibers of g. medius approach the tip of the greater trochanter, and attach via a relatively thin tendon (Fig. 6.3). By contrast, in vervets, the muscle attaches further caudad on the greater trochanter by a thicker, more substantial tendon that encroaches onto the proximal posterior surface of the trochanter (Fig. 6.3), thus implying a more powerful capability for extending the thigh.

Gluteus minimus lies beneath g. medius, converging distally from the distal dorsal and ventral margins of the caudal ilium to attach to the greater trochanter. The anterior fibers, often referred to as scansorius, when more distinctly separated from pars posterior (although this is not the case in either vervets or red-tailed monkeys), are thought to medially rotate and flex the thigh (Stern, 1971; Vangor, 1979) in cebid monkeys, while the posterior fibers are considered to laterally rotate and abduct the thigh.

The posterior flexors of the thigh, or hamstrings, consist of biceps femoris, semimembranosus accessorius, semimembranosus proprius, and semitendinosus. Proximally, they attach, more or less in common, to the ischial tuberosity. The action of the biarticular hamstrings (biceps femoris, semimembranosus proprius, and semitendinosus) differs from that of the gluteals and semimembranosus accessorius by an added flexor (of the knee) component.

Table 6.1. *Means ± standard deviations of most proximal (PROX) and most distal (DIST) extents, midpoints (MDPT), and total lengths (SPRD) of the distal attachments (ATT) of the muscles of the extensors of the hip and the hamstrings. Each measurement is expressed as a proportion of the length of the long bone (femoral shaft for gluteus superficialis posterior and semimembranosus accessorius, and tibia for biceps femoris, semimembranosus proprius, and semitendinosus). Boldface indicates statistically significant interspecific differences (P < 0.10) with the larger value underscored.*

Muscle	ATT	n	Vervets	n	Red-tails
Gluteus superficialis posterior	PROX	6	0.351 ± 0.103	6	0.310 ± 0.051
	MDPT	6	**<u>0.500 ± 0.072</u>**	6	**0.434 ± 0.039**
	DIST	6	**<u>0.649 ± 0.068</u>**	6	**0.558 ± 0.056**
	SPRD	6	0.298 ± 0.100	6	0.247 ± 0.074
Biceps femoris	PROX	5	0.000 ± 0.000	6	0.000 ± 0.000
	MDPT	5	0.190 ± 0.033	6	0.209 ± 0.027
	DIST	5	0.380 ± 0.066	6	0.417 ± 0.053
	SPRD	5	0.380 ± 0.066	6	0.417 ± 0.053
Semimembranosus accessorius	PROX	6	0.000 ± 0.000	6	0.000 ± 0.000
	MDPT	6	**0.010 ± 0.002**	6	**<u>0.015 ± 0.006</u>**
	DIST	6	**0.019 ± 0.005**	6	**<u>0.030 ± 0.012</u>**
	SPRD	6	**0.019 ± 0.005**	6	**<u>0.030 ± 0.012</u>**
Semimembranosus proprius	PROX	6	0.034 ± 0.019	5	0.039 ± 0.015
	MDPT	6	0.047 ± 0.020	5	0.052 ± 0.013
	DIST	6	0.059 ± 0.021	5	0.066 ± 0.014
	SPRD	6	0.025 ± 0.005	5	0.026 ± 0.013
Semitendinosus	PROX	6	0.216 ± 0.041	6	0.218 ± 0.047
	MDPT	6	0.260 ± 0.036	6	0.279 ± 0.063
	DIST	6	0.304 ± 0.035	6	0.339 ± 0.080
	SPRD	6	**0.087 ± 0.024**	6	**<u>0.121 ± 0.036</u>**

Biceps femoris (flexor cruris lateralis) consists only of a single belly in guenons. From a narrow attachment on the ischial tuberosity, its longitudinal parallel fibers spread to a relatively wide distal attachment. The anterior-most fibers attach to the fascia lata with the remainder continuing into the crural fascia of the tibia. Biceps femoris is poised to extend the hip, flex the knee, and laterally rotate the leg. No significant differences distinguish vervets from red-tailed monkeys with respect to the distal attachment site of b. femoris.

Semimembranosus attaches proximally to the ischial tuberosity, from below the origin of semitendinosus, and splits along its length into two bellies, semimembranosus accessorius and semimembranosus proprius, which are treated separately here. Like all of the hamstrings, semimembranosus consists

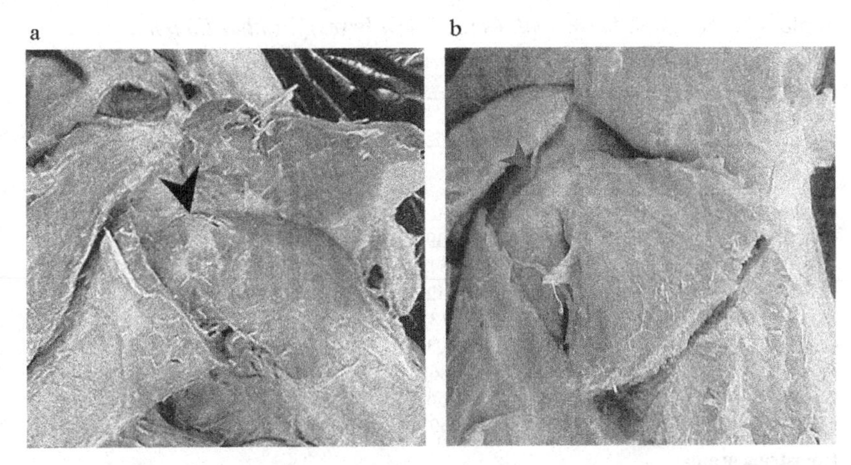

Figure 6.3. The attachment of gluteus medius on the greater trochanter is shown (arrowhead) for (a) *Chlorocebus aethiops* and (b) *Cercopithecus ascanius*. The tendon in (a) is more substantial than in (b) and wraps caudally over the trochanter.

of long, parallel fibers whose lengths vary with distal attachment. The fleshy attachment of semimembranosus accessorius is on the femoral shaft, medial to the linea aspera as far down as the level of the medial condyle. Although the proximal margin is identical in both species, the distal attachment of semimembranosus accessorius is slightly more extensive ($P = 0.067$) in red-tailed monkeys, its distal extent located somewhat further ($P = 0.067$) distally on the femoral shaft (Table 6.1). Semimembranosus proprius attaches via a small, cord-like tendon slightly ($\sim$ 2 cm) distal to the medial epicondyle of the tibia. No significant differences distinguish vervets from red-tailed monkeys with respect to the distal attachment site of semimembranosus proprius. Semimembranosus is poised to resist protraction of the femur (semimembranosus accessorius) and tibia (semimembranosus proprius) and contribute to propulsion.

Semitendinosus also attaches to the ischial tuberosity, but separately, beneath the attachment of biceps femoris, and, by a long broad tendon, attaches distally to the medial-to-posterior surface of the tibia, 2–3 cm below the level of the tuberosity. Although it may medially rotate the hind limb in some species (Howell and Straus, 1933), its more posterior attachment in vervets and red-tailed monkeys favors flexion of the leg as its primary function. Situated distal to that of semimembranosus on the medial surface of the tibia, the proximal extent of the distal attachment of semitendinosus begins just slightly more proximal in red-tailed monkeys than in vervets (Table 6.1). The distal extent of the attachment, however, is sufficiently further down the shaft in red-tailed monkeys to generate an interspecific difference ($P = 0.095$) in the total extent of the attachment.

Table 6.2. *Means and error values (in parentheses) for hind-limb muscle measurements in vervets (V) and red-tailed monkeys (R).*

	W		l_f		θ		l_b	
	V	R	V	R	V	R	V	R
Gluteal group								
GMed	18.167	14.258	44.730	39.939	3.804	4.863	74.914	70.916
	(2.370)	(6.552)	(3.782)	(4.054)	(0.558)	(0.559)	(4.452)	(5.115)
GMin	1.880	1.788	20.828	13.912	3.438	8.823	39.762	37.322
	(0.706)	(0.342)	(1.636)	(1.113)	(0.409)	(0.902)	(3.285)	(2.011)
GSuA	3.170	2.945	47.962	44.903	–	–	52.914	45.913
	(0.661)	(1.496)	(6.084)	(7.486)	–	–	(5.525)	(7.333)
GSuP	4.314	3.545	96.545	65.756	–	–	92.161	70.888
	(1.102)	(1.403)	(14.270)	(6.060)	–	–	(7.842)	(4.625)
Hamstring group								
BF	22.000	23.355	134.378	138.907	–	–	132.310	137.251
	(3.865)	(12.034)	(6.819)	(12.108)	–	–	(5.629)	(12.744)
SmA	16.993	18.208	117.458	134.116	–	–	129.723	121.902
	(4.569)	(9.627)	(12.079)	(13.004)	–	–	(9.695)	(11.410)
SmP	8.005	7.970	155.118	137.557	–	–	155.006	137.057
	(1.911)	(3.876)	(14.036)	(11.569)	–	–	(15.300)	(11.424)
St	5.405	6.508	150.200	139.295	–	–	149.900	138.555
	(0.775)	(3.155)	(13.082)	(9.899)	–	–	(13.235)	(9.535)

W: wet muscle weight
l_f: adjusted fasciculus length
θ: adjusted angle of pinnation
l_b: adjusted muscle belly length
GMed: gluteus medius
GMin: g. minimus
GSuA: g. superficialis pars anterior
GSuP: g. superficialis pars posterior
BF: biceps femoris
SmA: semimembranosus accessorius
SmP: semimembranosus proprius
St: semitendinosus
Error values are standard deviations for wet weight and standard errors for the remaining variables
$n = 6$, except for vervet $n = 5$ for GSuA and GSuP

Internal architectural morphology

In Table 6.2, means and error values are presented for raw muscle weight, adjusted fasciculus length, adjusted angle of pinnation, and adjusted belly length. Although each mean gluteal muscle of vervets outweighs its homolog in red-tailed monkeys, the reverse is generally true for each hamstring muscle. The single exception is semimembranosus proprius, which is more or less of equal weight in both species.

In Table 6.3, the results are presented for the calculated architectural variables. Significant differences were determined by using the student's *t*-test on species means. Because each muscle is represented by a mean of several samples, the use of the standard errors in this test is justified.

For each variable, the coefficient of variation adjusted for small-sample bias (V^*) is shown for gluteal and hamstring muscles separately. The muscles of the gluteal group have substantially higher values for V^* for all variables except $l_t/l_f + l_t$. The broadest inter-group disparities are found in both measures of velocity W/P_0 and h/l_b. No pronounced interspecific differences occur. The sample size ($n = 4$ muscles in each group) is too small to test for significant interspecific differences. Despite the absence of statistically significant differences between gluteal and hamstring muscle groups, differences in V^* likely indicate biological differences.

To account for interspecific body-size differences, the wet weight of each muscle of the gluteal or hamstring group is expressed as a percentage of the total weight of all muscles of its group (%W). For %W, no interspecific differences are statistically significant ($P < 0.05$).

Percentage reduced physiologic cross-sectional area (%RPCA) is the reduced physiologic cross-sectional area (RPCA) of a muscle expressed as a percentage of the total RPCA of the muscle group (gluteal or hamstring) to which it belongs (Table 6.3). For the gluteal group, statistically significant ($P < 0.05$) interspecific differences occur for both gluteus medius and g. minimus. G. medius shows greater %RPCA in vervets than in red-tailed monkeys and gluteus minimus shows greater %RPCA in red-tailed monkeys. Among the hamstrings, semimembranosus shows significantly greater %RPCA ($P < 0.05$) in vervets, while semitendinosus has a larger value ($P < 0.05$) in red-tailed monkeys.

The variable W/P_0 compares muscle weight to maximum tetanic tension, as calculated from RPCA (non-percentage), and reflects whether an individual muscle is better suited for power or for excursion/velocity. Higher values indicate that increased muscle weight is due to longer fiber length, rather than increasing muscle girth. The hamstrings all showed greater values for W/P_0 than did any of the gluteal muscles. Significantly higher values for W/P_0 for vervets ($P < 0.05$) occur in g. medius, g. minimus and the posterior portion of g. superficialis, with a higher (but not significantly higher) value in semimembranosus proprius. A significantly higher value for semimembranosus accessorius occurs in red-tailed monkeys.

All four gluteal muscles show interspecific significant differences ($P < 0.05$) for h/l_b, a variable that estimates potential whole-muscle excursion for an animal

Table 6.3. *Means and standard errors (in parentheses) for calculated architectural variables of hind-limb muscles in vervets (V) and red-tailed monkeys (R). Boldface indicates statistically significant interspecific differences ($P < 0.05$) with the larger value underscored.*

	%W		%RPCA		W/P_0		h/l_b		$l_t/l_f + l_t$	
	V	R	V	R	V	R	V	R	V	R
Gluteal group										
GMed	0.661	0.624	**0.677**	**0.587**	2.060	1.841	**0.138**	**0.128**	0.321	0.280
	(0.057)	(0.041)	**(0.031)**	**(0.031)**	(0.173)	(0.186)	**(0.007)**	**(0.004)**	(0.043)	(0.028)
GMin	0.068	0.086	**0.134**	**0.217**	**0.959**	**0.647**	**0.121**	**0.088**	**0.272**	**0.474**
	(0.023)	(0.024)	**(0.011)**	**(0.019)**	**(0.075)**	**(0.050)**	**(0.003)**	**(0.008)**	**(0.028)**	**(0.047)**
GSuA	0.115	0.132	0.110	0.112	2.203	2.062	**0.271**	**0.294**	0.646	0.675
	(0.023)	(0.032)	(0.022)	(0.019)	(0.279)	(0.344)	**(0.015)**	**(0.012)**	(0.030)	(0.028)
GSuP	0.155	0.158	0.078	0.085	**4.434**	**3.020**	**0.308**	**0.276**	0.272	0.303
	(0.036)	(0.020)	(0.021)	(0.006)	**(0.655)**	**(0.278)**	**(0.029)**	**(0.007)**	(0.042)	(0.034)
V*	118	107	122	98	**64**	**55**	**48**	**56**	51	45
Hamstring group										
BF	0.422	0.413	0.404	0.409	6.172	6.380	0.304	0.304	0.068	0.082
	(0.036)	(0.048)	(0.031)	(0.022)	(0.313)	(0.556)	(0.005)	(0.002)	(0.009)	(0.011)
SmA	0.321	0.324	**0.376**	**0.331**	**5.395**	**6.160**	**0.269**	**0.331**	–	–
	(0.035)	(0.077)	**(0.032)**	**(0.023)**	**(0.555)**	**(0.597)**	**(0.015)**	**(0.015)**	–	–
SmP	0.153	0.142	0.130	0.143	**7.125**	**6.318**	0.301	0.301	0.049	0.050
	(0.028)	(0.012)	(0.018)	(0.012)	**(0.645)**	**(0.531)**	(0.003)	(0.003)	(0.004)	(0.002)
St	0.104	0.121	**0.090**	**0.116**	6.899	6.398	0.301	0.301	**0.053**	**0.076**
	(0.010)	(0.053)	**(0.007)**	**(0.016)**	(0.601)	(0.455)	(0.001)	(0.001)	**(0.014)**	**(0.018)**
V*	63	60	69	61	**13**	**2**	**6**	**5**	72	75

Muscle abbreviations as in Table 6.2
See text for further details of calculated variables
$n = 6$, except for vervet $n = 5$ for GSuA and GSuP (all calculated variables) and GMed and GMin (%RPCA only)

in postural position (see Materials and methods). Generally, but not always, h/l_b tracks W/P_0. Similar to the results for W/P_0, greater values occur in vervets for g. minimus and the posterior portion of g. superficialis. Red-tailed monkeys have the higher value ($P < 0.05$) for the anterior portion of g. superficialis. Among the hamstring muscles, only semimembranosus accessorius exhibits a significant ($P < 0.05$) interspecific difference: it is better suited for excursion/velocity in red-tailed monkeys than in vervets. Expressed as whole-muscle length, the design of the hamstrings and g. superficialis is consistent with that fitting our model of a muscle producing large-excursion/high-velocity by contrast to g. medius or g. minimus in either species.

Both of these interspecific comparisons for potential velocity appear somewhat enigmatic. For g. superficialis anterior, red-tailed monkeys have the advantage ($P < 0.05$) over vervets for h/l_b, while for W/P_0 vervets have a sizable but not statistically significant advantage over red-tailed monkeys. Not only are the fasciculi relatively longer in red-tailed monkeys, accounting for higher h/l_b, but its weight as a proportion of the group total is slightly higher, thus yielding a muscle suited for both force and velocity. In semimembranosus proprius, vervets show a statistically significant ($P < 0.05$) advantage over red-tailed monkeys for W/P_0 yet both species have the same value for h/l_b. The explanation for this is similar to that for anterior g. superficialis, in that vervets have a proportionally heavier semimembranosus proprius with the additional weight dedicated to increasing fiber length. Thus, while the potential excursion/velocity *per se* may not distinguish these two species, semimembranosus proprius favors excursion/velocity over strength in vervets, more so than in red-tailed monkeys.

Interspecific differences in $l_t/l_f + l_t$ are found only in g. minimus and semitendinosus. For both muscles, red-tailed monkeys had longer tendons of attachment per muscle fasciculus.

The results of the randomization tests are presented in Table 6.4. In only six cases are differences significant ($P < 0.05$), all of which substantiate the statistical results shown in Table 6.3 and discussed above. No statistical differences are found for %RPCA. For h/l_b, statistically significant differences show g. minimus to have a greater value in vervets, and semimembranosus accessorius to have a greater value for red-tailed monkeys. Vervets also have significantly higher values for W/P_0 in g. minimus and in the posterior portion of g. superficialis, and show a significantly higher value for $l_t/l_f + l_t$ in g. minimus.

The predictability values derived from the logistic regression equations showed a grand mean of 64.90 for eight muscles, four calculated variables each. The grand mean for the six bone variables was 77.3. This suggests that the measurements taken of the bone of the hip region (distal index, dorsal index, crural index ($P < 0.05$), femoral robusticity ($P < 0.05$), height of the greater trochanter, and ischium length/ilium length) distinguish between vervets and

Table 6.4. *Randomization test results. The values in the table are the number of the 10 000 samples where the difference between the species equaled or exceeded the value of the observed value of the vervet mean minus the red-tail mean. P values are calculated by dividing the number in the table by 10 000. Values under 500 (vervet mean smaller than red-tail mean) and over 9,500 (vervet mean greater than red-tail mean) reach conventional significance levels of 0.05 and are indicated in bold.*

Muscle	%RPCA	W/P_0	h/l_b	$l_t/l_f + l_t$
GMed	5526	2672	2375	3145
GMin	4929	**195**	**94**	**9899**
GSuA	4808	3873	8686	7404
GSuP	3425	**327**	1424	6407
BF	5604	6124	4947	8347
SmA	4414	8042	**9802**	NA
SmP	5433	1521	**1**	6749
St	5201	2332	6824	8612

Abbreviations as in Tables 6.2. and 6.3.

red-tail monkeys better than do the gluteal muscles and posterior flexors of the thigh.

Discussion

An analysis of the organization of the constituent components of whole muscle provides for the "fine-tuning" of locomotor studies such that new, and potentially more accurate, interpretations of the functional "division of labor" of muscles in behavior can be considered. To this end, physiologically significant morphological variables are used to evaluate each muscle according to its potential maximum force output, potential maximum velocity, propensity for force or excursion, cost of tension transmission, and whether it is more suitable for isometric or isotonic use. The enhancement of functional interpretation bestowed by this approach over traditional means by which muscles are compared, e.g., muscle weight as a representative of force output, has been demonstrated in previous studies (Anapol and Jungers, 1986) and also is exemplified in the current study: e.g., for gluteus minimus, %RPCA is twice that of %W in vervets and almost thrice that of %W in red-tailed monkeys; for posterior g. superficialis, %RPCA is half that of %W in both species. The difference between these results relies on the simple, but often overlooked, premise, that most of the weight of any muscle results from fiber length, rather than cross-sectional area.

With respect to the calculated variables of these two muscle groups of the hip functional complex, fewer statistically significant differences separated these species than previously observed in their anti-gravity hind limb musculature (Anapol and Barry, 1996) and intrinsic muscles of the shoulder and arm (Anapol and Gray, 2003). Although the randomization test eliminated several of the statistically significant interspecific differences, those substantiated by student's *t*-test, and upon which the ensuing discussion will be based, likely represent biological differences suffering from sample size deprivation, due to the paucity of available material. Since both species primarily practice arboreality, any significant differences statistically demonstrated here can best be understood in terms of vervet terrestriality and their transition between ground and canopy.

In both of these guenon species, the muscles of the gluteal group are more different from one another (higher V*) for each of the four calculated variables (%RPCA, W/P_0, h/l_b, and $l_t/l_f + l_t$) than reported in previously studied anti-gravity muscle groups – quadriceps femoris of these same guenons (Anapol and Barry, 1996), *Lemur fulvus* (Anapol and Jungers, 1986), and *Felis domesticus* (Anapol, 1984) – as well as the hamstrings (for all but $l_t/l_f + l_t$) reported here. This is particularly true with respect to the two excursion/velocity variables. These results are consistent with those previously published by Anapol and Barry (1996) that showed higher values for V* among the bellies of more proximally oriented anti-gravity muscles groups. This also appears to be the case in the shoulder region, i.e., the more proximally located intrinsic shoulder muscles are more disparate than the intrinsic arm muscles (Anapol and Gray, 2003).

Interspecific comparisons

To facilitate the ensuing discussion, Table 6.5 shows the means from Table 6.3 (plus wet weight) normalized to the smallest value, for gluteal and hamstring functional groups separately. For each variable, the magnitude of interspecific differences between the relative distributions among muscles of a group can more easily be visualized.

Gluteal group

Anterior and posterior portions of gluteus superficialis exhibit some functional diversity in terrestrial, but not arboreal, species (Vangor, 1979). In the arboreal cebid monkeys, g. superficialis abducts and laterally rotates the thigh during climbing (Vangor, 1979). In the Old World terrestrial patas monkey, however, g. superficialis functions more like that of a dog (Tokuriki, 1973) or cat

Table 6.5. *For each calculated variable shown in Tables 6.2 and 6.3, the relative value of each muscle of a synergistic group (see Discussion)*

	W		%W		%RPCA		W/P_0		h/l_b		$l_t/l_f + l_t$	
	V	R	V	R	V	R	V	R	V	R	V	R
Gluteal group												
GMed	9.7	8	9.7	7.3	8.7	6.9	2.1	2.8	1.1	1.5	1.2	1
GMin	1	1	1	1	1.7	2.6	1	1	1	1	1	1.7
GSuA	1.7	1.6	1.7	1.5	1.4	1.3	2.3	3.2	2.2	3.3	2.4	2.4
GSuP	2.3	2.0	2.3	1.8	1	1	4.6	4.7	2.5	3.1	1	1.1
Hamstring group												
BF	4.1	3.6	4.1	3.4	4.5	3.5	1.1	1	1.1	1	1.4	1.6
SmA	3.1	2.8	3.1	2.7	4.2	2.9	1	1	1	1.1	–	–
SmP	1.5	1.2	1.5	1.2	1.4	1.2	1.3	1	1.1	1	1	1
St	1	1	1	1	1	1	1.3	1	1.1	1	1.1	1.5

Abbreviations as in Tables 6.2 and 6.3

(Rasmussen *et al.*, 1978): the anterior portion flexes the hip while the posterior portion extends the hip and has a greater role in lateral rotation of the thigh (Vangor, 1979). In red-tailed monkeys, higher excursion/velocity in the anterior portion during hip flexion would be advantageous in the rapid recapturing of purchase at end-swing phase in an animal ascending diagonal branches: greater excursion would increase stride length, and hence speed. That the posterior portion of g. superficialis both attaches further distally and has a statistically significant advantage for excursion/velocity in vervets, by contrast to red-tailed monkeys, implies a better mechanical advantage for extension of the thigh, when the hind limb is protracted, either for rapid terrestrial running (galloping) or during ascension from ground to tree.

In gluteus medius, which is by far the largest and strongest muscle of the group for either species, vervets hold an advantage over red-tailed monkeys in both potential force and potential velocity. The force advantage for vervets occurs at the expense of g. minimus, since force is shifted to a more superficial location, thus providing a somewhat longer lever arm. Apparently joint stability, needed by red-tailed monkeys in the arboreal habitat, is traded for increased force at a better moment arm for vervets running on the ground or possibly for ascension into the canopy. This interpretation is consistent with two general principles of muscle recruitment in anti-gravity muscles: (1) muscles situated closer to the bone tend to be more postural in their propensity to stabilize the joint, than more superficial muscles; (2) as the intensity of activity increases, muscle tissue is recruited from deep to more superficial muscles, and in more superficial fibers within whole muscles (Walmsley *et al.*, 1978; Armstrong,

1980); when force recruitment is exceptionally rapid, the deeper muscle may not be active at all, with the entire burden assumed by the more superficial muscle (Gillespie *et al.*, 1974; Smith *et al.*, 1977, 1980; Jungers *et al.*, 1980; Anapol and Jungers, 1987).

In cebids, g. medius inserts at the same level as the axis for extension of the hip joint, leading Stern (1971) to suggest that in these exclusively arboreal monkeys this muscle functions primarily as a medial rotator of the thigh. He noted, however, citing Gregory (1912) and Smith and Savage 1955), that raising the trochanteric attachment further dorsal to the femoral head would justify assigning the role of g. medius as a thigh extender. No statistically significant ($P < 0.05$) interspecific differences in trochanteric height distinguish these two guenon species (Anapol and Bischoff, 1992; Gebo and Chapman, 1995). However, as noted above (see Results), the attachment of g. medius wraps over the tip of the trochanter in vervets to attach more on its posterior surface than in red-tailed monkeys (Fig. 6.3). In this light, the compositionally more substantial and more posteriorly positioned attachment of g. medius on the trochanter in vervets indicates a greater potential for strong hind-limb extension.

Another point that bears mentioning is that, although g. medius is positioned to medially rotate the thigh, its effective function when the hind limb is stabilized would be to rotate the remainder of the body toward the side of the stable limb, e.g., when the right limb is in support phase, g. medius would redirect the animal toward the right. This would effectively increase lateral undulation, which would be somewhat less wieldy in the arboreal habitat. However, when running on the ground, the increased moment arm, coupled with increased force potential and increased excursion/velocity potential, would be significantly advantageous. Substrate reaction forces are directed medially when monkeys travel on a horizontal pole and laterally when traveling on the ground (Schmitt, 2003).

The relevance of this to the current study is that, if the orientation of g. medius for medial rotation without extension is standard for arboreal monkeys such as cebids, then the repositioning of the distal attachment of g. medius to a more posterior and more stable position on the trochanter indicates a muscular, rather than skeletal (e.g., increased trochanteric height), adaptation to accommodate a shift in locomotor modality. Electromyographic activity in g. medius increases with increased load in orthograde propulsive activities, such as climbing and bipedality, in otherwise quadrupedal primates (Vangor, 1979). Thus, the statistically significant advantages bestowed upon vervets for excursion/velocity for most muscles of the gluteal group may not necessarily be attributable exclusively to terrestrial vis-à-vis arboreal locomotion, but perhaps to their transition between ground and trees, while retaining an essentially terrestrial bony morphology (Gebo and Chapman, 1995).

In these two guenon species, g. minimus is not divisible into anterior (often known as scansorius) and posterior (pars posterior) portions (Stern, 1971), which is not to say that function does not vary across the breadth of the muscle. In cebid monkeys, the anterior fibers are thought to medially rotate and flex the thigh with some slight extension and abduction capability when the thigh is flexed, while the posterior fibers are thought to laterally rotate and abduct the thigh (Stern, 1971; Vangor, 1979). Having a greater force potential in red-tailed monkeys than in vervets, g. minimus provides a greater proportion of force located closer to the bone. This enables increased joint stability in the more arboreal species during a wider variety of hind-limb positions among irregularly spaced support structures of the three-dimensional substrate. As expected for terrestrial locomotion, the excursion/velocity variables favor the semiterrestrial vervets, while the muscle is used more isometrically in red-tailed monkeys.

Hamstring group

The hamstrings are the primary flexors of the knee and also retract the thigh when the knee joint is fixed (Howell and Straus, 1933). Vangor (1979), however, disputes the contribution of biceps femoris to knee flexion because of its broad, expansive distal attachment across the joint which may stabilize, rather than flex. Fewer interspecific differences occur in the hamstrings than in the gluteal muscles. Composed entirely of parallel-fibered muscles with primarily fleshy attachments, the hamstrings are all about speed, with values for W/P_0, and for the most part h/l_b, generally much higher than in the gluteals (above) and all hind-limb extensors for which comparable classes of results have been published (Anapol, 1984; Anapol and Jungers, 1986; Anapol and Barry, 1996; Anapol, 2003).

Irrespective of statistical significance, the distal attachment sites of all hamstring muscles are situated further proximally on the tibia in vervets, resulting in a smaller group moment arm caudal to the knee joint. A smaller moment arm produces less acceleration, but of longer duration, resulting in greater speeds, i.e., the limb would be traveling faster at the end of movement (Stern, 1974) – a beneficial adaptation for the rapid terrestrial quadrupedalism practiced by this species.

With respect to potential velocity, except for semimembranosus accessorius, the hamstrings of these two guenon species are generally indistinguishable. This is similar to the constituent bellies of the antagonistic muscle group, quadriceps femoris (Anapol and Barry, 1996). Thus, both flexion and extension of the crus are comparable in both species, regardless of locomotor preference. This is by contrast to the more proximal (hip) and distal (ankle) joints: when on the ground, the gluteal group and triceps surae (Anapol and Barry, 1996) provide greater excursion/velocity in vervets. The lack of interspecific differences across the knee joint likely results from a general requirement both to resist hyperextension at the knee joint and to coordinate support- and swing-phase activity in both these, and probably most other, quadrupeds.

Semimembranosus accessorius is active monophasically from late swing through early support phase, functioning both to brake the forward momentum of the hind limb and to contribute to propulsion from touchdown to the end of support (Vangor, 1979). In both the suspensory arboreal quadruped *Ateles* and the terrestrial quadruped *Erythrocebus*, semimembranosus in its entirety functions similarly to semimembranosus accessorius of dogs and cats as a hip extender with no effect on the knee, while in the arboreal quadrupedal climber *Lagothrix* it functions like semimembranosus proprius of these nonprimates, i.e., as a knee flexor and adductor of the hind limb (Vangor, 1979). Semimembranosus proprius, however, is active out of phase with semimembranosus accessorius, sometimes firing biphasically (Vangor, 1979).

As in the gluteals, an intermuscular redistribution occurs among the hamstrings, differentially positioning the concentration of force: vervets have a greater potential force output for semimembranosus accessorius at the expense of semitendinosus, the latter of which is proportionally greater in red-tailed monkeys. Since semimembranosus accessorius attaches to the distal femur and semitendinosus to the proximal tibia, this reflects a redistribution of power to a more proximal position, as is found in ungulates (Hildebrand, 1959), for terrestrial running often practiced by vervets, and further distally for the more committed arborealist, the red-tailed monkey. Its position in the latter also may indicate a greater dependence upon crural-based propulsion while in the semi-crouched position necessary for stability on arboreal supports (Schmitt, 1998). In cats, dogs, and primates, semitendinosus fires biphasically, relating it to both end-swing deceleration and hip extension, in preparation of toe-off during swing phase (Vangor, 1979).

Semimembranosus accessorius seems to be more suited for excursion/velocity ($P < 0.05$) than force in red-tailed monkeys, while the more distally attaching semimembranosus proprius is more inclined toward excursion/velocity in vervets ($P < 0.05$). This may, in vervets, reflect an increase in a medially directed vector of "braking" to modulate termination of protraction during rapid terrestrial progression.

Implications for evolutionary morphology

Statistically significant differences between vervets and red-tailed monkeys in the morphology of the gluteal and hamstring muscles are interpretable with respect to their locomotor preferences. However, predictability of class membership using logistic regression equations suggests that interspecific differences are much less pronounced in these muscles than are bony measurements taken of the hip region (Anapol and Bischoff, 1992). Using logistic regression equations, Anapol and Gray (2003) present comparable results between intrinsic muscles of the shoulder and arm and related bony measurements. Anapol and Barry (1996) initially alluded to this, more generally, by comparing interspecific overlap in error bars between similar data from hind-limb extensors and measurements on related long bones (Anapol and Bischoff, 1992).

Similar to what is observed in the skeleton, gross muscle morphology, e.g., attachment positions, innervation and compartmentalization patterns, clearly distinguishes between species to the extent that phylogenetic (Woods and Hermanson, 1985) as well as functional (English, 1984) hypotheses can be generated. However, the overall implication from the current, and other architecture studies (Anapol and Barry, 1996; Anapol and Gray, 2003), is that physiologically related morphological features of muscle distinguish less clearly between closely related species than do the topographical features of their bones. Within an evolutionary context, the physiologic underpinnings of muscles can provide a variably broad range of adaptive potential, allowing a species to "experiment" facultatively with one or more alternative locomotor niches during any behavioral transitions that may result from environmental instability. When one alternative becomes more behaviorally constitutive, biomechanical advantages for functional complexes most adaptive for the new behavior would be proffered by selection for alterations in bone morphology in a manner which supports new muscle orientations (Anapol and Barry, 1996; Anapol and Gray, 2003).

Several authors (e.g., Groves, 1989, 2000; Fleagle, 1999; Disotell, 2000) remove vervets from the genus *Cerocopithecus* to *Chlorocebus*, a taxonomic position between the arboreal *C. diana* and the terrestrial *Erythrocebus*. Groves (1989) further suggests that *Erythrocebus* and *C. aethiops* might constitute a sister group to all other cercopithecine species. The morphological results presented here support this in that they are consistent with a concept of "semiterrestriality" as a unique locomotor category, rather than one that implies simply "sometimes arboreal" and "sometimes terrestrial." Semiterrestrial locomotion as practiced by vervets, therefore, can be perceived as a *bone fide* behavioral mode, rather than that of a terrestrial or arboreal animal that occasionally occupies the alternative habitat.

Conclusions

Fewer interspecific differences in the gluteal and hamstring muscle groups distinguish vervets from red-tailed monkeys than in the anti-gravity hind-limb muscle groups, quadriceps femoris and triceps surae. The most pronounced differences are that the gluteal muscles are generally constructed more for velocity in the semiterrestrial vervets than in the more arboreal red-tailed monkeys, a finding consistent with previous studies (Anapol and Barry, 1996; Anapol and Gray, 2003). In both species, the gluteal and hamstring muscles differ from the previously studied muscles in having dramatically less energy-efficient, more isotonically utilized construction. The calculated variables collected from the gluteal and hamstring muscles do not distinguish between these species nearly as much as do variables previously collected from their associated skeletal structures. This implies that the inherent plasticity of muscle tissue may be preadaptive for facultative behavioral modifications necessitated by a shift in environmental niche, prior to long-term selection for constitutive changes in bone proportions and topography.

Acknowledgments

We are grateful to Dr. Neil Tappen, who generously provided most of the cadavers used in this study, and to Jean Baulu of the Barbados Primate Research Center and Wildlife Reserve and Dr. Gary Baskin of the Tulane Regional Primate Center, who provided the remainder. We also appreciate the diligent work of the reviewers who greatly improved the quality of the manuscript. The work was funded by a grant from the National Science Foundation (DBS-9221795).

References

Alexander, R. McN. (1974). The mechanics of jumping by a dog (*Canis familiaris*). *J. Zool. Lond.*, **173**, 549–573.

Alexander, R. McN. and Bennet-Clark, H. C. (1977). Storage of elastic strain energy in muscle and other tissue. *Nature*, **265**, 114–117.

Anapol, F. (1984). *Morphological and functional diversity within the quadriceps femoris in* Lemur fulvus: *architectural, histochemical, and electromyographic considerations.* Ph.D. thesis, State University of New York at Stony Brook.

(1988). Morphological and videofluorographic study of the hyoid apparatus and its function in the rabbit (*Oryctolagus cuniculus*). *J. Morphol.*, **195**, 141–157.

(2003). Fiber architecture in primate limb muscles with new data for *triceps surae* in *Eulemur fulvus*. *Amer. J. Phys. Anthropol.* (Suppl. 36), 59.

Anapol, F. and Barry, K. (1996). Fiber architecture of the extensors of the hindlimb in semiterrestrial and arboreal guenons. *Amer. J. Phys. Anthropol.*, **99**, 429–447.

Anapol, F. and Bischoff, L. K. (1992). Comparative skeletal adaptations of *Cercopithecus ascanius* and *C. aethiops* to locomotor behavior. *Amer. J. Phys. Anthropol.* (Suppl. 14), 43.

Anapol, F. and Gray, J. P. (2003). Fiber architecture of the intrinsic muscles of the shoulder and arm in semiterrestrial and arboreal guenons. *Amer. J. Phys. Anthropol.*, **122**, 51–65.

Anapol, F. and Herring, S. W. (1989). Length–tension relationships of masseter and digastric muscles of miniature swine during ontogeny. *J. Exp. Biol.*, **143**, 1–16.

Anapol, F. and Jungers, W. L. (1986). Architectural and histochemical diversity within the quadriceps femoris of the brown lemur (*Lemur fulvus*). *Amer. J. Phys. Anthropol.*, **69**, 355–375.

(1987). Telemetered electromyography of the fast and slow extensors of the leg of the brown lemur (*Lemur fulvus*). *J. Exp. Biol.*, **130**, 341–358.

Anapol, F., Muhl, Z. F., and Fuller, J. H. (1987). The force–velocity relation of the rabbit digastric muscle. *Arch. Oral Biol.*, **32**, 93–99.

Appleton, A. B. (1921). The gluteal region of *Tarsius spectrum*. *Proc. Camb. Philos. Soc.*, **20**, 466–474.

Armstrong, R. B. (1980). Properties and distributions of the fiber types in the locomotory muscles of mammals. In: *Comparative Physiology: Primitive Mammals*, ed. K. Schmidt-Nielsen, L. Bolis, and C. R. Taylor. Cambridge: Cambridge University Press. pp. 243–254.

Armstrong, R. B. and Phelps, R. O. (1984). Muscle fiber type composition of the rat hindlimb. *Amer. J. Anat.*, **171**, 259–272.

Armstrong, R. B., Sauber, C. W., Seeherman, H. J., and Taylor, C. R. (1982). Distribution of fiber types in locomotory muscles of dogs. *Amer. J. Anat.*, **163**, 87–98.

Ashton, E. H. and Oxnard, C. E. (1963). The musculature of the primate shoulder. *Trans. Zool. Soc. Lond.*, **29**, 553–650.

Benninghoff, A. and Rollhäuser, H. (1952). Zur inneren Mechanik des gefiederten Muskels. *Pflugers Arch. Ges. Physiol.*, **254**, 527–548.

Biewener, A. (1998). Muscle function *in vivo*: a comparison of muscles used for elastic energy savings versus muscles used to generate mechanical power. *Amer. Zool.*, **38**, 703–717.

Biewener, A., Alexander, R. McN., and Heglund, N. C. (1981). Elastic energy storage in the hopping of kangaroo rats (*Dipodomys spectabilis*). *J. Zool. Lond.*, **195**, 369–383.

Biewener, A. and Baudinette, R. V. (1998). In vivo muscle force-length behavior during steady-speed hopping of tammar wallabies. *J. Exp. Biol.*, **201**, 1681–1694.

Buchner, H. (1877). Kritische und experimentelle Studien über den Zusammenhalt des Hüftgelenks während des Lebens in allen normalen Fallen. *Arch. Anat. Physiol.*, **1877**, 22–45.

Burke, R. E. and Edgerton, V. R. (1975). Motor unit properties and selective involvement in movement. *Exerc. Spt. Sci. Rev.*, **3**, 31–81.

Cavagna, G. A. (1969). Travail mécanique dans la marche et la course. *J. Physiol., Paris*, **61**, 3–42.

Cavagna, G. A. and Kaneko, M. (1977). Mechanical work and efficiency in level walking and running. *J. Physiol.*, **268**, 467–481.

Cavagna, G. A., Citterio, G., and Jacini P. (1980). Elastic storage: role of tendons and muscles. In: *Comparative Physiology: Primitive Mammals*, ed. K. Schmidt-Nielsen, L. Bolis, and C. R. Taylor. Cambridge: Cambridge University Press. pp. 231–242.

Disotell, T. R. (2000). Molecular systematics of the Cercopithecidae In: *Old World Monkeys*, ed. P. F. Whitehead and C. J. Jolly. Cambridge: Cambridge University Press. pp. 29–56.

Edgington, E. (1995). *Randomization Tests*. 3rd edn. New York, NY: Marcel Dekker.
English, A. W. (1984). An electromyographic analysis of compartments in cat lateral gastrocnemius muscle during unrestrained locomotion. *Exp. Neurol.*, **87**, 96–108.
Fawcett, D. W. (1986). *Bloom and Fawcett, a Textbook of Histology*. 11th edn. Philadelphia, PA: Saunders.
Fick, A. (1860). Über die Längenverhältnisse der Skelettmuskelfasern. Aus der Inaug.-Diss. des Herrn Dr. GUBLER, mitgeteilt von A. FICK. Moleschtt Us, **8**, 253–263 (1860) und GS, **1**, 445–455 (1903).
Fick, R. (1910). Anatomie und Mechanik der Gelenke unter Berücksichtigung der bewegenden Muskeln. In: *Handbuch der Anatomie des Menschen*, ed. K. V. Bardeleben. Jena: Gustav Fischer. Vol. 2, sect. 1, part 2, pp. 1–363.
Fleagle, J. G. (1977). Locomotor behavior and muscular anatomy of sympatric Malaysian leaf monkeys (*Presbytis obscura* and *Presbytis melalophos*). *Amer. J. Phys. Anthropol.*, **46**, 297–308.
(1999). *Primate Adaptation and Evolution*. 2nd edn. San Diego, CA: Academic Press.
Fleagle, J. G. and Anapol, F. (1992). The indriid ischium and the hominid hip. *J. Hum. Evol.*, **22**, 285–305.
Fleagle, J. G., Stern, J. T., Jr., Jungers, W. L., and Susman, R. L. (1981). Climbing: a biomechanical link with brachiation and with bipedalism. In: *Vertebrate Locomotion*, ed. M. Day. Symp. Zool. Soc. Lond. 48. London: Academic Press. pp. 359–375.
Gans, C. (1982). Fiber architecture and muscle function. *Exerc. Spt. Sci. Rev.*, **10**, 160–207.
Gans, C. and Bock, W. F. (1965). The functional significance of muscle architecture – a theoretical analysis. *Ergeb. Anat. Entwicklungsgesch.*, **38**, 115–142.
Gans, C. and de Vree, F. (1989). Functional bases of fiber length and angulation in muscle. *J. Morphol.*, **192**, 63–85.
Gaunt, A. S. and Gans, C. (1992). Serially arranged myofibers: an unappreciated variant in muscle architecture. *Experientia*, **48**, 864–868.
Gebo, D. L. and Chapman, C. A. (1995). Positional behavior in five sympatric Old World monkeys. *Amer. J. Phys. Anthropol.*, **97**, 49–76.
Gillespie, C. A., Simpson, D. R., and Edgerton, V. R. (1974). Motor unit recruitment as reflected by muscle fiber glycogen loss in a prosimian (bushbaby) after running and jumping. *J. Neurol. Neurosurg. Psych.*, **37**, 817–824.
Gonyea, W. J. and Ericson, G. C. (1977). Morphological and histochemical organization of the flexor carpi radialis muscle in the cat. *Amer. J. Anat.*, **148**, 329–344.
Good, P. I. (1999). *Resampling Methods: a Practical Guide to Data Analysis*. Boston, MA: Birkhäuser.
Gordon, A. M., Huxley, A. F., and Julian, F. J. (1966). The variation in isometric tension with sarcomere length in vertebrate muscle fibres. *J. Physiol.*, **184**, 170–192.
Goslow, G. E., Jr., Cameron, W. E., and Stuart, D. G. (1977). Ankle flexor muscles in the cat: length-active tension and muscle unit properties as related to locomotion. *J. Morphol.*, **15**, 23–38.
Grand, T. I. (1967). The functional anatomy of the ankle and foot of the slow loris (*Nycticebus coucang*). *Amer. J. Phys. Anthropol.*, **26**, 207–218.

(1968a). The functional anatomy of the lower limb of the Howler monkey (*Alouatta caraya*). *Amer. J. Phys. Anthropol.*, **28**, 163–183.

(1968b). Functional anatomy of the upper limb. In: *Biology of the Howler Monkey (Alouatta caraya)*, ed. M. R. Malinow. Bibl. Primatol., 7. Basel: Karger. pp. 104–125.

(1977). Body weight: its relation to tissue composition, segment distribution, and motor function. I. Interspecific comparisons. *Amer. J. Phys. Anthropol.*, **47**, 211–239.

Grand, T. I. and Lorenz, R. (1968). Functional analysis of the hip joint in *Tarsius bancanus* (Horsfield, 1821) and *Tarsius syrichta* (Linnaeus, 1758). *Folia Primatol.*, **9**, 161–181.

Gray, J. E. (1870). *Catalogue of Monkeys, Lemurs and Fruit-eating Bats in the Collection of the British Museum*. London: Trustees of the British Museum.

Gregory, W. K. (1912). Notes on the principles of quadrupedal locomotion and on the mechanism of the limbs in hoofed animals. *Ann. N. Y. Acad. Sci.*, **22**, 267–294.

Groves, C. P. (1989). *A Theory of Human and Primate Evolution*. Oxford: Oxford University Press.

(2000). The phylogeny of the Cercopithecoidea. In: *Old World Monkeys*, ed. P. F. Whitehead and C. J. Jolly. Cambridge: Cambridge University Press. pp. 77–98.

Haines, R. W. (1932). The laws of muscle and tendon growth. *J. Anat.*, **66**, 578–585.

Haxton, H. A. (1944). Absolute muscle force in the ankle flexors of man. *J. Physiol. Lond.*, **103**, 267–273.

Herring, S. W., Wineski, L. E., and Anapol, F. C. (1989). Neural organization of the masseter muscle in the pig. *J. Comp. Neurol.*, **280**, 563–576.

Herring, S. W., Anapol, F. C., and Wineski, L. E. (1991). Motor unit territories in the masseter muscle of infant pigs. *Archiv. Oral Biol.*, **36**, 867–873.

Hildebrand, M. (1959). Motions of the running cheetah and horse. *J. Mammal.*, **40**, 481–495.

(1974). *Analysis of Vertebrate Structure*. New York, NY: Wiley.

Hill, A. V. (1938). The heat of shortening and the dynamic constants of muscle. *Proc. Roy. Soc. B*, **126**, 136–194.

Howell, A. B. and Straus, W. L., Jr. (1933). The muscular system. In: *The Anatomy of the Rhesus Monkey*, ed. C. G. Hartman and W. L. Straus. New York, NY: Hafner. pp. 89–175.

Hoyle, G. and Smyth, T., Jr. (1963). Giant muscle fibers in a barnacle, *Balanus nubilus* (Darwin). *Sci.*, **139**, 49–50.

Huxley, A. F. (1957). Muscle structure and theories of contraction. *Prog. Biophys. Biophys. Chem.*, **7**, 255–318.

Jenkins, F. A. and Weijs, W. A. (1979). The functional anatomy of the shoulder in the Virginia opossum (*Didelphis virginiana*). *J. Zool. Lond.*, **188**, 379–410.

Jungers, W. L., Jouffroy, F. K., and Stern, J. T., Jr. (1980). Gross structure and function of the quadriceps femoris in *Lemur fulvus*: an analysis based on telemetered electromyography. *J. Morphol.*, **164**, 287–299.

Jungers, W. L., Stern, J. T., Jr., and Jouffroy, F. K. (1983). Functional morphology of the quadriceps femoris in primates: a comparative anatomical and experimental analysis. *Ann. Sci. Nat., Zool., Paris*, 13 Serie, **5**, 101–116.

Kimura, K., Takahashi, Y., Konishi, M., and Iwamoto, S. (1983). Extensor muscles of the thigh of crab-eating monkeys (*Macaca fascicularis*). *Primates*, **24**, 86–93.

Larson, S. G. and Stern, J. T., Jr. (1986). EMG of scapulohumeral muscles in the chimpanzee during reaching and "arboreal" locomotion. *Amer. J. Anat.*, **176**, 171–190.

(1987). EMG of chimpanzee shoulder muscles during knuckle-walking: problems of terrestrial locomotion in a suspensory adapted primate. *J. Zool. Lond.*, **212**, 629–655.

Lewontin, R. C. (1966). On the measurement of relative variability. *Syst. Zool.*, **15**, 141–142.

Lieber, R. L. and Blevins, F. T. (1989). Skeletal muscle architecture of the rabbit hindlimb: functional implications of muscle design. *J. Morphol.*, **199**, 93–101.

Manly, B. F. J. (1997). *Randomization, Bootstrap and Monte Carlo Methods in Biology*. 2nd edn. New York, NY: Chapman Hall.

McClearn, D. (1985). Anatomy of racoon (*Procyon lotor*) and coati (*Nasua narica* and *N. nasua*) forearm and leg muscles: relations between fiber length, moment-arm length, and joint-angle excursion. *J. Morphol.*, **183**, 87–115.

Muhl, Z. F. (1982). Active length–tension relation and the effect of muscle pinnation on fiber lengthening. *J. Morphol.* **173**, 285–292.

Murphy, R. A. and Beardsley, A. C. (1974). Mechanical properties of the cat soleus muscle *in situ*. *Amer. J. Physiol.*, **227**, 1008–1013.

Oxnard, C. E. (1975). *Uniqueness and Diversity in Human Evolution: Morphometric studies of Australopithecines*. Chicago, IL: University of Chicago Press.

Paul, R. J. (1983). Physical and biochemical energy balance during an isometric tetanus and steady state recovery in frog sartorius at 0 °C. *J. Gen. Physiol.*, **81**, 337–354.

Peters, S. E. and Rick, C. (1977). The actions of three hamstring muscles of the cat: a mechanical analysis. *J. Morphol.*, **152**, 315–328.

Pfuhl, W. (1937). Die gefiederten Muskeln, ihre Form und ihre Wirkungsewise. *Z. Anat. Enwickl.-Gesch.*, **196**, 749–769.

Prophet, E. B., Mills, B., Arrington, J. B., and Sobin, L. H. (1992). *Laboratory Methods in Histotechnology*. Washington, DC: Armed Forces Institute of Pathology.

Quest, W. J. (1987). *Estimation of Intrinsic Strength of Rabbit Digastric Muscles from Physiological Cross-Sectional Area*. M.S. thesis, University of Illinois at Chicago.

Ramsey, R. W. and Street, S. F. (1940). The isometric length tension diagram of isolated skeletal muscle fibers of the frog. *J. Cell. Comp. Physiol.*, **15**, 11–34.

Rasmussen, S., Chan, A. K., and Goslow, G. E., Jr. (1978). The cat step cycle: electromyographic patterns for hind limb muscles during posture and unrestrained locomotion. *J. Morphol.*, **155**, 253–270.

Reiser, P. J., Moss, R. L., Giulian, G. G., and Greaser, M. L. (1985). Shortening velocity in single fibers from adult rabbit soleus muscle is correlated with myosin heavy chain composition. *J. Biol. Chem.*, **260**, 9077–9080.

Richmond, F. J. R. (1998). Elements of style in neuromuscular architecture. *Amer. Zool.*, **38**, 729–742.

Sacks, R. D. and Roy, R. R. (1982). Architecture of the hind limb muscles of cats: Functional significance. *J. Morphol.*, **173**, 185–195.

Scapino, R. P. (1968). *Biomechanics of Feeding in Carnivora*. Ph.D. thesis, University of Illinois at Chicago.

Schmitt, D. (1998). Forelimb mechanics during arboreal and terrestrial quadrupedalism in Old World monkeys. In: *Primate Locomotion: Recent Advances*, ed. E. Strasser, J. G. Fleagle, H. McHenry, and A. Rosenberger. New York, NY: Plenum Press. pp. 175–200.

(2003). Mediolateral reaction forces and forelimb anatomy in quadrupedal primates: implications for interpreting locomotor behavior in fossil primates. *J. Hum. Evol.*, **44**, 47–58.

Schumacher, G. H. (1961). *Funktionelle Morphologie der Kaumuskulatur*. Jena: Gustav Fischer. (Trans. by Z. Muhl).

Shapiro, S. S. and Wilk, M. B. (1965). An analysis of variance test for normality (complete samples). *Biometrika*, **52**, 591–611.

Smith, J. L., Edgerton, V. R., Betts, B., and Collatos, T. C. (1977). EMG of slow and fast ankle extensors of cat during posture, locomotion, and jumping. *J. Neurophysiol.*, **40**, 503–513.

Smith, J. L., Betts, B., Edgerton, V. R., and Zernicke R. F. (1980). Rapid ankle extension during paw shakes: selective recruitment of fast ankle extensors. *J. Neurophysiol.*, **43**, 612–620.

Smith, J. M. and Savage, R. J. G. (1955). Some locomotory adaptations in mammals. *J. Linn. Soc. Lond. Zool.*, **42**, 603–622.

Sokal, R. R. and Rohlf, F. J. (1981). *Biometry.* 2nd edn. San Francisco, CA: W. H. Freeman.

Spector, S. A., Gardiner, P. F., Zernicke, R. F., Roy, R. R., and Edgerton, V. R. (1980). Muscle architecture and force–velocity characteristics of cat soleus and medial gastrocnemius: implications for motor control. *J. Neurophysiol.*, **44**, 951–960.

Stern, J. T., Jr. (1971). *Functional Myology of the Hip and Thigh of Cebid Monkeys and its Implications for the Evolution of Erect Posture.* Bibl. Primatol, **14**. Basel: Karger.

(1974). Computer modeling of gross muscle dynamics. *J. Biomech.*, **7**, 4111–28.

Stern, J. T. Jr., Wells, J. P., Vangor, A. K., and Fleagle, J. G. (1977). Electromyography of some muscles of the upper limb in *Ateles* and *Lagothrix*. *Yrbk. Phys. Anthropol.*, **20**, 98–507.

Swartz, S. M. and Tuttle, R. H. (1990). Allometric patterns in the limb musculature of catarrhine primates. *Amer. J. Phys. Anthropol.*, **81**, 304.

Tappen, N. C. (1991). Present and future status of the Tappen collection of primate specimens at the University of Wisconsin–Milwaukee. In: *Primatology Today*, ed. A. Ehara and T. Kimura. Amsterdam: Elsevier. pp. 555–558.

Tokuriki, M. (1973). Electromyographic and joint-mechanical studies in quadrupedal locomotion. I. Walking. *Jpn. J. Vet. Sci.*, **35**, 433–446.

Tuttle, R. H. and Basmajian, J. V. (1978a). Electromyography of pongid shoulder muscles. II. Deltoid, rhomboid and "rotator cuff." *Amer. J. Phys. Anthropol.*, **49**, 47–56.

(1978b). Electromyography of pongid shoulder muscles. III. Quadrupedal positional behavior. *Amer. J. Phys. Anthropol.*, **49**, 57–70.

Vangor, A. K. (1979). *Electromyography of Gait in Non-Human Primates and its Significance for the Evolution of Bipedality.* Ph.D. thesis, State University of New York at Stony Brook.

Walmsley, B. and Proske, U. (1981). Comparison of stiffness of soleus and medial gastrocnemius muscles in cats. *J. Neurophysiol.*, **46**, 250–259.
Walmsley, B., Hodgson, J. A., and Burke, R. E. (1978). Forces produced by medial gastrocnemius and soleus muscles during locomotion in freely moving cats. *J. Neurophysiol.*, **41**, 1203–1216.
Weber, E. F. (1851). Uber die Langenverhaltnisse der Fleischfasern der Muskeln im allgemeinen. *Ber. K. sachs. Ges. Wiss. nat. phys.*, **K1**, 64–86.
Wilkie, D. R. (1968). Heat work and phosphoryl creatine breakdown in muscle. *J. Physiol.*, **195**, 157–183.
Willemse, J. J. (1963). Some characteristics of muscle fibers in a pinnate muscle. *Proc. Kon. Ned. Akad. Wet.* Ser. C, **66**, 162–171.
(1977). Morphological and functional aspects of the arrangement of connective tissue and muscle fibres in the tail of the Mexican axolotl, *Siredon mexicanum* (Shaw) (Amphibia, Urodela). *Acta Anat.*, **97**, 266–285.
Wineski, L. and Gans, C. (1984). Morphological basis of the feeding mechanics in the shingleback lizard *Trachydosaurus rugosus* (Scincidae: Reptilia). *J. Morphol.*, **181**, 271–295.
Woods, C. A. and Hermanson, J. W. (1985). Myology of hystricognath rodents: an analysis of form, function and phylogeny. In: *Evolutionary Relationships Among Rodents: a Multidisciplinary Analysis*, ed. W. P. Luckett and J. L. Hartenberger. New York, NY: Plenum Press. pp. 515–548.
Zar, J. H. (1984). *Biostatistical Analysis*. 2nd edn. Englewood Cliffs, NJ: Prentice-Hall.

7 Comparative fiber-type composition and size in the antigravity muscles of primate limbs

FRANÇOISE K. JOUFFROY, MONIQUE F. MÉDINA
Muséum National d'Histoire Naturelle

Gravity: the most compelling factor in the evolution of primate locomotor systems

Inflationary increase in the number of space flights and length of time spent in space stations – likely to become casual trips for wealthy tourists – has brought to light how much the structure and function of the locomotor system depend upon gravitational force. The sort of "swimming" movements of astronauts within space stations have become familiar images on our TV screens. Adaptation to weightlessness for an easy life in space has required scientists to analyze and overcome the alterations generated in muscles by the absence of weight constraints. Astronauts are well aware of the changes occurring within their body in the microgravitational environment, as well as their reversal during the recovery phase following return to earth. Astronauts' sensations are in accordance with experimental results drawn from flown rats and macaques. Changes in muscle structure were reported first in rats, after only two weeks in space (Desplanches *et al.*, 1987; Riley *et al.*, 1990; Desplanches, 1997; Fitts *et al.*, 2001). Significant advances in our understanding of the effect of gravity on the primate muscle system resulted from two series of studies of flown rhesus monkeys (*Macaca mulatta*) during 14- and 12-day space flights: (1) COSMOS 2044, COSMOS 2229 (Bodine-Fowler *et al.*, 1995; Roy *et al.*, 1996) and (2) BION 11 (Roy *et al.*, 1999, 2000; Belozerova *et al.*, 2000; Mounier *et al.*, 2000; Chopard *et al.*, 2000; Fitts *et al.*, 2000a, 2000b, 2001; Kischel *et al.*, 2001).

Shaping Primate Evolution, ed. F. Anapol, R. Z. German, and N. G. Jablonski. Published by Cambridge University Press.

In addition to global metabolic protein deficit, a series of significant changes were observed, mainly in slow fibers:

(1) decrease of the overall volume and muscular force;
(2) modifications in the relative EMG recruitment of slow and fast extensors of the ankle (Hodgson *et al.*, 1991; Roy *et al.*, 1996);
(3) decrease in percentage of type I fibers in postural muscles (Chopard *et al.*, 2000);
(4) decrease in size of type I and hybrid fibers (Belozerova *et al.*, 2000; Mounier *et al.*, 2000);
(5) increase in percentage of hybrid fibers in postural muscles (Mounier *et al.*, 2000; Kischel *et al.*, 2001).

Amazingly, adaptive changes to the microgravitational environment of spacecraft shed new light on the adaptation and evolution of vertebrate locomotor systems. According to Newton's first law of motion, the resultant of the forces acting on a motionless body is nil. Within the earth's gravitational field this implies that when an animal stands still a system of forces generated by muscles counteracts gravity, i.e., the muscular forces. Hence, the specialization of some "postural" muscles for antigravity function. Differences between the muscle characteristics of ground-dwelling and arboreal mammals illustrate conspicuously opposite trends to overcome and master gravity (Jouffroy *et al.*, 1990). With ground-dwelling species, whether quadrupeds or bipeds, the travel direction as well as the body position keep a constant relationship with the gravity vector. Muscles are highly specialized for either postural or locomotor function: postural function is mainly performed by one extensor head at each limb joint (Jouffroy *et al.*, 1979; Jungers *et al.*, 1980, 1983; Jouffroy and Stern, 1990). Unlike terrestrial animals, arboreal ones move in every direction within their three-dimensional, discontinuous-support environment. The direction of travel as well as body position are at varied, changing angles in relation to the direction of gravity. Antigravity muscle specialization is minimal (see below).

Both types of specialization are the extant stages of a lengthy evolutionary process that began some 350 million years ago. When our fish ancestors emerged from their aquatic habitat to settle in continental dry lands, they gave up the benefits of buoyancy, which roughly neutralizes body weight, and they had to deal with the effect of gravitational acceleration pressing them down on the ground. The evolutionary onset of the quadrupedal pattern of locomotion, and correlative antigravity activity of limb muscles as well, was raising the body on four limbs above the ground, as happened with the paleozoic early amphibians Stegocephalia, and is still evidenced in extant urodeles. The distal parts of the limbs play the role of supporting pillars away from the body center of mass, whereas propulsion results from the transverse arms and thighs sweeping horizontally to-and-fro in phase with the lateral undulation of the spine (inherited from fish ancestry). Such "sprawling posture" (Romer, 1964, p. 201) is wasteful of energy for much muscular effort is used up in merely keeping the body off the ground and preventing collapse. Ever since, gravity has been the prevailing factor of selective pressure accounting for the evolution of land vertebrates insofar as it is the only force that acts uninterruptedly on the body, although one is rarely aware of it – except when falling down!

In the course of subsequent evolutionary events, biomechanical improvements resulted from the whole of the forelimbs and hind limbs having moved beneath the body into vertical, parasagittal planes close to the center of mass. As a result, most of the burden of body weight became supported directly by the limb bones. Hinge-like limb joints are so molded to confine limb movements in parasagittal planes in which the system of forces can be reduced into only two components: a vertical (antigravity) component and a horizontal one in the direction of locomotion (accelerating/braking). At each joint, one limb extensor specialized for antigravity function, as the other extensors acting more specifically as motors generated the propulsive force component (Jouffroy *et al.*, 1999). Simultaneously, complementary changes in skeletal and muscle morphology of the spine, including energy-saving mechanisms, favored dorsoventral movements at the expense of the transverse ones, as well as efficient transmission of the impulse of the hind limbs to the body center of mass (Alexander *et al.*, 1985; Jouffroy, 1992). Such trends were perfected in those animals that keep a constant relationship with the invariable direction of the gravitational force. For example, the ground-dwelling species do not merely resist gravity but also take advantage of it. By utilizing the gravity-related changes of potential energy resulting from up-and-down movements of the body center of mass, and benefiting from elastic properties of muscles and tendons, cursorial mammals such as horses and antelopes (Alexander, 1984; Alexander *et al.*, 1985; Dimery and Alexander, 1985; Dimery *et al.*, 1985), and terrestrial leapers such as kangaroos (Ker *et al.*, 1986) can speed up despite heavy body mass. Alexander and Jayes (1983) and Alexander (1990) demonstrated that locomotor patterns can be predicted on the basis of the interaction between size, speed, and gravity ("dynamic similarity hypothesis").

As opposed to ground-dwelling mammals, nonhuman primates are marked in their morphology by the changing relationships to gravity that characterize the arboreal locomotion and posture of most species. The direction of travel, as well as the body position, make varied and continuously changing angles with the line of gravity: horizontal (prone and upside down), oblique, vertical (up and down) (Jouffroy *et al.*, 1990). Three-dimensional components make up the force system: a significant transverse component, which hardly exists in cursorial mammals, supplements the vertical and axial ones. Limb joints with several degrees of freedom, and the presence of a clavicle, allow the animal to rotate, shorten, and lengthen its limbs for grasping branches in all directions and for bridging gaps. According to the fickle relationship between the body and its support, gravity may tend either to press the palms and soles against the support, as it does with terrestrial animals, or to pull them away, requiring prehensile mechanisms of the hands and feet (Ishida *et al.*, 1990). Bones, joints, and muscles must continually adapt to resist and overcome the varying biomechanical stresses (pressure or tension) to which they are subjected by gravity. When an animal stands still, some muscular force must be produced to counteract omnipresent gravity. The vertebrate body being made up of series of articulated elements, it would collapse under the effect of the gravitational force, if the different limb and spine joints were not equipped with appropriate muscles. Because the antigravity function of the various muscles depends on the transient position of the primate body, it can be fulfilled either by extensors or by flexors as shown by EMG records (Jouffroy and Stern, 1990). Comparative analysis of the antigravity muscle machinery in arboreal and terrestrial locomotor systems offers a fine clue to deciphering primate adaptation and evolution.

Muscles have a dual function: they are mechanical elements, like the bones they are attached to, and they constitute the unique source of energy that generates animal movements. By analogy with man-designed vehicles, muscles are at once the equivalent of the motor and an integral part of the transmission.

Muscles as mechanical parts of the locomotor system

From a mechanical point of view, movements (or stillness) are generated by variation (or constancy) of muscle length. When a muscle fiber contracts, it tends to approximate its points of attachment, but whether it does so depends on the balance between the force it generates and the forces which oppose its shortening. These forces are the elastic and viscous properties of the muscle itself, of the articular tissues and other soft tissues, activity of the antagonist musculature, inertia of the body segment to be moved, and finally the effect of gravity upon all the foregoing. Biomechanical characteristics of each muscle depend on its morphology (size, weight, shape, fascicular architecture) and relation to the bones (distance between muscle attachments and joint axes, length of lever arms). Muscle mechanical adjustments to gravity effects are included in general studies of comparative morphology, locomotor biomechanics, and experimental approaches to the different types of locomotion (see for example, as regards the primates: Jouffroy, 1968, 1971, 1989; Jenkins, 1974; Jouffroy and Gasc, 1974; Stern *et al.*, 1977, 1980; Preuschoft *et al.*, 1979; Preuschoft and Demes, 1984; Emerson, 1985; Hildebrand, 1985; Reynolds, 1985; Larson and Stern, 1989; Preuschoft, 1990; Anapol and Barry, 1996).

In this paper, we will focus on the energetic aspect of muscle involvement. First, we will review the recent advances of histoenzymology and immunocytochemistry for the study of muscles as energy-producing units. Second, we will summarize the story (1966–1996) of research that has used the histoenzymological approach to provide information on cytochemical characteristics of primate muscles. Third, we will discuss recent contributions of immuncytochemical techniques in furthering our knowledge of the structure-function relationships in primate limb muscles, with particular emphasis on postural muscles.

Muscles as energy-producing units

As energy-producing elements, skeletal muscles differ from each other in the types of fibers they are composed of. Muscle mechanical work is converted from chemical energy, in the form of ATP, which is derived either from the anaerobic glycolysis of glucose stored as glycogen, or from oxidation. Muscle fibers are specialized and differ in their content of enzymes and substrates of metabolism. As opposed to the glycolytic metabolism that exhausts glycogen stores, the oxidative metabolism provides energy for virtually unlimited periods of time as long as a sufficient blood supply is provided. It suits particularly the requirements of antigravity (postural) muscles (Armstrong *et al.*, 1982). Biochemical approaches to the main function of locomotor muscles are based on the relationships between their contractile and biochemical properties (substrates of metabolism, enzymic and proteinic characteristics).

The concept of histologic and functional heterogeneity of mammalian skeletal muscles (mosaic pattern) dates back to the late nineteenth century, as Ranvier (1873, 1874) observed that the red and well-vascularized muscles of rabbits contracted slowly whereas the pale and poorly vascularized fibers of the white muscles were fast. As early as 1885, Ehrlich observed that the oxidative capacity of muscles destined for continuous work was higher than that of muscles destined for rapid movements or short-lasting activities. Further advance did not occur before the 1960s, with improvements in histochemical procedures for demonstrating enzymic activity (oxidative enzymes, ATPase, and phophorylase). Two categories of fibers – 1 (= red) and 2 (= white) – were recognized

(Dubowitz and Pearse, 1960; Engel, 1962; Stein and Padykula, 1962; Ogata and Mori, 1964). However, for lack of adequate biochemical tools, the study of the relationships between muscle structure and function relied almost entirely on what could be inferred from physiological measurements of whole muscles in which a particular type of fiber predominates.

The "glycogen depletion method" described by Edström and Kugelberg (1968) and Kugelberg and Edström (1968) was a major contribution to our understanding of the relationships between functional and biochemical characteristics of individual muscle fibers. It demonstrated that: (1) all fibers composing a motor unit (innervated by a single motoneuron) are of the same metabolic type; and (2) the maximum speed of contraction and resistance to fatigue of a motor unit are related to the histochemical properties of its muscle fibers. It provided a basis for correlating physiological and biochemical investigation (Burke *et al.*, 1973; Burke and Tsairis, 1974; Dum and Kennedy, 1980; Burke, 1981). According to their physiological properties, motor units have been divided into three main categories: (1) slow (S) units, with slow maximum speed of contraction and high resistance to fatigue; (2) fast fatigable (FF) units, with fast maximum time of contraction and very readily fatigable; and (3) fast fatigue-resistant (FR) units, with intermediate contraction times and resistance to fatigue. In addition, a small group of motor units with intermediate resistance to fatigue between FF and FR were named F(int) (Burke and Tsairis, 1974, 1977; Burke *et al.*, 1976). Owing to the multiplicity of intermediate types of fast fibers, it has been advocated that FF and FR are the extremes of a spectrum rather than discrete categories (Stephens and Stuart, 1974). In view of such a complexity, it is lucky that the study of antigravity muscles rests mainly on the identification of the well-defined S type, as maximum resistance to fatigue is the essential property required to oppose the gravitational force. Moreover, the relations between physiological and biochemical characteristics are simpler to establish for the S type than for the different categories of the F type.

Identification of myofiber types is based on either enzymic or antigenic protein specificity.

Enzymic identification of myofibers

Twitch fibers of adult mammalian skeletal muscles can be classified into two main types based on the stability of their myosin ATPase (mATPase) to alkaline (pH 10.4) conditions: low activity or high activity. The pattern is reversed when the fibers are preincubated at acid pH (4.35). By varying slightly the pH within the acid range (from 4.35 to 4.6), differences in staining intensity reveal two categories of acid-stable fibers. Brooke and Kaiser (1970) used these results to classify myosin ATPase into two primary types: I and II, with type II having a less acid-stable form IIA and a more acid-stable form IIB. The results were corroborated by associating the alkaline ATPase method with the reaction to oxidative enzymes (NADH and SDH) to determine the mitochondrial content (an indicator of oxidative metabolism). Peter *et al.* (1972), by staining serial sections alternately for myosin ATPase and the metabolic enzymes, found that myosin ATPase type I had high oxidative capacity. For this reason they called them "slow-oxidative" fibers (SO). Myosin ATPase type IIA stained darkly for both oxidative and glycolytic enzymes: they were named "fast-oxidative/glygolytic" fibers (FOG). Lastly, myosin ATPase type IIB fibers had low oxidative and high glycolytic capacity: they were dubbed "fast-glycolytic" fibers (FG). Later investigations based on mATPase histochemistry delineated four major subtypes: three fast (IIB, IID/X, and IIA) and one slow (I) (Schiaffino *et al.*, 1989; Gorza, 1990; Hämäläinen and Pette, 1993). The classification based on myosin ATPase and oxidative enzymes cannot be strictly equated. However, Pette and Staron (1997) pointed out that a good correlation between types I and SO fibers remains undisputed, and this is the basis for the study of the antigravity muscle system.

The relationships between the functional categories of motor units S, FR, and FF and the enzymic categories of fibers were pointed out by Burke *et al.* (1971), Burke and Tsairis (1974), and Burke (1986). They showed that the fatigue-resistant, slow-twitch motor units consist of fibers that stain only weakly for alkaline ATPase and intensely for mitochondrial enzymes (SDH), and conversely for the fast-twitch motor units. Further subtypes of myofibers have been described, mainly fiber types IIC (intermediate between I and IIA: Brooke and Kaiser, 1970) and IIX/D (intermediate between IIA and IIB; previously called IIAB). Similarly, as regards resistance to fatigue and force output, Edström and Kugelberg (1968) and Burke (1975, 1981) found a small number of motor units (F-int) with intermediate characteristics between FR and FF ("FI" type: McDonagh *et al.*, 1980a, 1980b). These intermediate types are found in small numbers: they are considered as transitional forms expressing fiber plasticity (see review in Gauthier, 1986; Peters, 1989). Their existence does not modify the basic classification of the myofibers into the three main categories, I, IIA, and IIB. It is necessary to keep in mind that the relationships between enzymic reactivity of muscle fibers and functional characteristics of motor units are relative: they vary between animals or between muscles of the same animal. However, it has been found in most mammals investigated that type I myosin ATPase correlated well with fatigue-resistant SO fibers, which legitimates using enzymic methods for the study of the antigravity muscle system (Pette and Staron, 2000).

Antigenic protein identification of myofibers

As regards the proteinic diversity of myofibers, immunocytochemistry techniques supplement data drawn from enzyme cytochemistry. Myosin is the most abundant protein in muscles and makes up the primary component of the thick filament. Its molecule is made up of two heavy chains (MHCs) and four light chains (MLCs) (Gauthier, 1986; Peters, 1989). These chains occur as several distinct isoforms (isomyosins). Adult skeletal muscle fibers are composed of different isoforms (slow and fast) of the myosin heavy chains (Gauthier and Lowey, 1977). Myofibers defined as slow or fast by histoenzymic reactions have fundamentally different MHC (Staron and Pette, 1986). It has been shown that the distinct slow and fast MHC isoforms correlate with the distinct enzyme patterns of mATPase staining: molecular techniques have revealed a single slow MHC and three fast MHCs which correspond to the IIA, IIX/D, and IIB enzyme categories (Gauthier and Lowey, 1977, 1979; Bär and Pette, 1988; Schiaffino *et al.*, 1989; Gorza, 1990; Bottinelli *et al.*, 1994; Graziotti *et al.*, 2001). Antibodies specific for fast MHC react with the fibers that have high alkali-stable mATPase activiy (type II; Brooke and Kaiser, 1970), and conversely antibodies specific for slow MHC react with type I fibers that have a very low alkali-stable mATPase activity. Opposite responses to slow and fast antibodies can be clearly visualized with markers such as immunoperoxidase (ABC) or FITC immunofluorescence (Fig. 7.1).

According to Pette and Staron (2000), MHC isoforms represent the most appropriate markers for fiber type delineation. They show an excellent correlation with the classical fiber types I (= SO) and II (= FOG and FG) as assessed by myosin ATPase histoenzymology (Brooke and Kaiser, 1970; Peter *et al.*, 1972). As investigations about muscle immunology have advanced during the last decade, the number of intermediate fiber types that have been discovered has increased, suggesting a continuum rather than discrete categories. Recent studies have shown that, in addition to "pure" fiber types (Hämäläinen and Pette, 1995) that are characterized by the expression of a single MHC isoform, there are "hybrid" fibers that express two or more MHC isoforms (Gorza, 1990; Hämäläinen and Pette, 1993; Staron and Pette, 1993; Pette and Staron, 1997, 2000, 2001). These hybrid fibers bridge the gaps between the pure fiber types (I/IIA, IIA/IIX, IIX/IIB), hence the appearance of a continuum. Experimental studies have shown that electrical stimuli, mechanical factors (immobilization, loading and unloading, space weightlessness), variations in thyroid hormone levels, etc. induce changes in MHC expression heading in the direction of either fast-to-slow or the reverse (Hämäläinen & Pette, 1995; Pette & Staron, 1997, 2000, 2001; Sartorius *et al.*, 1998; Kischel *et al.*, 2001; Serrano *et al.*, 2001).

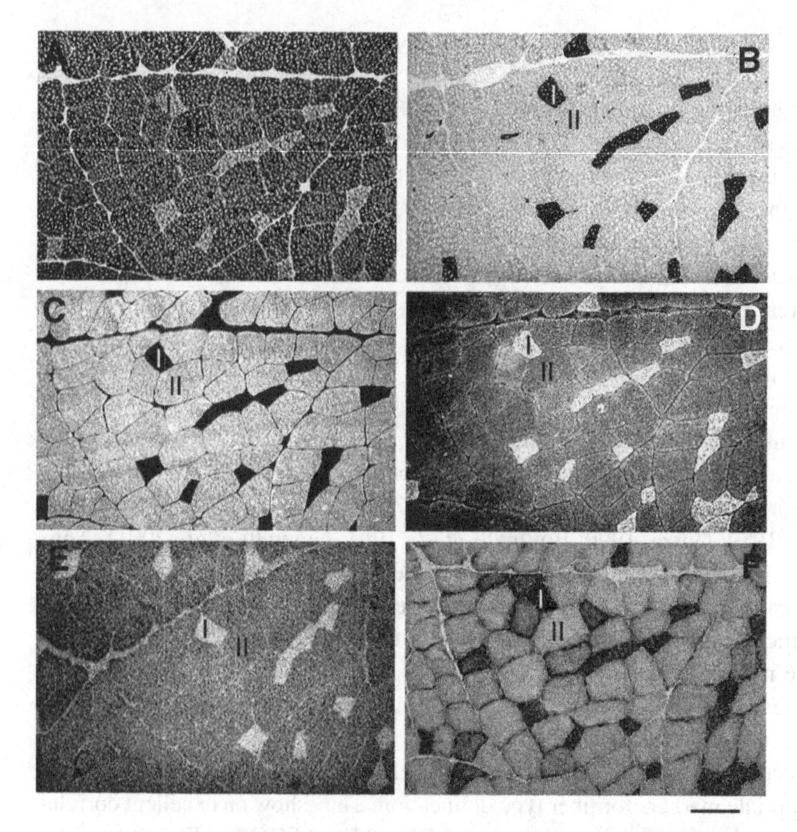

Figure 7.1. *Macaca mulatta* (adult). Gracilis (frozen specimen). Comparison of myofiber reactivity to myosin heavy chain antibodies (MHC I and II) with intensity of histochemical staining for myosin-ATPase and NADH. A: mATPase pH 9.4; B: mATPase pH 4.3; C: immunofluorescence (FITC) anti-MHC-II (monoclonal antibody); D: immunofluorescence (FITC) anti-MHC-I (monoclonal antibody); E: immunoperoxydase (ABC) anti-MHC-II (monoclonal antibody); F: NADH. I: type I fibers; II: type II fibers. Scale bar = 100 μm.

The number of hybrid fibers depends on external factors. As regards the study of fatigue-resistant fibers in relation to antigravity muscle function, the most important group of hybrid fibers is the one that bridges the gap between type I and IIA. They can be easily identified because they react to both categories of antibodies: against slow and fast myosin. In the muscles of adult primates investigated, they were scarce (Fig. 7.2).

The myosin isoforms depend also on developmental stages. Embryonic, neonatal, and adult MHC are sequentially synthesized (Whalen *et al.*, 1981). It has been shown in infant rats that adult slow myosin is present much earlier than fast myosin: for example, at birth, the soleus contains a large proportion

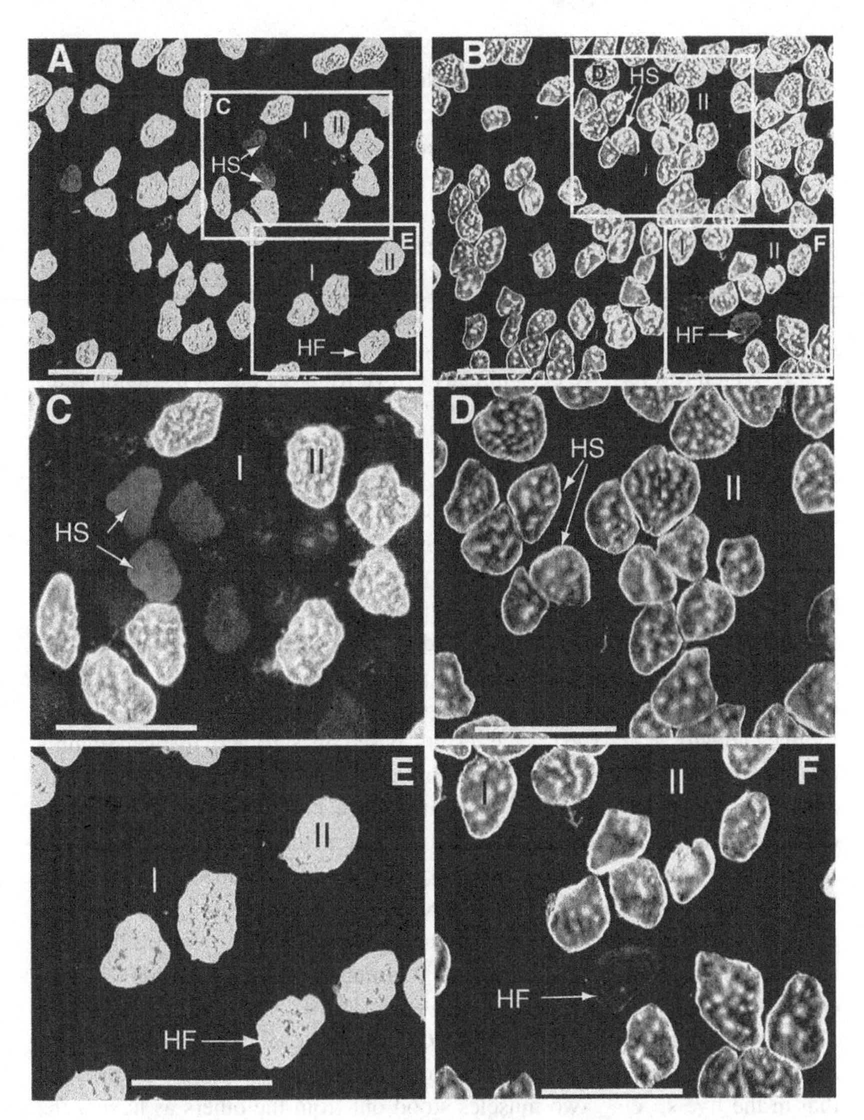

Figure 7.2. *Microcebus murinus* (adult). Soleus (frozen specimen). Hybrid fibers (coexpressing both I and II isoforms) detected by the immunofluorescence method. A, C, and E: anti-MHC-II (Sigma's clone MY-32). B, D, and F: anti-MHC-I (Sigma's clone NOQ7.5.4D). Middle and bottom: enlargements of framed areas C, D, E, and F from pictures A and B. The weak reactivity to MHC-II of slow hybrid fibers HS (expressing mainly the isoform I) is visible in A and C. The weak reactivity to MHC-I of fast hybrid fibers (expressing mainly the isoform II) is visible in B and F. I: type I fibers; II: type II fibers. Scale bar = 100 μm.

of adult slow myosin (Albis *et al.*, 1989, 1990). Nowadays, monoclonal antibodies against either the adult unique slow isoform MHC-I or all fast MHC-II and against neonatal myosin are commercially available. It makes immunocytochemistry procedures relatively easy and very reliable for detecting (relative number and size) those fibers that are most capable of sustained activity, as required for antigravity muscle activity.

In short, both enzymocytochemistry and immunocytochemistry appear to be alternative approaches to the analysis of fiber-type population and size of postural muscles. Are they equally suitable for the study of primate locomotor systems and are they equally used? Which assets, which hindrances characterize each of them? Which results have been already drawn from the one or the other in the domain of primate adaptation to gravity? These questions are addressed below.

The story (1966–1996) of histoenzymological approaches to nonhuman primate limb muscles

The first histochemical information concerning nonhuman primates appeared in the mid 1960s with a comparative study of the red and white voluntary skeletal muscles of several species of adult primates (Beatty *et al.*, 1966) and the differentiation of different types of fibers during the development of the rhesus monkey (Beatty *et al.*, 1967). The results, based on SDH activity, were mainly qualitative. In the following decade, four papers used enzyme cytochemistry as a complementary technique to seek the muscle biochemical bases of the ability of some prosimians to perform very specialized modes of locomotor behavior: mainly vertical clinging-and-leaping by the galagines and slow crawling by the lorisines (Jouffroy, 1989). These two modes of locomotion entail opposite physiological demands: short and powerful acceleration for triggering off the leaps as opposed to sustained activity for maintaining long-lasting motionless periods (Charles-Dominique, 1971). Ariano *et al.* (1973) compared the fiber type of 33 hind-limb muscles in the bushbaby *Galago senegalensis*, the slow climber *Nycticebus coucang* and three small nonprimate mammals. They pointed out that in the five species, two muscles stood out from the others as having the highest proportion of slow oxidative type I fibers: vastus intermedius (deepest head of knee extensors) and soleus (deepest head of ankle extensors). These results demonstrated the enzymic diversity of the various heads of the knee and ankle extensors, suggesting a functional specialization.

In a pioneering study associating oxidative enzyme SDH technique with morphological, biomechanical, and physiological approaches, Hall-Craggs (1974) compared muscle fiber types of four hind-limb muscles in *Galago senegalensis*, *Arctocebus calabarensis*, and *Loris tardigradus*, as well as two forelimb muscles in the last species. Even though the author specified the poor quality of his preparation "owing to the time that elapsed between death and removal of the muscles," a major restraint in enzyme techniques, the three main types of fibers were identifiable (p. 841). Qualitative results showed the presence of a high proportion of "possibly fast fibers" in the quadriceps femoris of *Galago*, "where speed of shortening is a paramount need," and the predominance of fatigue-resistant fibers in the muscles of "more slowly moving species that are known to maintain a posture for long periods without movement" (p. 844). These studies launched the use of enzyme cytochemistry in the field of primate locomotion. The results concerning lorisine muscles were confirmed by a study of limb muscles in another "slow-mover," *Perodicticus potto*. Using several oxidative enzymes and mATPase at pH 9.4 and 3.5, Marechal *et al.* (1976) found a chequerboard pattern of type I and type II myofibers in about equal proportions in both the ankle extensor gastrocnemius and the ankle flexor tibialis anterior. In addition to the exceptionally high percentage of type I fiber in the gastrocnemius (50%, as compared to less than 25% in most primates). Marechal *et al.* (1976) also observed an irregular distribution of the two types of fibers within the muscle, type I fibers being more numerous in the deeper part of the muscle than in the superficial part, and the reverse for type II. It turned out from subsequent studies that such a concentration of slow fibers in the deepest part of heterogeneous muscles is the most general condition in mammals (Armstrong, 1980; Armstrong *et al.*, 1982; Anapol and Jungers, 1986; Acosta and Roy, 1987).

It is to be mentioned that the bushbaby *Galago senegalensis* was also the subject of the first experimental investigation of primate myofiber-type plasticity depending on physical constraints, such as immobilization and overloading (Edgerton *et al.*, 1972, 1975). Such plasticity accounts for the changes that were observed subsequently in the fiber-type population of rat hind limb when muscles are subjected to training or rest, or freed from gravitational constraints (e.g., hind-limb suspension, space microgravity) (Booth and Kelso, 1973; Booth and Seider, 1979; Booth, 1982; Desplanches *et al.*, 1987, 1990; Desplanches, 1997).

Major improvements in enzyme cytochemistry relied on the study of laboratory animals, easily sacrificed for immediate treatment without reservation (mice, rats, guinea-pigs, rabbits, etc). However the technique was also used to investigate the relationships between muscle fiber-type population and primate locomotor behavior. The comparative study by Sickles and Pinkstaff (1980, 1981) of 30 hind-limb muscles in the leaper *Galago* and the slow-mover *Nycticebus*, both compared to the tree-shrew *Tupaia*, was a major step in this research. It showed that the various muscle heads that make up the knee and ankle extensors (quadriceps femoris, triceps surae) present quite different fiber-type populations. At each one of these joints, one head (vastus intermedius, soleus) contains a very high percentage of fatigue-resistant fibers (up to 95% in the soleus of the bushbaby) while the other heads are richer in glycolytic IIB fibers (up to 85%, in the vastus lateralis of the bushbaby). This corresponds to functional specialization of each head of *Galago* extensors either for sustained activity or for short and powerful acceleration. Moreover, the absence of glycolytic fibers IIB in all muscles of the slow loris suggested a certain ability of all muscles to resist gravity in turn. These functional implications were confirmed by the results of subsequent telemetered-EMG studies of these muscles in different species of primates in which vastus intermedius and soleus appeared to be the only knee and ankle extensors to be active during still posture (Jungers *et al.*, 1980, 1983; Anapol and Jungers, 1987; Jouffroy and Stern, 1990).

Other histoenzymatic studies of primates focused on a single genus showing specialized modes of locomotion: e.g., the slow-mover *Nycticebus* (Sickles and Pinkstaff, 1980; Kimura *et al.*, 1987), the leapers *Aotus* (Plaghki *et al*; 1981) and *Lemur* (Anapol and Jungers, 1986), the climber *Microcebus* (Petter and Jouffroy, 1993), and the armswinger *Hylobates* (Kimura and Inokuchi, 1985). In their study of the brown lemur, Anapol and Jungers (1986) studied the regional variation of fiber-type population within distal/proximal and deep/superficial regions of heterogeneous muscles. A main interest of this study was also to take into account not only the fiber-type population, but also the total cross-sectional areas of each type of fibers. Since the force and resistance to fatigue of a muscle depend on the myosin content, they depend on the fiber cross-sectional areas as well as on the percentages of each type of fibers. No wonder that the macaques, the primates most commonly found in laboratories, have also been the subject of many histoenzymological studies (Moriyama, 1983; Roy *et al.*, 1984; McIntosh *et al.*, 1985; Acosta and Roy, 1987; Roy *et al.*, 1991; Suzuki and Hayama, 1991, 1994; Kojima and Okada, 1996).

All these studies – and especially the comparison of the nimblest leapers with the sluggish, upside-down-moving lorisines – contributed to show that the enzyme cytochemical characteristics of primate limb muscles correlate with the predominating modes of locomotion and posture. The relationships between enzyme data and the precise function of a muscle during a well-defined movement came out of a series of telemetered-EMG studies (coupled with video-recordings of the subject) which have been carried out since 1977 by the Primate Locomotion Laboratory at Stony Brook (Stern *et al.*, 1977; Jouffroy *et al.*, 1979; Jungers *et al.*, 1980, 1983; Anapol and Jungers, 1987; Larson and Stern, 1989; Jouffroy and Stern, 1990; Jouffroy *et al.*, 1999). The concept of "antigravity muscles," explicitly formulated at first in an histoenzymatic study of dogs (Armstrong *et al.*, 1982), materialized in electromyograms corroborating histoenzymatic data ("postural muscles": Jungers *et al.*, 1980; Anapol and

Jungers, 1986; Jouffroy and Stern, 1990). Significance of fiber-type population as an indicator of adaptation to physiological and environmental factors, including prevailing gravity, was also confirmed by numerous experimental investigations in rats and humans. They showed the plasticity of fiber-type population under the effects of varied factors: training, loading, rest, immobilization, hind-limb suspension, space microgravitational environment, levels of thyroid hormone, changes in nerve supply, cross-reinnervation (Edgerton *et al.*, 1972, 1975).

As a matter of fact, after two decades of flourishing and fruitful investigations, enzymocytochemistry fell into disuse in the field of evolutionary primatology. Various reasons account for such a decline. First, the increase in the number of fiber types successively discovered, suggesting a continuum rather than well-defined categories (see above), hampered easy interpretation of cytochemical data in terms of muscle specialization and function. Second, even using the simplest mATPase method for distinguishing slow/fatigue-resistant fibers from fast/fatigable ones was problematic for comparative studies because the final staining depends very strictly upon the experimental protocol. For example, slight differences in the pH of preincubation, duration of the successive stages, and temperature, result in different mosaic patterns that impede taking into account and comparing quantitative results obtained in different laboratories. Last but not least, enzymocytochemistry requires fresh samples of muscles being frozen in isopentane cooled in liquid nitrogen immediately after death, and stored at −80 °C. Such requirements are difficult to fulfill in the case of wild animals and protected species, such as primates.

Immunocytochemical approaches to the postural limb muscles of primates

As opposed to the myosin enzymatic activity, the antigenicity (antibody epitopes) of the myosin persists in the protein after death. Muscles preserved in formaldehyde react to antibodies against MHCs (Fig. 7.3). Such persistence of immunoreactivity makes immunocytochemical fiber-typing convenient for the study of preserved cadavers, a useful procedure for comparative zoological investigations (Jouffroy and Médina, 1996; Médina and Jouffroy, 1998). Using immunocytochemical methods for comparative studies of skeletal muscles necessitates appropriate antibodies to be easily available, which is the case for standardized, marketed monoclonal antibodies against slow and fast myosin as well as neonatal myosin. It is also possible to cross-check the results of immunoreactivity to MHC against the result of immunoreactivity to troponin I (Dhoot and Perry, 1979). The regulatory protein troponin I inhibits

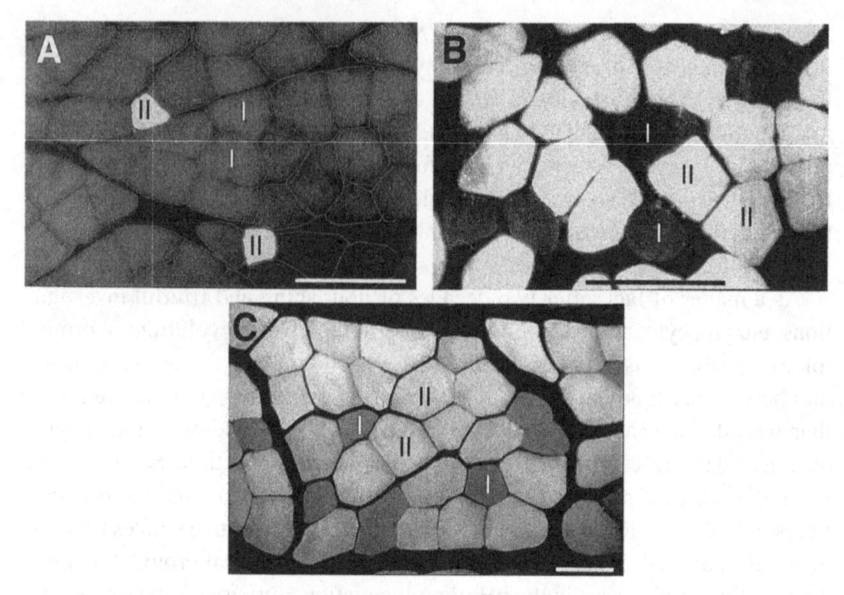

Figure 7.3. *Lemur catta* (adult). A: soleus; B: gastrocnemius medialis. *Macaca mulatta* (adult). C: triceps brachii caput laterale. All formaldehyde-preserved specimens. Note the relatively greater cross-sectional areas of type I fibers (dark) in the postural muscle soleus (A) and conversely the greater cross-sectional areas of type II fibers (white) in the nonpostural extensor muscles (B and C) I: type I fibers; II: type II fibers. Scale bar = 100 μm.

the connection between actin and myosin, and consequently mATPase activity. Reactivity to troponin I and MHCs (either fast or slow isoforms) is very similar with the exception of IIX fibers which react to fast troponin I but not to fast MHC (clone MY-32) antibodies (Jouffroy *et al.*, 2003). Antibodies against fast and slow troponin I are nowadays commercially available. It is to be noted that most muscles (soleus excepted) are heterogenous, with the deeper part richer in type I fibers than the more superficial part (Fig. 7.4) and with proximal/distal variations (Anapol and Jungers, 1986). Therefore, accurate comparisons between various individuals or species must be based on equivalent parts of the muscles, generally fibers located mid-belly and close to the bone, unless otherwise stated.

Immunocytochemical methods (e.g., using Sigma's clones MY-32 and NOQ7.5.4D) have provided clear and reliable information on the type I and II population and cell size of primate postural muscles. A study associating telemetered-EMG and immunocytochemical approaches has demonstrated the relationships between the postural (antigravity) function and cytochemical characteristics of the rhesus monkey's limb muscles (Jouffroy *et al.*, 1999). Each

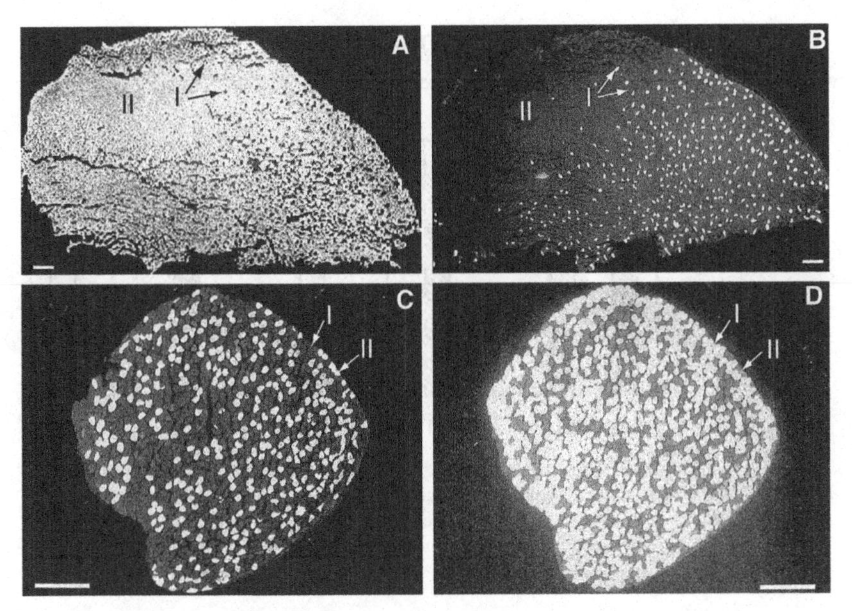

Figure 7.4. *Microcebus murinus* (adult). Heterogenous fiber-type distribution within the gastrocnemius lateralis (A and B), a nonpostural extensor, as compared to the homogenous distribution across the entire soleus (C and D), a postural extensor muscle of the ankle. Left (A and C): anti-MHC-II (monoclonal antibody). Right (B and D): anti-MHC-I (monoclonal antibody). I: type I fibers (dark in A and C, white in B and D); II: type II fibers (white in A and C, dark in B and D). Scale bar = 500 μm.

of the elbow, knee, and ankle joints is endowed with a multi-headed extensor muscle group (triceps brachii, quadriceps femoris, and triceps surae, respectively). EMG records have shown that one head at each joint fulfills postural function (triceps brachii caput mediale, vastus intermedius, and soleus, respectively). Recruitment at low levels during posture characterizes these postural heads, which prevent collapse of the body by stabilizing joint angles against the effect of gravity. They are also active, near maximum levels, during locomotion, together with the non-postural extensor heads that are recruited at levels correlated to the strenuousness of the effort. As regards myofiber types, the postural heads differ from the non-postural ones by high content (more than 85%) and relatively large cross-sectional areas of type I fibers (Figs. 7.5 and 7.6).

Because most primate modes of locomotion (bipedalism and brachiating excepted) bring into play both pairs of limbs, generic diversity of locomotor adaptation is noticeable not only in the hind limb – the most investigated – but also in the forelimb. In the rhesus monkey, triceps brachii caput mediale presents a higher percentage of type I fibers than the postural hind-limb muscles.

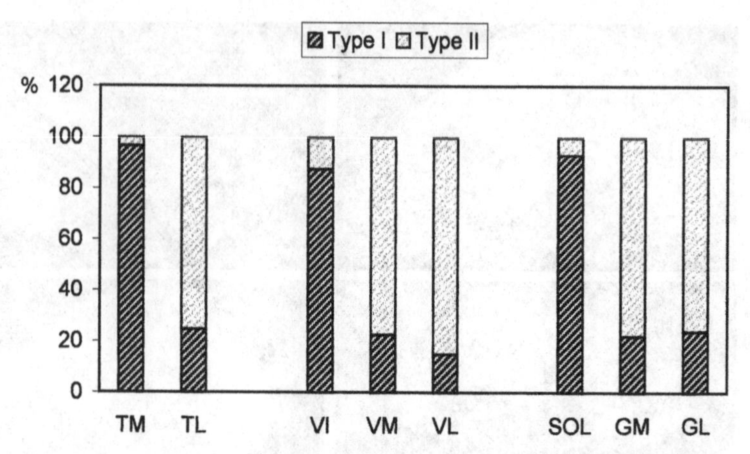

Figure 7.5. *Macaca mulatta* (adult). Stacked percentages of types I and II fibers in elbow, knee, and ankle extensor muscles: triceps brachii caput mediale (TM); triceps brachii caput laterale (TL); vastus intermedius (VI); vastus medialis (VM); vastus lateralis (VL); soleus (SOL); gastrocnemius medialis (GM); gastrocnemius lateralis (GL).

The 97% figure is similar to that of the dog (Armstrong *et al.*, 1982). It is only about 85% in the lemur (unpublished data). Gravitational constraints vary in relation to body weight: percentages of slow fibers are lower in light species like *Microcebus* (Petter and Jouffroy, 1993). Their quasi-absence in the forelimbs of the saltatorial jerboa (*Allactaga*, 5%: Jouffroy *et al.*, 2003) is very likely related to the forelimbs being completely excluded from locomotion.

The postural extensor head of the primate elbow is also characterized by the great size (cross-sectional area) of type I fibers compared to type II (Fig. 7.6). The higher percentage and greater size of type I fibers in the postural head of the elbow, as compared to knee and ankle, is in accord with the observation that in the Japanese macaque "the forelimb muscles have a higher resistance to fatigue" (Moriyama, 1983, p.107). We see that immunocytochemical data can be an indication of the postural behavior: macaques in general, and especially the rhesus monkey, rarely stand quadrupedally for long. Rather they rest in a sitting attitude, trunk sloping forwards, as do cats. EMG records have shown that in this posture the support of head and trunk weight by the arms entails isometric contraction of the elbow postural head, while hind-limb muscles remain silent (Jouffroy *et al.*, 1999). The average cross-sectional area of type I fibers in relation to type II in hind-limb postural muscles seems to depend on the locomotor behavior. While both types are of the same size in the rhesus monkey, cross-sectional areas of type I fibers are greater than type II in *Lemur* (Fig. 7.3). On the contrary in non-postural extensor muscles of *Lemur*,

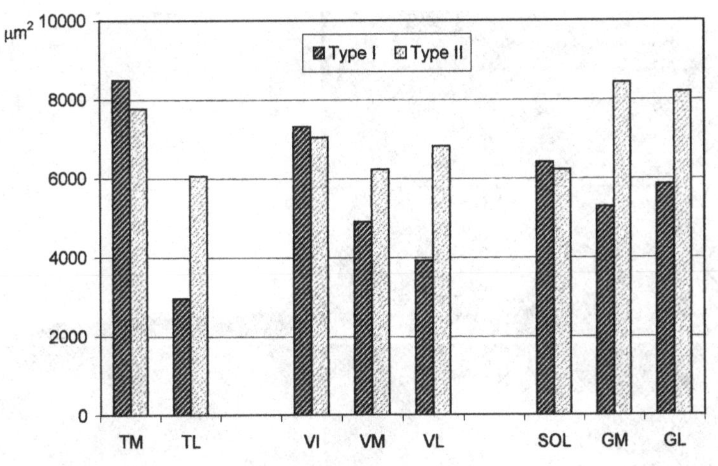

Figure 7.6. *Macaca mulatta* (adult). Average cross-sectional areas of type I and type II fibers in elbow, knee, and ankle extensor muscles: triceps brachii caput mediale (TM); triceps brachii caput laterale (TL). vastus intermedius (VI); vastus medialis (VM); vastus lateralis (VL); soleus (SOL); gastrocnemius medialis (GM); gastrocnemius lateralis (GL).

and especially in those involved in triggering the leap (EMG data: Jungers *et al.*, 1983) cross-sectional areas of type II fibers are greater than those of type I. Such opposite characteristics of the two types of muscles are possibly related to the leaping ability of *Lemur*, requiring some extensor muscles to produce a very high acceleration at takeoff, as has also been observed to be the case in the five-toed jerboa (*Allactaga*; Jouffroy *et al.*, 2003). The fact that in the jerboas the cross-sectional areas of type I fibers are much greater than those of type II in postural muscles (and conversely for type II in propulsive muscles) as in *Lemur* calls for further comparative investigation of the size of fibers in a range of primate species diversely adapted to leaping. The study of fiber-type population cannot be dissociated from the study of fiber size (cross-sectional areas), nor the study of postural muscles from the study of propulsive muscles.

Finally, imunocytochemical approach to muscle fiber types and size provides information about the developmental process that ends up in the adult configuration. In newborn *Microcebus murinus*, neonatal myosin coexists with the adult types of slow and fast myosin isoforms and disappears at about eight days (work in progress). The existence of adult myosin as early as birth makes the immunofluorescence technique an accurate marker tool for the study of posnatal development of limb muscles. At birth, the fiber-type population in non-postural extensor heads of elbow, knee, and ankle is very similar to the adult condition: about 20–25% of type I fibers. On the contrary, the fiber-type

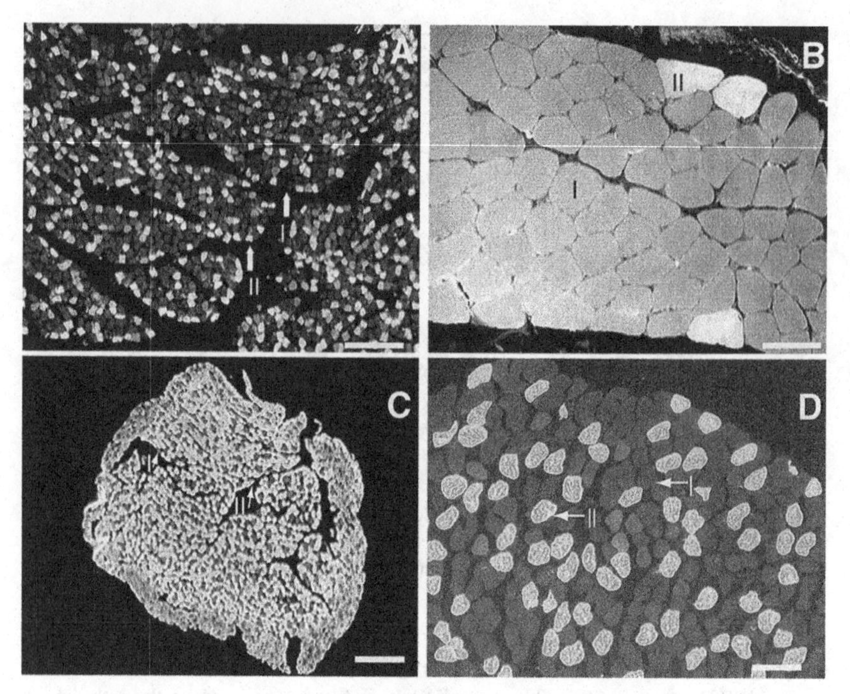

Figure 7.7. Changes in the fiber-type population and size from birth to adulthood. *Macaca mulatta* (top): newborn (A) and adult (B). *Microcebus murinus* (bottom): newborn (C) and adult (D). Soleus muscle, anti-MHC-II (monoclonal antibody) I: type I fibers (grey); II: type II fibers (white). Scale bar = 100 μm.

population of postural muscles changes from birth to adulthood, with an increase in the percentage of type I fibers: from 37% to 97% at the elbow, from 46% to 88% at the knee, and from 34% to 93% at the ankle in the rhesus monkey (Jouffroy and Médina, 1996). Although the percentage of type I fibers is already slightly greater in postural muscles than in non-postural ones at birth, the setting up of the muscle antigravity system results from a postnatal developmental process (Fig. 7.7). Very likely, increase in body weight and EMG activity generates the conditions of conversion from fast to slow fiber types (Streter *et al.*, 1973; Kugelberg, 1976; Salmons and Henriksson, 1981). Such developmental increase in number of type I fibers is the opposite of their experimental decrease that has been observed in rats freed from gravitational constraints, either after hind-limb suspension or space flight, and in flown macaques (Chopard *et al.*, 2000). The results concerning the progressive setting-up of the muscle antigravity system call for complementary studies of developmental series of individuals in various primate species characterized by different locomotor behaviors.

This review shows that the primates, although not offering the advantages of common laboratory animals for the study of myofiber cytochemistry, are characterized by a locomotor diversity that indisputably makes them the best models to study the effect of gravity on locomotor systems. Such research requires that various primate species with different locomotor behaviors be comparatively studied. Ideally, EMG and cytochemical investigations should be carried out in parallel. But EMG and histoenzymology are, for different reasons, techniques not easily applicable to research on wild and rare mammals such as most nonhuman primates. While it might not be impossibly difficult to obtain muscle biopsies from such animals, biopsy analyses are not reliable for comparative studies due to the heterogeneity of many muscles. In practice, immunocytochemistry has opened a new area of investigation of the structure-function relationships by making the study of preserved muscles feasible. For those primatologists who wish to supplement the study of muscle gross morphology and biomechanics with data about muscle power capability and performance, immunocytochemistry has emerged as a reliable tool of investigation to obtain new insights into the effects of gravity on the development, adaptation, and evolution of primate locomotor systems.

Acknowledgments

We would like to express our appreciation to F. Anapol, R. Z. German, and N. G. Jablonski for their invitation to contribute to this book. We thank F. Anapol, W. L. Luckett, J. T. Stern, Jr., and the reviewers for their valuable comments on the manuscript and for editorial help. We are indebted to M. Geze (Laboratoire de Photobiologie, CEMIM-MNHN) for his assistance in initiating us in the use of the Nikon *Eclipse* TE 300 DV inverted microscope and the Metaview and Metamorph software. We also thank B. Jay for technical assistance. Data about *Microcebus* come from a collaborative study (in progress) with M. Perret (CNRS-UMR 8572). This work was carried out in the Muséum National d'Histoire Naturelle, Laboratoire d'Anatomie Comparée (Paris) and supported by the Centre National de la Recherche Scientifique (France), grant UMR 8570.

References

Acosta, L. and Roy, R. R. (1987). Fiber-type composition of selected hindlimb muscles of a primate (*Cynomolgus* monkey). *Anat. Rec.*, **218**, 136–141.

Albis, A. d', Couteaux, R., Janmot, C., and Roulet, A. (1989). Specific program of myosin expression in the postnatal development of rat muscles. *Eur. J. Biochem.*, **183**, 583–590.

Albis, A. d', Chanoine, C., Janmot, C., Mira, J. C., and Couteaux, R. (1990). Muscle-specific response to thyroid hormone of myosin isoform transitions during rat postnatal development. *Eur. J. Biochem.*, **193**, 155–161.

Alexander, R. McN. (1984). Elastic energy stores in running vertebrates. *Amer. Zool.*, **24**, 85–94.

(1990). The dependence of gait on size, speed and gravity. In: *Gravity, Posture, and Locomotion in Primates*, ed. F. K. Jouffroy, M. H. Stack, and C. Niemitz. Firenze: Sedicesimo. pp. 79–85.

Alexander, R. McN. and Jayes, A. S. (1983). A dynamic similarity hypothesis for the gaits of quadrupedal mammals. *J. Zool. Lond.*, **201**, 135–152.

Alexander, R. McN. Dimery, N. J., and Ker, R. F. (1985). Elastic structures in the back and their rôle in galloping in some mammals. *J. Zool. Lond.* (*A*), **207**, 467–482.

Anapol, F. and Barry, K. (1996). Fiber architecture of the extensors of the hindlimb in semiterrestrial and arboreal guenons. *Amer. J. Phys. Anthropol.*, **99**, 429–447.

Anapol, F. and Jungers, W. L. (1986). Architectural and histochemical diversity within the quadriceps femoris of the brown lemur (*Lemur fulvus*). *Amer. J. Phys. Anthropol.*, **69**, 355–375.

(1987). Telemetered electromyography of the fast and slow extensors of the leg of the brown lemur *(Lemur fulvus)*. *J. Exp. Biol.*, **130**, 341–358.

Ariano, M. A., Armstrong, R. B., and Edgerton, V. R. (1973). Hindlimb muscle fiber population of five mammals. *J. Histochem. Cytochem.*, **21**, 51–55.

Armstrong, R. B. (1980). Properties and distributions of the fiber types in the locomotory muscles of mammals. In: *Comparative Physiology: Primitive Mammals*, ed. K. Schmidt-Nielsen and C. R. Taylor. Cambridge: Cambridge University Press. pp. 243–254.

Armstrong, R. B., Saubert, C. W., Seeherman, H. J., and Taylor, C. R. (1982). Distribution of fiber types in locomotory muscles of dogs. *Amer. J. Anat.*, **163**, 87–98.

Bär, A. and Pette, D. (1988). Three fast myosin heavy chain in adult rat skeletal muscle. *FEBS Lett.*, **235**, 153–155.

Beatty, C. H., Basinger, G. M., Dully, C. C., and Bocek, R. M. (1966). Comparison of red and white voluntary skeletal muscles of several species of primates. *J. Histochem. Cytochem.*, **14**, 590–600.

Beatty, C. H., Basinger, G. M., and Bocek, R. M., (1967). Differentiation of red and white fibers in muscle from fetal, neonatal, and infant rhesus monkey. *J. Histochem. Cytochem.*, **15**, 93–103.

Belozerova, I. N., Nemirovskaya, T. L., and Shenkman, B. S. (2000). Structural and metabolic profile of rhesus monkey *M. Vastus lateralis* after spaceflight. *J. Gravit. Physiol.*, **7**, S55–S58.

Bodine-Fowler, S. C., Pierotti, D. J., and Talmadge, R. J. (1995). Function and cellular adaptation to weightlessness in primates. *J. Gravit. Physiol.*, **2**, P43–46.

Booth, F. W. (1982). Effect of limb immobilization on skeletal muscle. *J. Appl. Physiol.*, **52**, 1113–1118.

Booth, F. W. and Kelso J. R. (1973). Effect of hindimb immobilization on contractile and histochemical properties of skeletal muscles. *Pflügers Arch.*, **342**, 231–238.

Booth, F. W. and Seider, M. J. (1979). Recovery of skeletal muscle after three months of hindlimb immobilization in rats. *J. Appl. Physiol.*, **47**, 435–439.

Bottinelli, R., Betto, R., Schiaffino, S., and Reggiani, C. (1994). Maximum shortening velocity and coexistence of myosin heavy chains isoforms in single skinned fast fibres of rat skeletal muscles. *J. Muscle Res. Cell Motil.* **15**, 413–419.

Brooke, M. H. and Kaiser, K. K. (1970). Muscle fiber types. How many and what kind? *Arch. Neurol.*, **23**, 369–379.
Burke, R. E. (1975). A comment on the existence of motor unit types. In: *The Nervous System. The Basic Neuroscience*, ed. R. O. Brady. New York, NY: Raven. Vol. I, pp. 611–619.
(1981). Motor units: anatomy, physiology, and functional organization. In: *Motor Control. Handbook of Physiology. Sec. I, the Nervous System*, ed. V. B. Brooks. Washington, DC: American Physiological Society. Vol. II, part 1, pp. 345–422.
(1986). Physiology of motor units. In: *Myology*, ed. A. G. Engel and Q. B. Banker. New York, NY: McGraw-Hill. pp. 419–443.
Burke, R. E. and Tsairis, P. (1974). The correlation of physiological properties with histochemical characteristics in single muscle units. *Ann. N. Y. Acad. Sci.*, **228**, 145–159.
(1977). Histochemical and physiological profile of a skeletofusimotor (β) unit in cat soleus muscle. *Brain Res.*, **129**, 341–345.
Burke, R. E., Levine, D. N., and Zajac, F. E. (1971). Mammalian motor units: physiological–histochemical correlation in three types of cat gastrocnemius. *Science*, **174**, 709–712.
Burke, R. E., Levine, D. N., Tsairis, P., and Zajac, F. E. (1973). Physiological types and histochemical profiles in motor units of the cat gastrocnemius. *J. Physiol.*, **234**, 723–748.
Burke, R. E, Reymer, W. Z., and Walsh, J. V. (1976). Relative strength of synaptic input from short-latency pathways to motor units of defined type in cat medial gastrocnemius. *J. Neurophysiol.*, **39**, 447–458.
Charles-Dominique, P. (1971). Eco-éthologie des Prosimiens du Gabon. *Biologia Gabonica*, **7**, 121–228.
Chopard, A., Lecler, L., Pons, F., Leger, J. J., and Marini, J. F. (2000). Effects of 14-day spaceflight on myosin heavy chain expression in biceps and triceps muscles of the rhesus monkey. *J. Gravit. Physiol.*, **7**, S47–S49.
Desplanches, D. (1997). Structural and functional adaptation of skeletal muscles to weightlessness. *Int. J. Sports Med.*, **18**, 259–264.
Desplanches, D., Sempore, B., and Flandrois, R. (1987). Structural and functional responses to prolonged hindlimb suspension in rat muscle. *J. Appl. Physiol.*, **63**, 558–563.
Desplanches, D., Mayet, M. H., Ilyina-Kakueva, E. I., Sempore, B., and Flandrois, R. (1990). Skeletal muscles adaptation in rats flown on Cosmos 1667. *J. Appl. Physiol.*, **68**, 48–52.
Dhoot, G. K. and Perry, S. V. (1979). Distribution of polymorphic forms of troponin components and tropomyosin in skeletal muscle. *Nature*, **278**, 714–718.
Dimery, N. J. and Alexander, R. McN. (1985). Elastic properties of the hind foot of the donkey, *Equus asimus*. *J. Zool.* A, **207**, 9–20.
Dimery, N. J., Alexander, R. McN., and Deyst, K. A. (1985). Mechanics of the ligamentum nuchae of some artiodactyls. *J. Zool. Lond.* (A), **206**, 3341–3351.
Dubowitz, V. and Pearse, A. G. E. (1960). A comparative histochemical study of oxidative enzyme and phosphorylase activity in skeletal muscle. *Histochemie*, **2**, 105–117.

Dum, R. P. and Kennedy, T. T. (1980). Physiological and histochemical characteristics of motor units in cat tibialis anterior and extensor digitorum longus muscles. *J. Neurophysiol.*, **43**, 1615–1630.

Edgerton, V. R., Barnard, R. J., Peter, J. B., Gillepsie, C. A., and Simpson, D. R. (1972). Overloaded skeletal muscles of a nonhuman primate, *Galago senegalensis*. *Exp. Neurol.*, **37**, 322–339.

Edgerton, V. R., Barnard, R. J., Peter, J. B., Maier, P. A., and Simpson, D. R. (1975). Properties of immobilized hindlimb muscles of the *Galago senegalensis*. *Exp. Neurol.*, **46**, 115–131.

Edström, L. and Kugelberg, E. (1968). Histochemical composition, distribution of fibres, and fatigability of single motor units. Anterior tibial muscle of the rat. *J. Neurol. Neurosurg. Psychiatry*, **31**, 424–433.

Ehrlich, P. (1885). *Das Sauerstoff-Bedürfniss des Organismus*. Berlin: Hirschwald.

Emerson, S. B. (1985). Jumping and leaping. In: *Functional Vertebrate Morphology*, ed. M. Hildebrand, D. M. Bramble, K. F. Liem, and D. B. Wake. Cambridge, MA: Harvard University Press. pp. 58–72.

Engel, W. K. (1962). The essentiality of histo- and cytochemical studies of skeletal muscles in the investigation of neuromuscular disease. *Neurology*, **12**, 778–794.

Fitts, R. H., Desplanches, D., Romatowski, J. G., and Widrick, J. J. (2000a). Spaceflight effects on single skeletal muscle fiber function in the rhesus monkey. *Amer. J. Physiol. Regulatory Integrative Comp. Physiol.*, **279**, R1546–R1557.

Fitts, R. H., Romatowski, J. G., De la Cruz, L., Widrick, J. J., and Desplanches, D. (2000b). Effect of spaceflight on the maximal shortening velocity, morphology, and enzyme profile of fast- and slow-twitch skeletal muscle fibers in rhesus monkeys. *J. Gravit. Physiol.*, **7**, S37–S38.

Fitts, R. H., Reiley D. R., and Widrick J. J. (2001). Functional and structural adaptations of skeletal muscle to microgravity. *J. Exp. Biol.*, **204**, 3201–3208.

Gauthier, G. F. (1986). Skeletal muscle fiber types. In: *Myology*, ed. A. G. Engel and C. Franzini-Armstrong. New York, NY: McGraw-Hill. pp. 255–283.

Gauthier, G. F. and Lowey, S. (1977). Polymorphism of myosin among skeletal muscle fiber types. *J. Cell Biol.*, **74**, 760–779.

(1979). Distribution of myosin isoenzymes among skeletal muscle fiber types. *J. Cell Biol.*, **81**, 10–25.

Gorza, L. (1990). Identification of a novel type 2 fiber population in mammalian skeletal muscle by combined use of histochemical myosin ATPase and anti-myosin monoclonal antibodies. *J. Histochem. Cytochem.*, **38**, 257–265.

Graziotti, G. H., Rios, C. M., and Rivero, J. L. (2001). Evidence for three fast myosin heavy chain isoforms in type II skeletal muscle fibers in the adult llama (*Lama glama*). *J. Histochem. Cytochem.*, **49**, 1033–1044.

Hall-Craggs, E. C. B. (1974). Physiological and histochemical parameters in comparative locomotor studies. In: *Prosimian Biology*, ed. R. D. Martin, G. A. Doyle, and A. C. Walker. London: Duckworth. pp. 829–845.

Hämäläinen, N. and Pette, D. (1993). The histochemical profiles of fast fiber IIB, IID and IIA in skeletal muscle of mouse, rat, and rabbit. *J. Histochem. Cytochem.*, **41**, 733–743.

(1995). Patterns of myosin isoforms in mammalian skeletal muscles fibres. *Micros. Res. Tech.* **30**, 381–389.
Hildebrand, M. (1985). Walking and running. In: *Functional Vertebrate Morphology*, ed. M. Hildebrand, D. M. Bramble, K. F. Liem, and D. B. Wake. Cambridge, MA: Harvard University Press. pp. 38–57.
Hodgson, S. A., Bodine-Fowler, S. C., Roy, R. R., *et al.* (1991). Changes in recruitment of Rhesus soleus and gastrocnemius muscles following a 14 day spaceflight. *Physiologist*, **34** (suppl.), S102–S103.
Ishida, H., Jouffroy, F. K., and Nakano, Y. (1990). Comparative dynamics of pronograde and upside down horizontal quadrupedalism in the slow loris (Nycticebus coucang). In: *Gravity, Posture, and Locomotion in Primates*, ed. F. K. Jouffroy, M. H. Stack, and C. Niemitz. Firenze: Sedicesimo. pp. 209–220.
Jenkins, F. A., Jr. (ed.) (1974). *Primate Locomotion*. New York, NY: Academic Press.
Jouffroy, F. K. (1968). Musculature épisomatique. In: *Traité de Zoologie, Mammifères, XVI, 2*, ed. P.-P. Grassé. Paris: Masson. pp. 479–548.
(1971). Musculature des membres. In *Traité de Zoologie, Mammifères, XVI, 3*, ed. P.-P. Grassé. Paris: Masson. pp. 1–475.
(1989). Quantitative and experimental approaches to primate locomotion. A review or recent advances. In: *Perspectives in Primate Biology*, Vol. 2, ed. P. K. Seth and S. Seth. New Delhi: Today and Tomorrow. pp. 47–108.
(1992). Evolution of the dorsal muscles of the spine in light of their adaptation to gravity effects. In: *The Head–Neck Sensory Motor System*, ed. A. Berthoz, W. Graf, and P. P. Vidal. Oxford: Oxford University Press. pp. 22–35.
Jouffroy, F. K. and Gasc, J. P. (1974). A cineradiographical analysis of leaping in an African prosimian (*Galago alleni*). In: *Primate Locomotion*, ed. F. A. Jenkins, Jr. New York, NY: Academic Press. pp. 117–142.
Jouffroy, F. K. and Médina, M. F. (1996). Developmental changes in the fibre composition of elbow, knee, and ankle extensor muscles in cercopithecid monkeys. *Folia Primatol.*, **66**, 55–67.
Jouffroy, F. K. and Stern, J. T., Jr. (1990). Telemetered EMG study of the antigravity versus propulsive actions of knee and elbow muscles in the slow loris (*Nycticebus coucang*). In: *Gravity, Posture, and Locomotion in Primates*, ed. F. K. Jouffroy, M. H. Stack, and C. Niemitz. Firenze: Sedicesimo. pp. 221–236.
Jouffroy, F. K., Jungers, W. L., and Stern, J. T., Jr (1979). Télé-électromyographie des divers faisceaux du muscle quadriceps femoris au cours de la locomotion chez un Lémurien de Madagascar (*Lemur fulvus*). *C.R. Acad. Sci. Paris*, ser. D, **288**, 1627–1630.
Jouffroy, F. K., Stack M. H., and Niemitz, C. (1990). Nonhuman primates as a model to study the effect of gravity on human and nonhuman locomotor systems. In: *Gravity, Posture, and Locomotion in Primates*, ed. F. K. Jouffroy, M. H. Stack, and C. Niemitz. Firenze : Sedicesimo. pp. 11–18.
Jouffroy, F. K., Stern, J. T., Jr., Médina, F. M., and Larson S. G. (1999). Function and cytochemical characteristics of postural limb muscles of the rhesus monkey: a telemetered EMG and imunofluorescence study. *Folia Primatol.* **70**, 235–253.

Jouffroy, F. K., Médina, M. F., Renous S., and Gasc J. P. (2003). Immunocytochemical characteristics of elbow, knee, and ankle muscles of the five-toed jerboa (*Allactaga elater*). *J. Anat.*, **202**, 373–386.

Jungers, W. L., Jouffroy, F. K., and Stern, J. T., Jr. (1980). Gross structure and function of the quadriceps femoris in *Lemur fulvus*. An analysis based in telemetered electromyography. *J. Morphol.*, **164**, 287–299.

Jungers, W. L., Stern, J. T., Jr., and Jouffroy, F. K. (1983). Functional morphology of the quadriceps femoris in primates: a comparative anatomical and experimental analysis. *Ann. Sc. Nat. Zool. Biol. Anim.*, **5**, 101–116.

Ker, R. F., Dimery, N. J., and Alexander, R. McN. (1986). The role of tendon elasticity in hopping in a wallaby (*Macropus rufogriseus*). *J. Zool., Lond.*, **208**, 417–428.

Kischel, P., Stevens, V., Montel, F., Picquet, F., and Mounier, Y. (2001). Plasticity of monkey triceps muscles fibers in microgravity conditions. *J. Appl. Physiol.*, **90**, 1825–1832.

Kimura, T. and Inokuchi, S. (1985). Distribution pattern of muscle fiber type in muscles biceps brachii of white-handed gibbon. *J. Anthrop. Soc. Nippon*, **93**, 371–380.

Kimura, T., Kumakura, H., Inokuchi, S., and Ishida, H. (1987). Composition of muscle fibers in the slow loris, using the biceps brachii as an example. *Primates*, **28**, 525–532.

Kojima, R. and Okada, M. (1996). Distribution of muscle fibre types in thoracic and lumbar epaxial muscles of Japanese macaques (*Macaca fuscata*). *Folia Primatol.*, **66**, 38–43.

Kugelberg, E. (1976). Adaptative transformation of rat soleus motor units during growth. *J. Neurol. Sci.*, **27**, 269–289.

Kugelberg, E. and Edström, L. (1968). Differential histochemical effects of muscle contractions on phosphorylase and glycogen in various types of fibres: relation to fatigue. *J. Neurol. Neurosurg. Psychiatry*, **31**, 415–423.

Larson, S. G. and Stern, J. T., Jr. (1989). The rôle of propulsive muscles of the shoulder during quadrupedalism in vervet monkeys (*Cercopithecus aethiops*). *J. Mot. Behav.*, **21**, 457–472.

Marechal, G., Goffart, M., Reznik, M., and Gerebtzoff, M. A. (1976). The striated muscles in a slow-mover *Perodicticus potto* (Prosimii, Lorisidae, Lorisinae). *Comp. Biochem. Physiol.*, **54A**, 81–93.

McDonagh, J. C., Binder, M. D., Reinking, R. M., and Stuart, D. G. (1980a). Tetrapartite classification of motor units of cat tibialis posterior. *J. Neurophysiol.*, **44**, 696–712.

(1980b). A commentary on muscle unit properties in cat hindlimb muscles. *J. Morphol.*, **166**, 217–230.

McIntosh, J. S., Ringqvist, M., and Schmidt, E. M. (1985). Fiber type composition of monkey fore arm muscle. *Anat. Rec.*, **211**, 403–409.

Médina, M. F. and Jouffroy, F. K. (1998). Immunocytochemical approach to the study of postural limb muscles in reptiles. *Biona Rep.*, **13**, 252–253.

Moriyama, K. (1983). Functional differentiation of fore- and hindlimb muscles in *Macaca fuscata* determined on the basis of enzymatic activity. *Primates*, **24**, 94–108.

Mounier, Y., Stevens, L., Shenkman, B. S., *et al.* (2000). Effect of spaceflight on single fiber function of triceps and biceps muscles in rhesus monkeys. *J. Gravit. Physiol.*, **7**, S51–S52.

Ogata, T. and Mori, M. (1964). Histochemical study of oxidative enzymes in vertebrate muscles. *J. Histochem. Cytochem.*, **12**, 171–182.

Peter, J. B., Barnard, R. J., Edgerton, V. R., Gillepsie, C. A., and Stempel, K. E. (1972). Metabolic profiles of three fiber types of skeletal muscle in guinea pigs and rabbits. *Biochemistry*, **11**, 2627–2633.

Peters, S. E. (1989). Structure and function in vertebrate skeletal muscle. *Amer. Zool.*, **29**, 221–234.

Pette, D. and Staron, R. S. (1997). Mammalian skeletal muscle fiber type transitions. *Int. Rev. Cytol.*, **170**, 143–223.

(2000). Myosin isoforms, muscle fiber types, and transitions. *Microsc. Res. Tech.*, **50**, 500–509.

(2001). Transitions of muscle fiber phenotypic profiles. *Histochem. and Cell Biol.* **115**, 359–372.

Petter, A. and Jouffroy, F. K. (1993). Fiber type population in limb muscles of *Microcebus murinus*. *Primates*, **34**, 181–196.

Plaghki, L., Goffart, M., Beckers-Bleukx, G., and Moureau-Lebbe, A. (1981). Some characteristics of the hindlimb muscles in the leaping night monkey: *Aotus trivirgatus* (Primates, Anthropoidea, Cebidae). *Comp. Biochem. Physiol.*, **70A**, 341–349.

Preuschoft, H. (1990). Gravity in primates and its relation to body shape and locomotion. In: *Gravity, Posture, and Locomotion in Primates*, ed. F. K. Jouffroy, M. H. Stack, and C. Niemitz. Firenze: Sedicesimo. pp. 109–127.

Preuschoft, H. and Demes, B. (1984). Biomechanics of brachiation. In: *The Lesser Apes: Evolution and Behavioural Biology*, ed. H. Preuschoft, D. Chivers, W. Y. Brockelman, and N. Creel. Edinburgh: Edinburgh University Press. pp. 96–118.

Preuschoft, H., Fritz, M., and Niemitz, C. (1979). Biomechanics of the trunk in primates and problems of leaping in *Tarsius*. In: *Environment, Behaviour and Morphology: Dynamic Interactions in Primates*, ed. M. E. Morbeck, H. Preuschoft, and N. Gomberg. New York and Stuttgart: Gustav Fischer. pp. 327–345.

Ranvier, L. (1873). Propriétés et structures différentes des muscles rouges et des muscles blancs, chez les lapins et chez les raies. *Pr. Verb. S. Hebd. Ac. Sc. Paris*, **77**, 1030–1034.

(1874). De quelques faits relatifs à l'histologie et à la physiologie des muscles striés. *Arch. Physiol. Norm. Pathol.*, **1**, 5–15.

Reynolds, T. R. (1985). Mechanics of increased support of weight by hindlimbs in primates. *Amer. J. Phys. Anthrop.*, **67**, 335–349.

Riley, D. A., Ilyina-Kakueva, E. I., Ellis, S., Bain, J. L. W., Slocum, G. R., and Sedlak, F. R. (1990). Skeletal muscle fiber, nerve, and blood vessel breakdown in space flown rats. *FASEB J.*, **4**, 84–91.

Romer, A. S. (1964). *The Vertebrate Body.* 3rd edn. Philadelphia, PA: Saunders.

Roy, R. R., Powell, P., Kanim, P., and Simpson, D. R. (1984). Architectural design and fiber-type distribution of the major elbow flexors and extensors of the monkey (*Cynomolgus*). *Amer. J. Anat.*, **171**, 285–293.

Roy, R. R., Bodine-Fowler, F. C., Kim, J., *et al.* (1991). Architectural and fiber type distribution properties of selected *Rhesus* leg muscles: feasibility of multiple independent biopsies. *Acta Anat.*, **140**, 350–356.

Roy, R. R., Hodgson, J. A., Aragon, J., Kathleen Day, M., Kozlovskaya, I., and Edgerton, V. R. (1996). Recruitment of the rhesus soleus and medial gastrocnemius before, during and after spaceflight. *J. Gravit. Physiol.*, **3**, 11–15.

Roy, R. R., Bodine, S. C., Pierotti, D. J., *et al.* (1999). Fiber size and myosin phenotypes of selected rhesus hindlimb muscles after a 14-day spaceflight. *J. Gravit. Physiol.*, **6**, 55–62.

Roy, R. R., Zhong, H., Bodine, S. C., *et al.* (2000). Fiber size and myosin phenotypes of selected rhesus lower limb muscles after a 14-day spaceflight. *J. Gravit. Physiol.*, **7**, S-45.

Salmons, S. and Henriksson, J. (1981). The adaptive response of skeletal muscle to increased use. *Muscle Nerve*, **4**, 94–105.

Sartorius, C. A., Lu, B. D., Acakpo-Satchivi, L., Jacobsen, R. P., Byrnes, W. C., and Leinwand, L. A. (1998). Myosin heavy chains IIa and IId are functionally distinct in the mouse. *J. Cell Biol.* **141**, 943–953.

Schiaffino, S., Gorza, L., Sartore, S., *et al.* (1989). Three myosin heavy chain in type 2 skeletal muscle fibers. *J. Musc. Res. Cell Motil.*, **10**, 197–205.

Serrano, A. L., Perez, M., Lucia, A., Chicharro, J. L., Quiroz-Roth, E., and Rivero, J. L. (2001). Immunolabeling, histochemistry and in situ hybridization in human skeletal muscle fibres to detect myosin heavy chain expression at the protein and mRNA level. *J. Anat.* **199**, 329–337.

Sickles, D. W. and Pinkstaff, C. A. (1980). Are fast glycolytic fibers present in slow loris *Nycticebus coucang* hindlimb muscles? *J. Histochem. Cytochem.*, **28**, 57–59.

(1981). Comparative histochemical study of prosimian primate muscles. *Amer. J. Anat.*, **160**, 175–194.

Staron, R. S. and Pette, D. (1986). Correlation between myofibrillar ATPase activity and myosin heay chain composition in rabbit muscle fibers. *Histochemistry*, **86**, 19–23.

(1993). The continuum of pure and hybrid myosin heavy chain-based fiber types in rat skeletal muscle. *Histochemistry*, **100**, 149–153.

Stein, J. M. and Padykula, H. A. (1962). Histochemical classification of individual skeletal muscle fibers of the rat. *Amer. J. Anat.*, **110**, 103–123.

Stephens, J. A. and Stuart D. G. (1974). Proceedings: the classification of motor units in cat medial gastrocnemius muscle. *J. Physiol.*, **240**, 43P–44P.

Stern, J. T., Jr., Wells, J. P., Vangor, A. K., and Fleagle J. G. (1977) An electromyographic study of the pectoralis major in *Ateles* and *Lagothrix*. *Yearbk. Phys. Anthropol.*, **20**, 498–507.

Stern, J. T., Jr., Wells, J. P., Jungers, W. L., and Vangor, A. K. (1980). An electromyographic study of serratus anterior in atelines and *Alouatta*: implications for hominoid evolution. *Amer. J. Phys. Anthropol.*, **52**, 323–334.

Streter, F. A., Gergely, J., Salmons, S., and Romanul, F. (1973). Synthesis by fast muscle of myosin light chain characteristic of slow muscle in response to long-term stimulation. *Nature*, **241**, 17–19.

Suzuki, A. and Hayama, S. (1991). Histochemical classification of myofiber types in the triceps surae and flexor digitorum superficialis of Japanese macaques. *Acta Histochem. Cytochem.*, **24**, 323–328.

(1994). Individual variation in myofiber type composition in the triceps surae and flexor digitorium superficialis muscles of Japanese macaques. *Anthropol. Sci.*, **102**, 127–138.

Whalen, R. G., Sell, S. M., Butler-Browne, G. S., Schwartz, K., Bouveret, P., and Pinset-Harmström, I. (1981). Three myosin heavy-chains appear sequentially in rat muscle development. *Nature*, **292**, 805–809.

8 *On the nature of morphology: selected canonical variates analyses of the hominoid hindtarsus and their interpretation*

ROBERT S. KIDD
University of Western Sydney

Introduction

Large-scale distinguishing features of the larger tarsal bones between different species of hominoids may readily be discerned with the naked eye. For example the talus of gorillas is visually quite different from that of humans: characteristic differences in the detail of the body, trochlea, and head may easily be identified. Likewise the calcanei of orangutans may be differentiated unambiguously from those of humans, or indeed from other apes. A visual distinction between the smaller tarsal elements, such as cuneiform bones, in differing hominoid groups is not so obvious, but nevertheless possible.

However, many far more subtle and complex morphological features and patterns do exist, both within species as have been identified in humans, or between related species such as between hominoid groups. Such patterns of morphological variation are only identifiable with the aid of quantification and multivariate statistical techniques (e.g., Day and Wood, 1968; Lisowski *et al.*, 1974, 1976; Steele, 1976; Rhoads and Trinkaus, 1977; Pickering, 1986; Kidd, 1995, 2001; Kidd *et al.*, 1996; Kidd and Oxnard, 1997, 2002). Appropriate multivariate techniques, applied correctly and with prudence, provide the investigator with a far richer investigative methodology than simple univariate statistics. The choice of multivariate technique will depend largely upon the nature of the question being asked. If, as is the case in this series of studies, interest lies in differences in tarsal morphology between related groups and species, canonical variates analysis (CVA) is the most appropriate technique as it

Shaping Primate Evolution, ed. F. Anapol, R. Z. German, and N. G. Jablonski. Published by Cambridge University Press.

maximizes differences between those groups (Albrecht, 1980, 1992; Reyment *et al.*, 1984).

CVA produces discriminations based upon the original variable offered up for analysis. However, this raises an important question: upon what criteria has this discrimination been made? Early morphometric studies of primates such as those of the scapula (Ashton *et al.*, 1965a, 1965b; Oxnard, 1967; Oxnard and Flynn, 1971) or the talus (Day and Wood, 1968; Oxnard, 1972; Lisowski *et al.*, 1974, 1976; Rhoads and Trinkaus, 1977) are clearly of an isolated area of an organism. Thus it is axiomatic that this part of the organism's anatomy is adapted to a fairly narrow spectrum of functions. It is to be expected, therefore, that discriminations produced in these studies will reflect functional attributes. As an aside in the scapula study, Ashton *et al.* (1965b) undertook a secondary study utilizing dimensions specifically chosen, as far as possible, not to represent underlying function. The discrimination provided by this secondary set of dimensions was broadly the same as that provided by the first set, though to a lesser degree. Importantly, Ashton *et al.* went on to note that the discriminations produced are capable of separating the primate species studied on a taxonomic basis. Thus it would appear that not only is "functional" information contained within canonical variates, but also information of a taxonomic or genetic/phylogenetic nature. A succinct account of the interplay between these various bases of discrimination may be found in Oxnard (1997).

Further, the very nature of the subsequent analytical treatment of the dimensions under scrutiny may in itself provide insight as to the nature of the discriminations. In a functional study specific to the OH8 fossil assemblage (Kidd *et al.*, 1996) a series of indices was constructed from the linear data. The primary reason for this was to emphasize biomechanically important features. While the main reason for this was not to remove any effects of size, it is recognized that some gross effects may have been ameliorated, though the degree and nature remains intangible.

However, there is another perhaps not so obvious effect of indexing the data in this manner. As all dimensions used in this study were chosen to represent functionally important aspects of morphology, it is reasonable to expect the discriminations produced to reflect functional attributes of the bone's form. Nevertheless, it is possible that some non-adaptive, phylogenetic discrimination would also be represented, perhaps as a result of "residual information" in the manner described by Ashton *et al.* (1965b). The very act of indexing the data to emphasize functional attributes, however, is highly likely to de-emphasize any latent phylogenetic information. Thus, canonical plots produced from analyses of indexed data may be regarded as (largely) functional interpretations, while those from linear data are likely to contain elements of both, though the extent remains obscure.

If parts from various regions of the body are pooled and subjected to an integrated analysis, then several, perhaps contrasting, functional stories are likely to be represented, and to some extent may cancel each other out. The underlying apparently non-adaptive phylogenetic information, being the same for each part, is highly likely to be more readily seen; this was found to be the case by Oxnard (1983). However, there is another possibility. If the species or body parts chosen for analyses are very similar, then the functional information revealed should also be very similar; thus the latent phylogenetic information may be portrayed in a manner which is discernible.

Considerable insight as to the meaning of patterns of discrimination may be gained by referral back to the original univariate dimensions with the aid of the appropriate canonical coefficients; this may reveal the extent to which the original dimensions are supporting the canonical discrimination. However, for this step to be of any real benefit, patterns of biological discrimination must be broadly coincident with the statistical discrimination, the canonical variates. Frequently, patterns of discrimination are held jointly between two canonical variates; in this circumstance individual coefficients often do not reflect accurately the manner in which their dimensions are contributing to the overall pattern of discrimination.

Perhaps due to its favor in the fossil record, the talus has been studied more than any other pedal element (Day and Wood 1968; Oxnard, 1972, 1973; Lisowski *et al.*, 1974, 1976; Rhoads and Trinkaus, 1977; Kidd and Oxnard, 1997). Steele (1976) and Pickering (1986) studied the calcaneus with a view to identifying human subgroups, in the case of the latter for the express purpose of repatriation of American war dead. Kidd *et al.* (1996) undertook a study of the four hindmost tarsal bones specifically for the purpose of identifying functional affinities within the OH8 fossil foot assemblage. Clearly, while there has been an abundance of morphometric studies of the talus, there has been only minor investigation of other tarsal elements.

Thus, the first purpose of this study is to determine whether or not patterns of morphometric discrimination can be identified in other bones of the hominoid hindtarsus. For example, sex, size, functional, and genetic/phylogenetic differences could all be implicated in the discriminations between the four species chosen for analysis. The second purpose is to determine the degree to which information provided by any one bone is similar to, or separate from, that provided by the other bones. For example, how much information, if any, is lost in the examination of individual bones alone? How much new information may be provided by an integrated analysis, not discernible from analyses of individual bones? The third purpose is to understand how important it might be to examine as many elements in fossils as are available. For example, so far, multivariate studies of the Olduvai foot have been confined to individual bones, or a limited integrated study utilizing only indexed data (Kidd *et al.*, 1996). Is it possible that this might have obscured important information about the foot as a whole?

To these ends, a series of studies is presented which examine in some considerable detail the exact nature of discriminations produced by CVA of certain tarsal elements in hominoids, either as individual bones or as part of an integrated study.

Materials

Two sources of material were used in this study. The talus, calcaneus, navicular, and cuboid of four extant primate species, human (*Homo sapiens*), chimpanzee (*Pan troglodytes*), gorilla (*Gorilla gorilla*) and orangutan (*Pongo pygmaeus*) were utilized. In addition, accurate casts were used of the talus, navicular, and cuboid, and of the remaining calcaneal fragment, of the OH8 foot from Olduvai Gorge, attributed to the species *Homo habilis*. The sexes from the extant species were treated as groups in their own right, and, so far as was possible, were equally represented. The samples used consisted of approximately 20 males and females from each species. The African ape specimens were provided courtesy of the Powell-Cotton Museum, England and the orangutan samples by the Smithsonian Institution, Washington, DC. The human sample was made up of a mixture of four groups in roughly equal proportions, namely Victorian and Romano-British, Southern Chinese, and Zulu. These were provided by the British Museum of Natural History, The University of Hong Kong, and the University of Witwatersrand, South Africa.

Bony shape changes in the talus associated with ontogeny, particularly the early stages, have been noted (Straus, 1927; Lisowski, 1967). Accordingly, this study was restricted to adult specimens (as judged from epiphyseal closing in associated long bones). Measurements were taken from elements from the left foot whenever possible, previous statistical tests having shown no significant difference between sides (Kidd, 1995).

Methods

Choice of variables

The choice of measured variables in any morphometric study is of paramount importance as it may have significant effects upon the results. All dimensions utilized were chosen so as to reflect what are thought to be functionally important components, typified by articular dimensions, bone heights, or overall bone length. While most variables are linear, two from the talus and one from the calcaneus are indices in the form of angles. Space does not permit a full description of the reference positions utilized for the four bones or the actual linear and angular dimensions; full details of these are available in a report of the functional attributes of the OH8 hindtarsus (Kidd *et al.*, 1996).

From many of the linear dimensions, a series of indices was constructed from the linear dimensions for use in certain parts of the study. In those instances, the primary use of indices was not to reduce the gross effects of size, but rather to emphasize those aspects of the osseous morphology thought to be of particular biomechanical importance. The problems associated with the use of indices are well recognized (Albrecht, 1993; Atchley *et al.*, 1976; Corruccini, 1975); however, where the primary intention is to emphasize biomechanically important features rather than a deliberate attempt to remove the effects of size, their use is not only acceptable, it is essential. By way of example, the length of one lever arm is of no importance biomechanically: the ratio of two lever arms is the biomechanically critical variable. Details of the indices used in the relevant parts of this study, including their suggested mechanical implications, may be found in Kidd *et al.* (1996).

Methods of data collection

All linear dimensions were obtained using standard osteometric calipers with digital readout. Angular values were obtained with a geometric calculation from digitized photographs. All photographs utilized for data collection were taken in carefully referenced positions to minimize distortions resulting from camera or object orientation. In order to establish their validity, a reproducibility study was undertaken on all dimensions. A description of the tests used and results is provided in Kidd *et al.* (1996). Linear data were recorded to 0.1 mm and angles were recorded in degrees to one decimal place; both were entered directly into a spreadsheet.

Analytical methods

An initial univariate analysis was undertaken in which the spreads of individual values for each linear dimension, angle, or derived index were compared in each group. In addition, the standard univariate descriptors of mean, standard deviation, coefficients of variance, and distribution shape were closely examined. Plots of mean values against their standard deviations revealed a strong positive regression for most dimensions; thus the data were transformed to their natural logarithms. However, in the case of linear data, a subsequent series of plots revealed a notable negative correlation and regression; thus the transformation was too strong. Consequently, a square-root transformation was undertaken (Snedecor and Cochran, 1967) and subsequent plots did not reveal any obvious trends. In the case of the indexed data, the logarithmic transformation did satisfactorily remove the positive regression and no obvious trend was observed.

The primary objective of this study was to establish the basis of, and to seek biological explanation for, discriminations produced by canonical variates analysis of the hominoid tarsal elements under scrutiny, both as individual bones and as components of integrated studies. Thus a series of plots of the first, second and higher canonical variates was constructed. Some clue as to the nature of canonical plots of linear data may be gleaned from the inclusion of the OH8 fossil. The provenance of this assemblage is not in doubt: it is from East Africa and is thus genetically more closely related to African apes and humans than it is to Asian apes. Of course, this line of argument is only tenable once the functional attributes of the fossil specimens have been assigned by detailed analysis of the indexed data, specifically designed to emphasize function. Bearing this in mind, and given the desire to obtain as much insight as possible, three series of analyses were undertaken. First, a series of analyses was undertaken utilizing indexed data of apes-and-humans, including the isolated fossil. This was undertaken both as individual bones and as an integrated analysis. Second, a series of analyses using linear data was undertaken, of apes and humans and of apes only. This series was then repeated but with the inclusion of the isolated fossil. Third, an integrated analysis of all four tarsal bones was undertaken utilizing linear data, initially on their own and later including the OH8 fossil elements.

Though analyses of all four bones were undertaken in each series, only those results which best answer the questions outlined earlier are presented here. In all analyses including isolated fossil elements, the fossil was entered into the analysis directly with a sample size of unity, and not into a previously determined matrix of extant species (Oxnard, 1972). In all the analyses, whether of linear or indexed data, the majority of discrimination was contained within the first two variates, together accounting for at least 90% in each case. Thus, though they were all carefully examined, variates beyond the second are only considered further in those instances where they add biologically important detail.

Table 8.1. *The eigenvalues, coefficients, and percentages of the discrimination for the canonical variates analysis of hominoid tali, including OH8. See Figure 8.1 and text for details.*

		CA1	CA2
Eigenvalue		6.25	5.56
% of discrimination		48.06	9.08
Coefficients	Ind1	0.571	0.575
	Ind2	0.234	0.552
	Ind3	−0.069	0.195
	Ind4	0.186	−0.544
	Ind5	−0.077	−0.081
	Ind6	−0.224	0.515
	Ind7	0.218	0.198
	Ind8	−0.117	0.112
	Ind9	0.166	0.232
	Ind10	−0.688	0.163
	Ind11	0.410	0.340

Results

Indexed data and functional analyses: two examples from the OH8 hindtarsus

This report is not intended to be a detailed functional account of the OH8 fossil hindfoot. The findings and functional interpretation given here are not complete: for such an account of the OH8 hindtarsus, the reader is referred to Kidd *et al.* (1996).

The talus

The first canonical variate demonstrates a marked discrimination between apes as a whole and humans, there being slightly over three standard deviation units (SDU) between their means. The fossil talus occupies a position well within the overall ape group. The chief dimensions contributing towards this are the height index (ind1), the neck and head torsion angles (ind10 and ind11) and the trochlear groove index (ind2) in decreasing order of importance (Table 8.1).

The second canonical variate demonstrates considerably less discrimination between groups. Nevertheless there is still a clear division between African apes and orangutans with slightly over four SDU between them. Humans are positioned roughly equidistant between these two. The OH8 talus is located

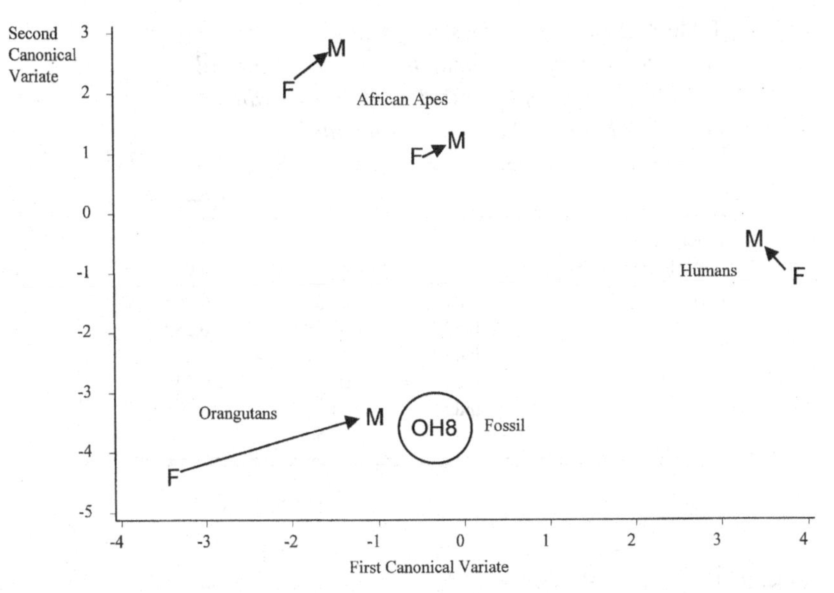

Figure 8.1. The talus (apes, humans, and fossil). Plot of the first and second canonical variates utilizing indexed variables. Note the large discrimination of the fossil from humans and African apes, and the close affinity to orangutans. Units for all plots are standard deviation units.

among the orangutans, slightly over two SDU from the nearest human group. The height index (ind1), the trochlear groove index (ind2), the trochlear breadth index (ind4), and the breadth index (ind6), in decreasing order of importance, are the most important dimensions in this pattern of discrimination (Table 8.1).

A plot of the first and second canonical variates together places the OH8 talus among orangutans, particularly the males (Fig. 8.1); thus the overall morphology of the fossil talus is markedly similar to that of orangutans.

The cuboid

The first canonical variate discriminates all apes from all humans with five SDU between them (Fig. 8.2). The fossil cuboid falls between apes and humans but is far closer to humans. Two dimensions contribute most to this discrimination, both of which reflect the presence (or absence) of the *process calcaneus*, ind3 and ind5 (Table 8.2).

The second canonical variate does not clearly separate any group, although there is a marginal discrimination between humans and African apes on one hand and orangutans on the other. The fossil cuboid falls among orangutans. A plot of the first two canonical variates provides an obvious visual display of the

Table 8.2. *The eigenvalues, coefficients, and percentages of the discrimination for the canonical variates analysis of hominoid cuboids, including OH8. See Figure 8.2 and text for details.*

		CA1	CA2
Eigenvalue		10.65	8.48
% of discrimination		84.90	8.03
Coefficients	Ind1	−0.338	0.903
	Ind2	−0.317	−0.282
	Ind3	1.068	0.203
	Ind4	−0.114	0.338
	Ind5	1.008	0.205
	Ind6	0.350	−0.038

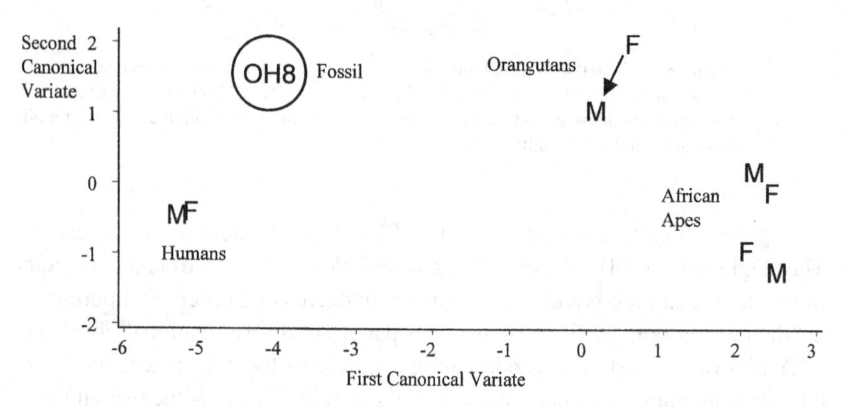

Figure 8.2. The cuboid (apes, humans, and fossil). Plot of the first and second canonical variates utilizing indexed variables. Note the large discrimination of the fossil from all apes, and the close affinity to humans.

affinity of the fossil cuboid with that of humans, and its distance from all apes (Fig. 8.2).

An integrated analysis of indexed data from all four bones

An examination of a plot of the first two variates reveals quite clearly the unique nature of this fossil assemblage, it being positioned more than 5 SDU from the nearest extant group (Fig. 8.3). Broadly, the three extant groups (humans,

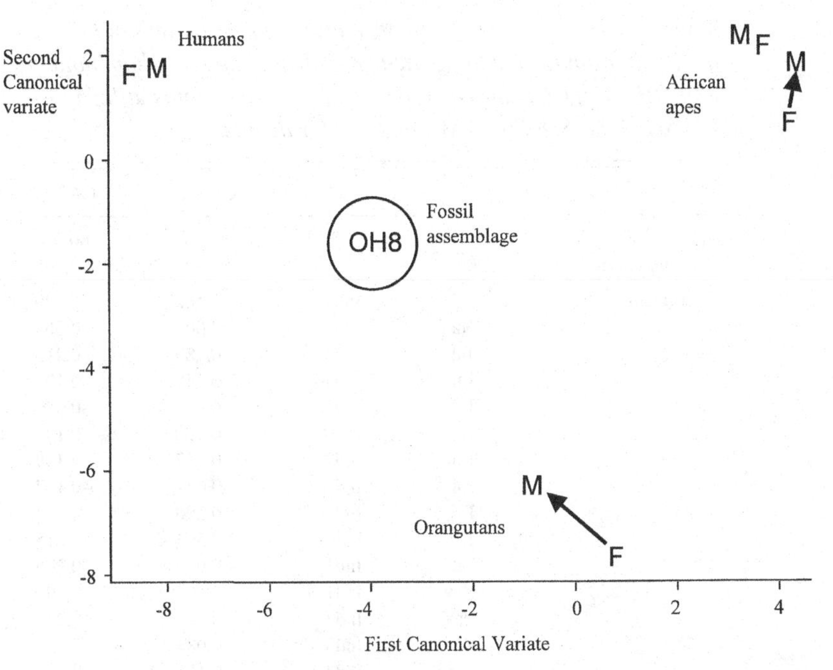

Figure 8.3. Integrated analysis (apes, humans, and fossil). Plot of the talus, navicular, cuboid, and (partial) calcaneus. Plot of the first and second canonical variates utilizing indexed variables. Note the unique position of the fossil assemblage when compared to extant groups.

African apes, and orangutans) occupy extremities of the plot, with the fossil assemblage being positioned centrally. When considered wholly, rather than piecemeal, the unique nature of the specimen is evident. An understanding of the key variables contributing to this discrimination may be gained from Table 8.3.

A functional interpretation of the OH8 fossil

In the case of the talus, with even a cursory glance one's eye is quickly drawn to the manner in which the extant species are distributed, occupying what may best be described as the apices of a triangle (Fig. 8.1). Functionally, this may neatly be interpreted as representing the three locomotor specialities of knuckle-walking in African apes, arboreality in orangutans, and terrestrial bipedalism in humans.

The position of the isolated fossil, therefore, gives important information as to its functional affinity. The position of the fossil talus, close to the orangutans,

Table 8.3. *The eigenvalues, coefficients, and percentages of the discrimination for the integrated canonical analysis of hominoid and OH8 tali, (partial) calcanei, naviculi, and cuboids utilizing indexed data. See Figure 8.3 and text for details.*

			CA1	CA2
Eigenvalue			28.13	60.02
% of discrimination			60.00	28.90
Coefficients	Tal	Ind1	−0.102	0.539
	Tal	Ind2	0.005	0.363
	Tal	Ind3	0.289	0.213
	Tal	Ind4	−0.249	−0.192
	Tal	Ind5	0.229	−0.326
	Tal	Ind6	0.120	0.179
	Tal	Ind7	−0.127	0.149
	Tal	Ind8	0.031	−0.117
	Tal	Ind9	−0.200	0.214
	Tal	Ind10	0.594	−0.215
	Tal	Ind11	−0.070	0.266
	Nav	Ind1	−0.509	0.195
	Nav	Ind2	−0.292	0.246
	Nav	Ind3	−0.084	0.437
	Nav	Ind4	0.105	−0.310
	Nav	Ind5	−0.216	−0.327
	Nav	Ind6	0.226	0.245
	Calc	Ind1	−0.238	−0.232
	Calc	Ind2	−0.214	−0.183
	Calc	Ind3	0.998	−0.110
	Calc	Ind4	−0.044	−0.172
	Cub	Ind1	0.911	−0.154
	Cub	Ind2	0.414	−0.194
	Cub	Ind3	0.222	0.097
	Cub	Ind4	0.074	−0.252
	Cub	Ind5	−0.065	−0.004
	Cub	Ind6	0.166	−0.274

strongly implies an arboreal function in a manner similar to that of the modern-day orangutan. While it is accepted that a minor component of genetic information may be being expressed in these analyses, it is reasonable to suppose that the majority of the discrimination is describing functional information for the reasons associated with the use of indexed data in the manner discussed earlier.

The position of the fossil cuboid, close to humans and distant from all the apes, is strong evidence for the human-like nature of this bone (Fig. 8.2). This is clearly in contrast to that found with the talus. A re-examination of the univariate data reveals that this discrimination is afforded almost entirely by those variables that describe the presence (or absence) of the process calcaneus. This process is a uniquely human feature and is absent in all the apes, though it is present in the fossil cuboid. The process has an important role, articulating intimately with the opposing calcaneal facet, providing stability to the lateral longitudinal arch of the foot; it is thus manifest that this analysis reveals a decidedly functional discrimination. Although much further information is revealed from similar analyses of bones not reported here, these two alone reveal quite differing functional attributes. The talus, from the medial side of the foot, clearly indicates an ape-like arboreal function and the cuboid, from the lateral aspect, a decidedly terrestrial function, perhaps analogous to modern humans.

Consideration of the integrated analysis of all four bones lends much evidence to the unique nature of the functional affinity of this organism. As in the talus, the position of the three extant groups at the apices of a triangle again describes the three known hominoid locomotor patterns of knuckle-walking, arborealism, and upright bipedal striding gait, and it is worth noting that the degree of discrimination is considerably greater than in the single bone (Fig. 8.3). The central position occupied by the fossil assemblage is clear evidence that this organism employed none of these, but another as yet unknown locomotor strategy.

Linear data: examples from the talus and navicular

The first variate provides a discrimination of gorillas and humans from chimpanzees and orangutans (Fig. 8.4). Within each species, there is a marked discrimination of males and females. Along the first canonical variate, the two most heavily weighted variables describe overall bone dimensions of breadth and height (var6 and var7). In contrast, the second variate provides discrimination of orangutans and humans from African apes. The heavily weighted coefficients seem to describe aspects of overall bone morphology, but in a more detailed and intricate fashion than along the first variate; again and in particular, the maximum breadth (var6), but also the cuneiform facet dimension (var3), overall height and sagittal thickness (var7 and var8) are represented here. Details of the canonical coefficients from this analysis may be found in Table 8.4.

It is clear that the two essential lines of biological discrimination of species are broadly concordant with the canonical variates that describe statistical discrimination (Fig. 8.4). As a consequence, examination of the heavily weighted coefficients for each variate is likely to reveal much functionally relevant information. The large discrimination described along the first variate seems to

Table 8.4. *The eigenvalues, coefficients, and percentages of the discrimination for the canonical analysis of hominoid naviculi. See Figure 8.4 and text for details.*

		CA1	CA2
Eigenvalue		13.31	4.37
% of discrimination		67.80	22.27
Coefficients	Var1	0.120	−0.080
	Var2	0.053	0.113
	Var3	−0.048	0.706
	Var4	0.054	−0.128
	Var5	0.113	0.124
	Var6	0.537	−1.257
	Var7	0.334	0.536
	Var8	0.160	0.519
	Var9	0.110	0.035

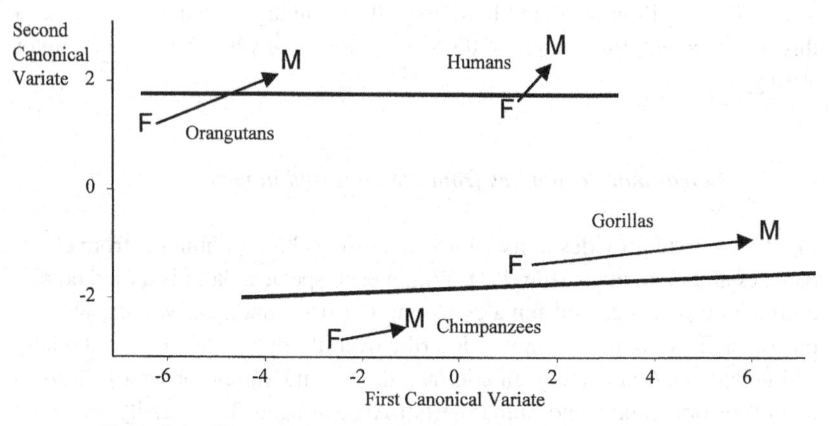

Figure 8.4. The navicular (apes and humans). Plot of the first and second canonical variates utilizing linear variables. Note the broadly concordant nature of the lines of biological discrimination and the canonical variates.

be essentially size-related, evidenced by the position of humans and gorillas at the positive extreme of the variate. In contrast, that found along the second variate does seem to be shape-related, at least in the broadest sense, providing a discrimination of African apes from orangutans and humans. Since orangutans have a major arboreal component to their lifestyle and humans are fully terrestrial it is implausible to relate this to a common functional affinity. The naviculi of the African apes differ from those of orangutans and humans in several fairly distinct areas. First, they are broader and rather more squat than those of orangutans and humans. This is partly due to their having a considerably larger tuberosity, projecting medially beyond the main component of the bone, but is also in part due to a broader and less tall profile of the main part of the bone itself. Second, the bone is relatively thinner (in the sagittal plane) than in orangutans and humans.

These patterns of morphology are reflected by the weighted coefficients along the second variate. The most negative coefficient is from the maximum breadth and the most positive from the long cuneiform dimension. Essentially, the latter dimension is a measure of the breadth of the main body of the bone while the former represents the main body plus the tuberosity projection. Effectively, therefore, what is being described is tuberosity magnitude or projection. Thus, humans and orangutans at the positive end of the variate are characterized by a small tuberosity while the African apes at the negative end of the variate are characterized by a large tuberosity projection.

Although morphological similarity of African apes on one hand and orangutans and humans on the other has been succinctly demonstrated, it is not immediately obvious what is the functional significance of this discrimination. The navicular tuberosity is primarily a site of insertion for the tibialis posterior tendon. Several authors (e.g., Lewis, 1964; Langdon, 1986) mention prolongation of this insertion to, among other sites, the medial (ento) cuneiform also.

Thus it would seem to have some importance in movements of the first ray, perhaps as a synergist (in this respect) to the long hallucial flexor and extensor and to the intrinsic muscles. In humans the function of tibialis posterior muscle has become very much modified as a powerful subtalar joint inverter and as a factor in the maintenance of the long and transverse arches; clearly, while the former of these functions may be shared with apes, the latter is not. In addition to the large tuberosity size in African apes indicating a major insertion, it will actually increase the lever arm of movement of the first ray in a medial direction, perhaps important as a part of grasping.

The first digit of orangutans has undergone significant modification, essentially shortening (Schultz, 1930). This is recognized as being a specialization in orangutans, presumably connected with their locomotor affinity. Humans have also undergone considerable shortening of the first digit, albeit as a component of a generalized decrease in digital length. Thus, though for quite different reasons and in response to the requirements of quite different locomotor strategies, the naviculi of orangutans and humans seem to have a parallel function, and have therefore a parallel morphology.

In the previous example, the biologically meaningful discriminations were broadly concordant with the statistical information, the canonical variates. In the next examples, the biologically meaningful discriminations are decidedly

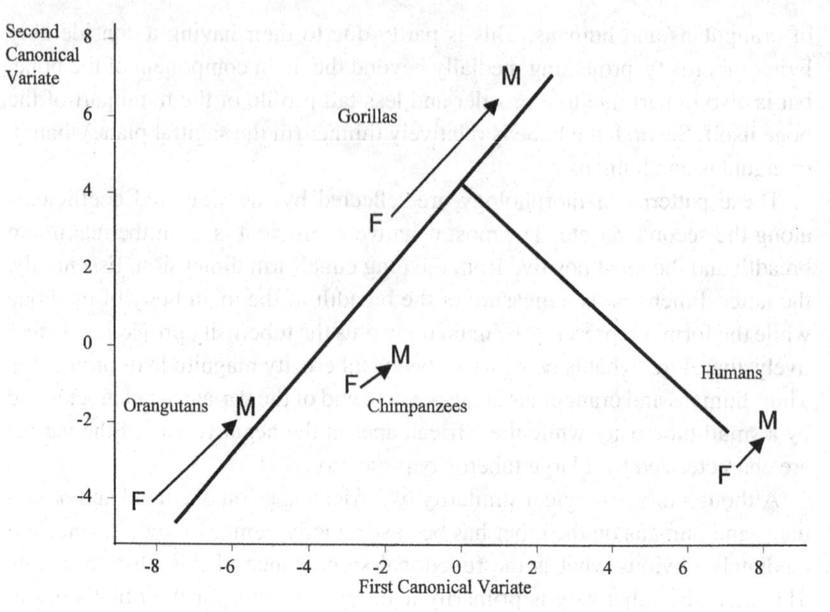

Figure 8.5. The talus (apes and humans). Plot of the first and second canonical variates utilizing linear variables. Note the decidedly oblique nature of the lines of biological discrimination compared to the canonical variates, and the separation of humans from apes generally.

oblique compared to the canonical variates. This alters somewhat the manner in which they may be interpreted.

Apes and humans

The first variate provides, with only minor overlap, discrimination on the grounds of both species and sex (Fig. 8.5). The human talus, while being smaller than that of gorillas, is positioned considerably more positively, thus demonstrating at least a component of shape and size-related-shape information along this variate. The most heavily weighted dimensions describe bone height (var3), a composite of height and breadth (var8), the medial facet length (var12), and the head torsion angle (var20). The second variate provides only a discrimination of gorillas from the other species. The heavily weighted variables being from bone height (var1 and var3) and the trochlear breadth (var4 and var5). It is noteworthy that one of each group of variables is positive and the other negative (Table 8.5); they are thus providing contrast with each other and describing key aspects of bone morphology. A plot of the first two variates provides clear separations by species and sex, held jointly between the two variates.

Table 8.5. *The eigenvalues, coefficients, and percentages of the discrimination for the canonical analysis of ape and human tali. See Figures 8.5 and 8.6 and text for details.*

		CA1	CA2	CA3
Eigenvalue		29.83	13.26	2.65
% of discrimination		63.80	28.35	5.67
Coefficients	Var1	0.182	−0.991	0.344
	Var2	0.010	−0.224	0.613
	Var3	0.541	1.103	−0.987
	Var4	0.088	−0.596	0.753
	Var5	−0.160	0.855	−0.461
	Var6	−0.102	0.061	0.206
	Var7	0.163	−0.317	−0.189
	Var8	0.416	0.121	−0.154
	Var9	−0.118	0.002	0.372
	Var10	−0.360	0.588	−0.019
	Var11	0.119	−0.075	−0.114
	Var12	−0.579	0.070	−0.074
	Var13	0.022	−0.030	0.070
	Var14	−0.305	0.318	−0.106
	Var15	−0.208	0.034	0.153
	Var16	0.379	−0.169	0.293
	Var17	0.161	−0.023	−0.277
	Var18	0.312	0.129	0.216
	Var19	−0.247	0.162	−0.059
	Var20	0.417	0.006	−0.415

While providing separations too small to be explained definitively in terms of contributing variables, the third canonical variate is still worthy of exploration. While alone it does not provide any convincing separations, a plot of the third and first canonical variates provides a clear separation by species and sex, held jointly between the two variates (Fig. 8.6).

Apes only

The first variate provides a discrimination of each species and sex, though to a lesser extent than in the previous analyses (Fig. 8.7). The second variate, to only a minor extent and with some overlap, provides a discrimination of species from each other. A plot of the first two variates again provides a clear separation by species and sex held jointly between the two variates. The canonical coefficients for this analysis are reported in Table 8.6.

It is clear that in both analyses and in all three plots, the two essential lines of discrimination are not concordant with either canonical variate, but are held

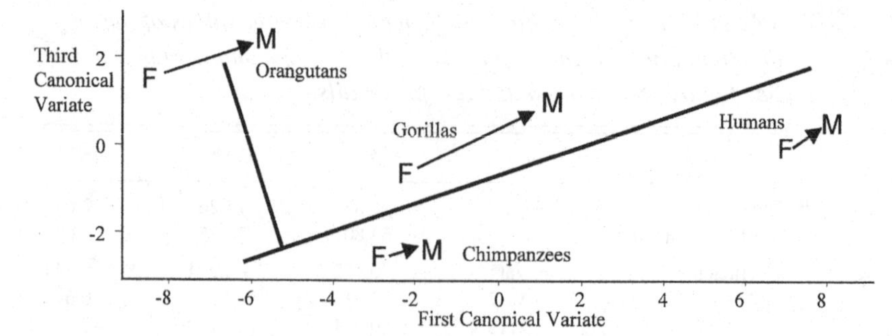

Figure 8.6. The talus (apes and humans). Plot of the first and third canonical variates utilizing linear variables. Note the oblique nature of the lines of biological discrimination compared to the canonical variates and the separation of orangutans from humans and African apes.

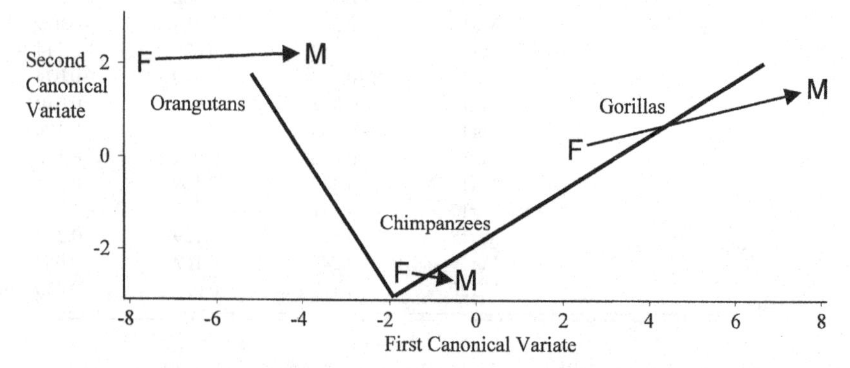

Figure 8.7. The talus (apes only). Plot of the first and second canonical variates utilizing linear variables. Note the oblique nature of the lines of biological discrimination and the separation of Asian and African apes.

obliquely between them. Thus though much may be learned from individual canonical coefficients, they are less meaningful than was found to be the case with the navicular. With even a cursory glance, it is clear that there are two essential lines of discrimination. In the case of the first and second variates of the apes-and-humans analyses, there is a clear line of discrimination describing the apes. In addition, there is a second line of discrimination separating apes as a whole from humans (Fig. 8.5). However a biological meaning for these discriminations is not immediately forthcoming. The variate along which the apes are positioned is characterized by orangutans at the negative end, gorillas at the positive, and chimpanzees between. Clearly this variate could, therefore, be a *functional* discrimination describing an increasing degree of terrestriality (or a decreasing degree of arboreality) towards the positive end from orangutans

Table 8.6. *The eigenvalues, coefficients, and percentages of the discrimination for the canonical analysis of apes-only tali. See Figure 8.7 and text for details.*

		CA1	CA2
Eigenvalue		25.24	4.22
% of discrimination		81.68	13.66
Coefficients	Var1	−0.235	−0.089
	Var2	−0.178	0.565
	Var3	0.967	−0.667
	Var4	−0.335	0.346
	Var5	0.743	0.044
	Var6	−0.048	0.345
	Var7	−0.226	−0.423
	Var8	0.424	−0.311
	Var9	0.147	0.430
	Var10	0.120	0.372
	Var11	−0.147	−0.311
	Var12	−0.248	0.071
	Var13	−0.264	0.001
	Var14	−0.058	0.248
	Var15	−0.328	0.079
	Var16	0.179	0.135
	Var17	−0.001	−0.401
	Var18	0.414	0.004
	Var19	0.042	0.150
	Var20	0.249	−0.487

through to gorillas. It could equally, however, be describing a *genetic* difference between apes of the Asian and African provinces.

The discrimination of humans from apes may also be viewed as being contained within an axis oblique to either of the canonical variates. Essentially, this is a perpendicular discrimination from apes as a whole. Is this discrimination one of a functional nature, separating bipedal humans from (broadly) quadrupedal or arboreal apes? Or is it perhaps of a phylogenetic, non-adaptive nature, separating apes and humans upon the basis of a genetic dissimilarity? Most likely, the answer is that there is a mixture of both components. However, for the reasons outlined earlier, that all dimensions are from a single bone, and that the majority of data are linear, it is most probable that significant functional discrimination is being expressed.

The plot of the third and first canonical variates again provides two clear lines of discrimination, though different from those described above (Fig. 8.6). There is a clear line of discrimination describing humans and African apes.

Equally, there is a clear separation of humans and African apes from orangutans. Does the African ape–human axis describe a functional trend with (broadly) quadrupedal apes at one end and upright bipedal humans at the other, or a genetic separation of humans and African apes? And, does the separation of orangutans from African hominoids describe a functional difference of arboreal and terrestrial locomotor strategies, or is it a genetic difference separating the Asian and African provinces?

The apes-only analysis again provides essentially two lines of discrimination, clearly separating African and Asian apes (Fig. 8.7). However, they present the same conundrum as outlined in the above scenarios. The separation of orangutans from African apes may describe a functional discrimination, of the essentially arboreal orangutan on one hand from the more terrestrial African apes on the other. However, it may also be describing an essentially phylogenetic discrimination, demonstrating the genetic distance of orangutans from African apes. In a similar manner, the African-ape axis could be a functional discrimination, describing a decreasing arborealism towards its more positive end, towards gorillas. Equally, however, it could also be describing a genetic separation of the two African ape species. What is noticeable is that the apes-only plot of the first and second variates provides the same discrimination as is found among the apes in the apes-and-humans plot of the first and third variates. Thus, the act of focusing on the ape groups seems to have relegated this discrimination to a lower variate.

Apes, humans, and fossil

A plot of the first two variates reveals the two (non-canonical) axes to be still prominent with the fossil bone positioned approximately midway along the first variate and at the extreme negative end of the second. Thus, its position is clearly neither on a line close to an extrapolation of the human nor ape groups (Fig. 8.8). However, examination of the third variate does reveal an interesting discrimination (Fig. 8.9). A plot of the first and third variates clearly reveals the fossil talus to be positioned on a line joining African apes and humans, close to the apes. Group mean positions for the plots in Figures 8.8 and 8.9 are given in Table 8.7.

Apes and fossil

A plot of the first two variates places unequivocally the fossil element near the negative end of both variates. In addition, the fossil talus is clearly positioned on an extrapolation of the non-canonical axis joining the African apes (Fig. 8.10). Group mean positions for this plot are recorded in Table 8.8.

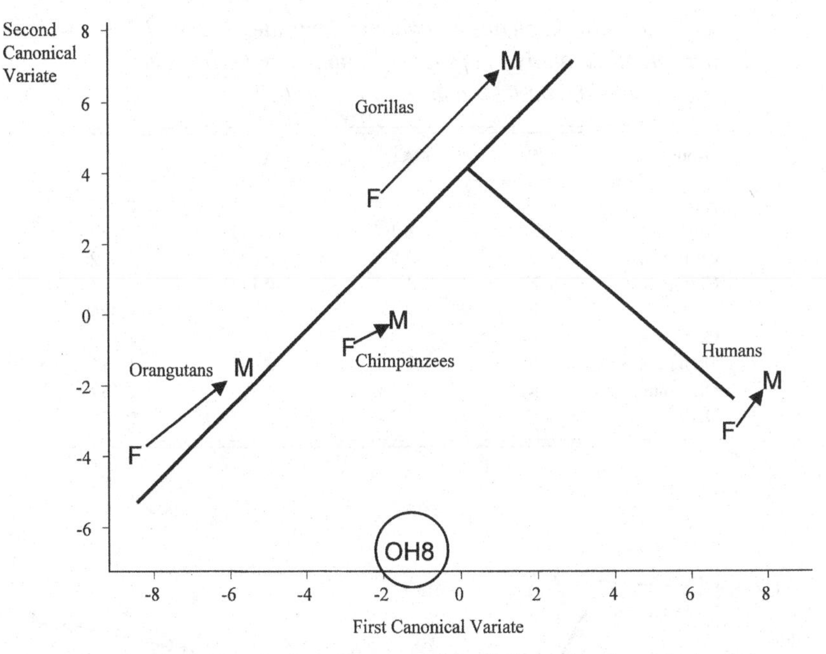

Figure 8.8. The talus (apes, humans, and fossil). Plot of the first and second canonical variates utilizing linear variables. Note the ambiguous position of the fossil talus, positioned on neither obvious line of discrimination.

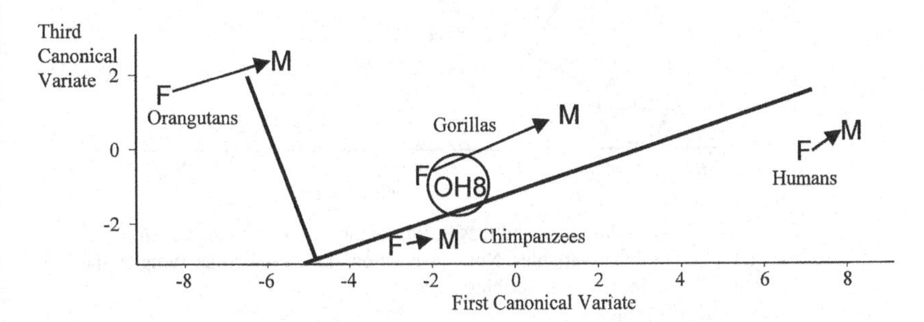

Figure 8.9. The talus (apes, humans, and fossil). Plot of the first and third canonical variates utilizing linear variables. Note the position of the fossil talus, lying on the clear line of discrimination describing humans and African apes.

The plot of the first and third canonical variates of the human, apes, and fossil clearly demonstrate the fossil to lie on the axis describing humans and African apes, and to be separated from the orangutans. Thus, tali with an unambiguous arboreal affinity are positioned towards both extremes of the third canonical

Table 8.7. *Group means on canonical variates 1, 2, and 3 for the canonical analysis of the ape, human, and OH8 talus. See Figures 8.8 and 8.9 and text for details.*

Group	Sex	CA1	CA2	CA3
Human	F	7.033	−3.274	0.193
Human	M	8.167	−1.841	0.739
Chimpanzee	F	−2.826	−0.886	−2.312
Chimpanzee	M	−1.547	−0.111	−2.144
Gorilla	F	−2.177	3.335	−0.425
Gorilla	M	1.372	7.193	1.190
Orangutan	F	−8.427	−3.950	1.737
Orangutan	M	−5.589	−1.453	2.659
OH8		−1.278	−6.631	−0.993

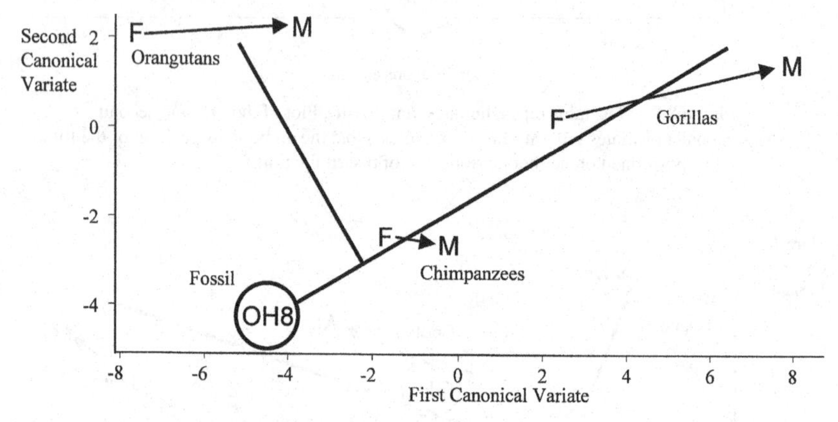

Figure 8.10. The talus (apes and fossil). Plot of the first and second canonical variates utilizing linear variables. Note the position of the fossil talus, lying on the clear line of discrimination describing African apes.

variate (Fig. 8.9). This is a clear indication that the discrimination between Asian and African apes is not one of a functional affinity, but one of a genetic nature, discriminating between Asian and African provinces. Moreover, the position of the fossil talus among the African apes lends tentative support to the notion of the African apes–human axis describing a functional trend, one of an increasing degree of terrestrialism, from chimpanzees through gorillas to humans.

Table 8.8. *Group means on canonical variates 1 and 2 for the canonical analysis of the ape and OH8 talus. See Figure 8.10 and text for details.*

Group	Sex	CA1	CA2
Chimpanzee	F	−1.66	−2.31
Chimpanzee	M	−0.19	−2.49
Gorilla	F	2.36	0.36
Gorilla	M	7.97	1.57
Orangutan	F	−7.57	2.21
Orangutan	M	−3.64	2.38
OH8		−4.38	−4.09

The apes-only plot of the first and second canonical variates provides a strikingly similar picture with the same arboreal tali at both ends of the second canonical variate (Fig. 8.10). Thus again there is described a genetic discrimination between the Asian and African provinces. In addition, there is again evidence for a functional trend being described along the African-ape axis with a decidedly arboreal talus at the lower end of the axis, through chimpanzees to gorillas, the least arboreal of the apes. The fact that in the apes-only study these trends are apparent on a lower variate lends convincing confirmation of the nature of these separations. Thus the notion of a genetic discrimination between the Asian and African apes is supported.

The final study: integrated analysis of linear data from all four bones

Apes and humans

The first canonical variate of the integrated analysis clearly separates humans from all apes and thus provides separation by species (Fig. 8.11). The second variate provides some separation by species and sex. A plot of the first two variates serves to reinforce these discriminations. A plot of the first and third canonical variate (not figured) reveals much the same story, though of a lesser discrimination. A plot of the third and second variates reveals an interesting discrimination, clearly present, but not obvious in first and higher variates (Fig. 8.12). Here there is a clear discrimination of orangutans from humans and African apes. And there is a clear line of discrimination defining humans and African apes. Details of the canonical coefficients for these plots are given in Table 8.9.

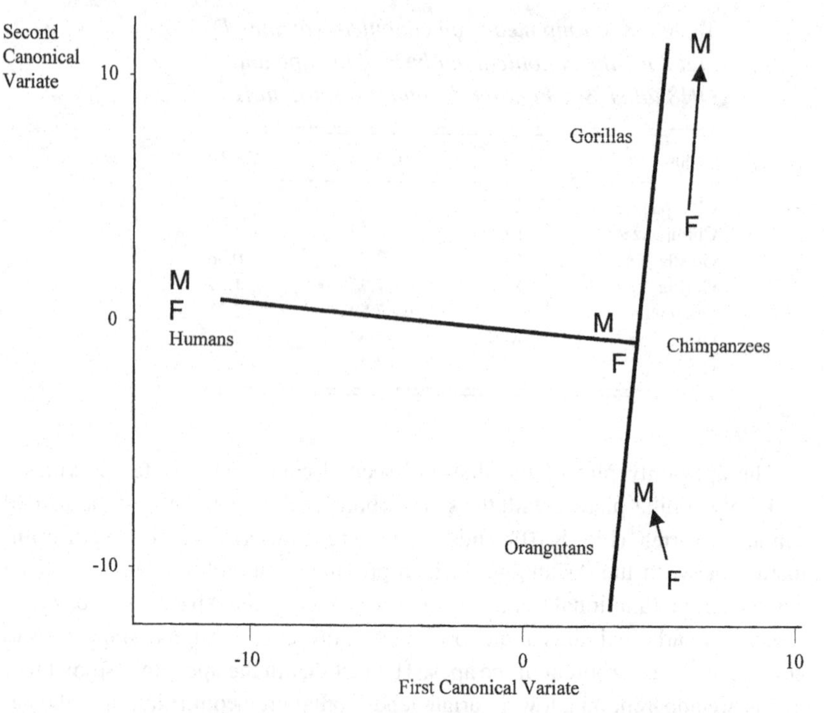

Figure 8.11. Integrated analysis (apes and humans). Plot of the first and second canonical variates of the talus, navicular, cuboid, and (partial) calcaneus utilizing linear variables. Note the large separation between apes as a whole and humans.

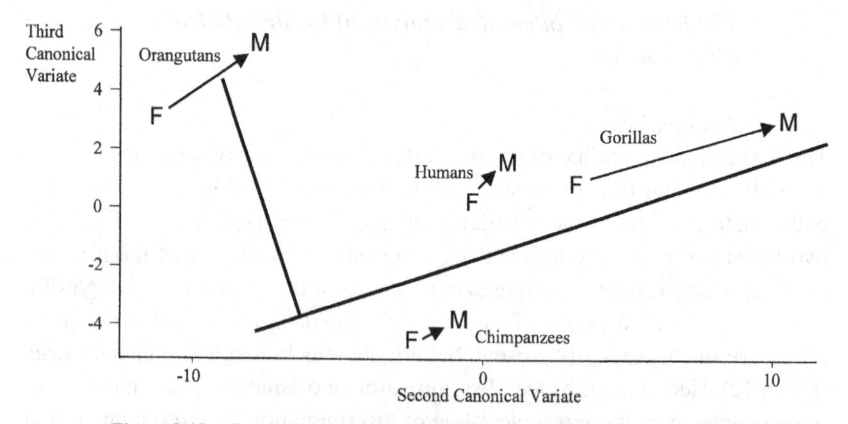

Figure 8.12. Integrated analysis (apes and humans). Plot of the second and third canonical variates of the talus, navicular, cuboid, and (partial) calcaneus utilizing linear variables. Note the oblique nature of the lines of biological discrimination compared to the canonical variates and the separation of orangutans from humans and African apes.

Table 8.9. *The eigenvalues, coefficients, and percentages of the discrimination for the integrated canonical analysis of hominoid tali, (partial) calcanei, naviculi, and cuboids utilizing transformed linear data. See Figures 8.11 and 8.12 and text for details.*

		CA1	CA2	CA3
Eigenvalue		66.09	36.94	10.14
% of discrimination		56.9	31.83	8.74
Coefficients	Tal 1	−0.239	−0.541	0.167
	Tal 2	0.281	0.238	0.684
	Tal 3	−0.471	1.004	−0.748
	Tal 4	−0.481	−0.799	0.285
	Tal 5	0.659	0.721	−0.066
	Tal 6	0.226	0.140	0.187
	Tal 7	−0.868	−0.487	−0.878
	Tal 8	−0.191	0.356	−0.212
	Tal 9	0.728	0.105	−0.060
	Tal 10	0.463	0.008	0.204
	Tal 11	−0.269	−0.034	0.166
	Tal 12	0.201	−0.518	−0.052
	Tal 13	0.012	0.091	0.248
	Tal 14	0.314	−0.123	−0.034
	Tal 15	0.050	−0.124	0.352
	Tal 16	0.015	0.254	0.230
	Tal 17	−0.317	−0.105	−0.011
	Tal 18	−0.188	0.518	−0.346
	Tal 19	0.284	0.016	−0.094
	Tal 20	−0.325	0.227	−0.223
	Cal 1	0.091	0.209	0.053
	Cal 2	−0.550	0.070	−0.030
	Cal 3	0.203	−0.405	0.130
	Cal 4	0.328	−0.037	0.569
	Cal 5	−0.207	0.233	0.269
	Cal 6	−0.057	0.545	0.291
	Cal 7	0.211	−0.121	0.367
	Nav 1	0.451	0.070	0.158
	Nav 2	−0.230	−0.104	−0.269
	Nav 3	0.057	−0.196	0.301
	Nav 4	−0.169	0.270	−0.271
	Nav 5	−0.328	0.154	0.117
	Nav 6	0.325	0.576	−0.431
	Nav 7	0.372	0.167	0.482
	Nav 8	0.299	0.142	0.420
	Nav 9	−0.099	0.124	−0.295
	Cub 1	0.353	−0.256	−0.336
	Cub 2	−0.278	−0.035	0.088
	Cub 3	0.153	−0.421	−0.572
	Cub 4	−0.125	−0.235	0.284
	Cub 5	−0.208	0.162	−0.220
	Cub 6	−0.770	−0.167	0.508
	Cub 7	−0.090	−0.556	−0.024
	Cub 8	−0.225	0.045	−0.299
	Cub 9	0.134	−0.201	−0.032

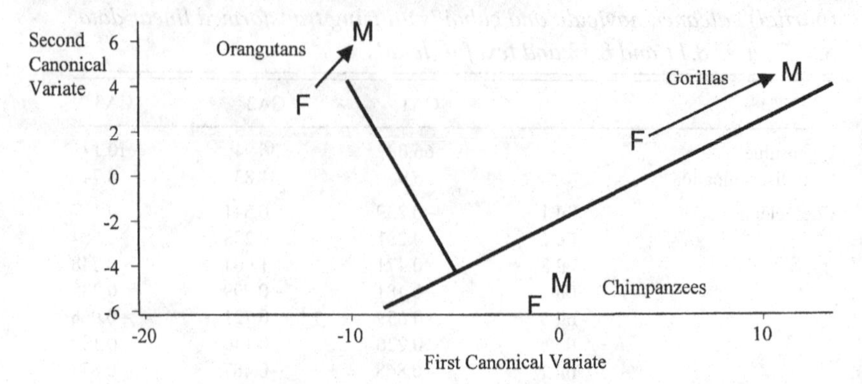

Figure 8.13. Integrated analysis (apes only). Plot of the first and second canonical variates of the talus, navicular, cuboid, and (partial) calcaneus utilizing linear variables. Note the large separation between Asian and African apes.

Apes only

The first canonical variate provides a clear separation by species and sex, while the second variate provides a partial separation (Fig. 8.13). A plot of the first two variates provides a clear separation, producing the same picture as that found in the talus alone, but to a greater degree: a clear line of discrimination describing chimpanzees and gorillas, and another line of discrimination separating orangutans and African apes. Details of the canonical coefficients for these plots are given in Table 8.10.

Apes, humans, and fossils

A plot of the first two variates shows that inclusion of the isolated fossils into the analysis alters the position of extant groups only marginally (not figured). The position of the fossil assemblage falls between apes and humans, not on either line of discrimination. The same pattern is found with a plot of the first and third variates (not figured): the position of the fossil is between apes and humans on neither line of discrimination. However, on the plot of the third and second variates, the fossil assemblage is unambiguously positioned on the line of discrimination describing African apes and humans (Fig. 8.14 and Table 8.11).

Apes and fossils

A plot of the first two variates shows that inclusion of the isolated fossils again alters the position of the extant groups only marginally. The fossil assemblage is clearly positioned on the African ape line, clearly separated from orangutans (Fig. 8.15 and Table 8.12).

Table 8.10. *The eigenvalues, coefficients, and percentages of the discrimination for the integrated canonical analysis of hominoid tali, (partial) calcanei, naviculi, and cuboids utilizing transformed linear data. See Figure 8.13 and text for details.*

		CA1	CA2
Eigenvalue		61.50	69.73
% of discrimination		21.63	24.53
Coefficients	Tal 1	−0.232	−0.269
	Tal 2	0.173	0.503
	Tal 3	0.771	−0.342
	Tal 4	−1.081	0.368
	Tal 5	0.965	0.294
	Tal 6	0.135	0.265
	Tal 7	−0.118	−1.041
	Tal 8	0.434	−0.272
	Tal 9	0.712	0.367
	Tal 10	−0.248	0.436
	Tal 11	−0.451	0.160
	Tal 12	−0.559	0.165
	Tal 13	−0.208	0.347
	Tal 14	−0.285	0.182
	Tal 15	−0.362	−0.098
	Tal 16	0.201	0.196
	Tal 17	−0.111	−0.224
	Tal 18	1.119	−0.543
	Tal 19	0.027	0.062
	Tal 20	0.159	−0.209
	Cal 1	−0.081	0.126
	Cal 2	−0.187	−0.160
	Cal 3	−0.508	0.027
	Cal 4	−0.267	0.727
	Cal 5	0.362	0.231
	Cal 6	0.330	0.543
	Cal 7	−0.242	0.395
	Nav 1	0.110	0.298
	Nav 2	0.134	−0.543
	Nav 3	−0.274	0.419
	Nav 4	0.187	0.042
	Nav 5	0.166	0.013
	Nav 6	0.770	−0.622
	Nav 7	0.032	0.830
	Nav 8	0.271	0.465
	Nav 9	0.157	−0.415
	Cub 1	0.553	−0.916
	Cub 2	−0.360	−0.169
	Cub 3	−0.112	−0.992
	Cub 4	−0.522	0.174
	Cub 5	0.229	−0.304
	Cub 6	−0.215	0.471
	Cub 7	−0.894	0.066
	Cub 8	−0.780	0.034
	Cub 9	0.299	0.131

Table 8.11. *Group means on canonical variates 2 and 3 for the integrated canonical analysis of apes, human, and OH8 utilizing transformed linear data. See Figure 8.14 and text for details.*

Group	Sex	CA2	CA3
Human	F	0.073	−0.046
Human	M	1.340	1.320
Chimpanzee	F	−2.133	−4.815
Chimpanzee	M	−0.458	−4.111
Gorilla	F	3.503	0.483
Gorilla	M	10.883	2.613
Orangutan	F	−10.906	2.813
Orangutan	M	−7.283	5.387
OH8		−5.955	−5.157

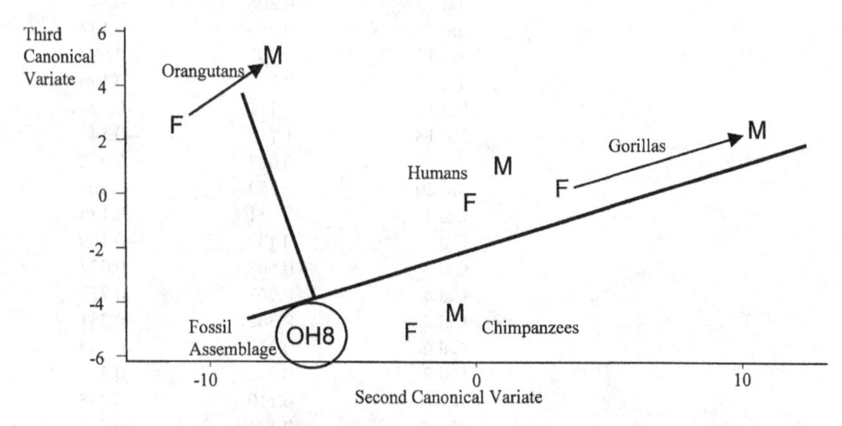

Figure 8.14. Integrated analysis (apes, humans, and fossil). Plot of the second and third canonical variates of the talus, navicular, cuboid, and (partial) calcaneus utilizing linear variables. Note the position of the fossil talus, lying on the clear line of discrimination describing humans and African apes.

The integrated analyses of extant species, whether of apes and humans, or of apes alone, provide a far greater degree of discrimination than that found in analyses of only one tarsal member. By way of example, the apes-and-humans integrated analysis provides nearly 20 SDU of separation between apes and humans. This may be read alongside only about 8 or 9 SDU in the talus alone. Thus, in an integrated analysis, information that must be present in the raw data, but which seems to be obscured in individual analyses by more dominant size and sex information, may be revealed additionally.

Table 8.12. *Group means on canonical variates 1 and 2 for the integrated canonical analysis of apes and OH8 utilizing transformed linear data. See Figure 8.15 and text for details.*

Group	Sex	CA1	CA2
Chimpanzee	F	−0.971	−5.895
Chimpanzee	M	0.322	−4.966
Gorilla	F	4.103	1.377
Gorilla	M	11.603	4.346
Orangutan	F	−12.061	3.119
Orangutan	M	−9.190	6.297
OH8		−4.660	−5.793

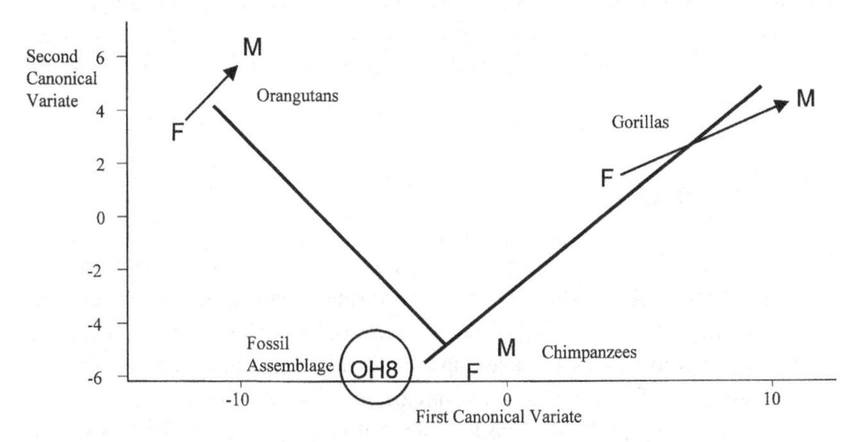

Figure 8.15. Integrated analysis (apes and fossil). Plot of the first and second canonical variates of the talus, navicular, cuboid, and (partial) calcaneus utilizing linear variables. Note the position of the fossil talus, lying on the clear line of discrimination describing African apes, and the separation from orangutans.

Plots of the first and second variates in the apes-and-humans analysis clearly reveal the by now familiar two lines of discrimination, one describing apes, from orangutans through chimpanzees to gorillas, the other describing a distance between apes as a whole and humans. It is noteworthy that the act of integrating bones for analysis has rendered these discriminations less oblique; broadly, the apes–humans discrimination is held within the first variate while the within-apes discrimination is held within the second variate.

There is another effect that has resulted from pooling of bones for analysis, best revealed in Figure 8.14. In an analysis of a single bone, the very slight discrimination held jointly between the second and third variates may be too small to be discernible; it is certainly too small in the case of these tarsal elements. Yet in the integrated analysis a clear and important discrimination is held between the two variates, though its nature remains elusive. However, inclusion of the OH8 fossil assemblage provides good evidence for the basis of this discrimination, the fossil assemblage lying unambiguously on an extrapolation of the human–African apes axis, clearly separated from orangutans. Thus, the separation between Asian and African provinces described earlier (Figs. 8.9 and 8.10) is supported by information from an integrated analysis, that is, information from more than one bone. This is clearly a discrimination with a genetic basis and, taken together with the information provided by Figures 8.9 and 8.10 from the talus, serves to remind us that *Homo sapiens* is of African descent.

This finding is further supported by the inclusion of the fossil assemblage into the integrated apes-only analysis (Fig. 8.15). Yet again one may clearly see its position on the African-ape axis, and the separation from orangutans. And as before with the talar study, this is present in the apes-only analysis on a lower variate and thus adds persuasive confirmation of the genetic nature of this discrimination.

Conclusion

It is clear from the various analyses reported here that both functional and genetic affinities may be revealed by canonical variates analysis, though frequently their basis may not be obvious. Under certain rather stringent circumstances, such as was demonstrated by indexing data to emphasize biomechanically important attributes, functional affinity may be emphasized. However, in the case of linear data, it is probably true to state that some portion of both types of information is being portrayed, though the extent of this remains uncertain. In the navicular study, for instance, it was found that functionally important inferences could be made quite easily, largely because the biological (non-canonical) discriminations lay broadly concordant with the statistical (canonical) discriminations. This was not found to be the case in any of the talar studies illustrated here, largely due to the oblique nature of the discrimination. In this situation, much may be revealed by the inclusion of a further group of known provenance such as the OH8 fossil assemblage.

With respect to the three purposes of this study outlined in the introduction, certain pieces of information have become abundantly clear. First, patterns of morphological discrimination may be identified in other tarsal elements: the indexed cuboid and linear navicular studies both provide succinct discriminations with a clear functional interpretation. Second, the need to undertake an integrated study is manifest. Third, the inclusion of appropriate fossil specimens into analyses, being careful to avoid any tautological leanings, may provide considerable insight into the biological meaning of discriminations.

It is thus clear that the type of analysis, the choice of variables, and indeed any subsequent manipulation of those variables, must depend critically upon the question or questions being asked. This study has demonstrated beyond doubt two points. First, in a study aimed specifically at functional attributes of a bone, or a suite of bones, it may be prudent to manipulate the data in such a manner that mechanically important features are emphasized. Second, in a study of a suite of bones such as this, it is important to utilize both individual and integrated analyses as they may each provide information not provided by the other.

References

Albrecht, G. (1980). Multivariate analysis and the study of form, with special reference to canonical variates analysis. *Amer. Zool.*, **20**, 679–693.

(1992). Assessing the affinities of fossils using canonical variates and generalised distances. *Hum. Evol.*, **7**, 49–69.

(1993). Ratios as a size adjustment in morphometrics. *Amer. J. Phys. Anthropol.*, **91**, 441–468.

Ashton, E. H., Healy, M. J. R., Oxnard, C. E., and Spence, T. F. (1965a). The combination of locomotor features of the primate shoulder girdle by canonical analysis. *J. Zool.*, **147**, 406–429.

Ashton, E. H., Oxnard, C. E., and Spence, T. F. (1965b). Scapular shape and primate classification. *Proc. Zool. Soc. Lond.*, **145**, 125–142.

Atchley, W. R., Gaskins, C. T., and Anderson, D. (1976). Statistical properties of ratios. 1. Empirical results. *Syst. Zool.*, **25**, 137–148.

Corruccini, R. S. (1975). Multivariate analyses in biological anthropology: some considerations. *J. Hum. Evol.*, **4**, 1–19.

Day, M. H. and Wood, B. A. (1968). Functional affinities of the Olduvai hominid 8 talus. *Man*, **3**, 440–455.

Kidd, R. (1995). *An Investigation into the Patterns of Morphological Variation in the Proximal Tarsus of Selected Human Groups, Apes and Fossils: a Morphometric Analysis*. Ph.D. thesis, University of Western Australia.

(2001). Individual and integrated analyses of tarsal morphology: a case study in humans. In: *Causes and Effects of Human Variation*, ed. M. Henneberg. Australasian Society for Human Biology, Department of Anatomical Sciences. Adelaide: University of Adelaide.

Kidd, R. S. and Oxnard, C. E. (1997). Patterns of morphological discrimination in the human talus: a consideration of the case for negative function. In: *Perspectives in Human Biology*, ed. C. E. Oxnard and L. Freedman. Perth: Centre for Human Biology; Singapore: World Scientific Publishing.

(2002). Patterns of morphological discrimination in selected human tarsal elements. *Amer. J. Phys. Anthropol.*, **117**, 169–181.

Kidd, R., O'Higgins, P., and Oxnard, C. E. (1996). The OH8 foot: a reappraisal of the functional morphology of the hindfoot utilizing a multivariate analysis. *J. Hum. Evol.*, **31**, 269–291.

Langdon, J. H. (1986). Functional morphology of the Miocene hominoid foot. *Contrib. Primatol.*, **22**, 1–225.

Lewis, O. J. (1964). The tibialis posterior tendon in the primate foot. *J. Anat.*, **93**, 209–218.

Lisowski, F. P. (1967). Angular growth changes and comparisons in the primate talus. *Folia Primatol.*, **7**, 81–97.

Lisowski, F. P., Albrecht, G. H., and Oxnard C. E. (1974). The form of the talus in some higher primates: a multivariate study. *Amer. J. Phys. Anthropol.*, **41**, 191–216.

(1976). African fossil tali: further multivariate studies. *Amer. J. Phys. Anthropol.*, **45**, 5–18.

Oxnard, C. E. (1967). The functional morphology of the primate shoulder as revealed by comparative anatomical, osteometric and discriminant function techniques. *Amer. J. Phys. Anthropol*, **26**, 219–240.

(1972). Some African fossil foot bones: a note on the interpolation of fossils into a matrix of extant species. *Amer. J. Phys. Anthropol.*, **37**, 3–12.

(1973). *Form and Pattern in Human Evolution: some Mathematical, Physical, and Engineering Approaches*. Chicago, IL: University of Chicago Press.

(1983). *The Order of Man: a Biomathematical Anatomy of the Primates*. Hong Kong: Hong Kong University Press.

(1997). The interface of function, genes, development and evolution: insights from primate morphometrics. In: *Perspectives in Human Biology*, ed. C. E. Oxnard and L. Freedman. Perth: Centre for Human Biology; Singapore: World Scientific Publishing.

Oxnard, C. E. and Flynn, R. M. (1971). The functional and classificatory significance of combined metrical features of the primate shoulder girdle. *J. Zool.*, **163**, 319–350.

Pickering, R. B. (1986). Population differences in the calcaneus as determined by discriminant function analysis. In: *Forensic Osteology: Advances in the Identification of Human Remains*, ed. K. J. Reichs. Springfield, IL: C. Thomas.

Reyment, R. A., Blackwith, R. E., and Campbell, N. A. (1984). *Multivariate Morphometrics.* London: Academic Press.

Rhoads, J. G. and Trinkaus, E. (1977). Morphometrics of the Neandertal talus. *Amer. J. Phys. Anthropol.*, **46**, 29–44.

Schultz, A. H. (1930). The skeleton of the trunk and limbs of higher primates. *Hum. Biol.*, **2**, 303–438.

Snedecor, G. W. and Cochran, W. G. (1967). *Statistical Methods.* Ames, IA: Iowa State University Press.

Steele, D. G. (1976). The estimation of sex on the basis of the talus and calcaneus. *Amer. J. Phys. Anthropol.*, **45**, 581–588.

Straus, W. L. (1927). The growth of the human foot and its significance. *Contrib. Embryol.*, **101**, 93–134.

9 *Plant mechanics and primate dental adaptations: an overview*

PETER W. LUCAS
University of Hong Kong

Introduction

Most mammals chew their food, a process generally involving the fracture and fragmentation of food particles. The efficiency of this process can be measured by the rate at which food particle size is reduced per chew and depends on the match of postcanine tooth shape to the mechanical properties of the food. The dentition of a mammal is presumably adapted to its diet so as to produce a rate of processing in the mouth sufficient for the gut to service its metabolic rate. Previously, I have offered very generalized models for the differentiation of molar shape in relation to food type (Lucas, 1979). The aim of this chapter is to suggest gains in understanding of tooth form that follow from more detailed analysis of food properties. The arguments presented here are snippets of those to be published in a forthcoming book (Lucas, in press).

For the dentition of a primate to be optimally adapted to its diet, the animal must either feed on a very restricted range of foods, with consequently limited variation in mechanical texture, or else actively select foods on a textural basis. It is difficult, at first glance, to imagine that primates employ the texture of plant parts as a primary criterion for food selection. Species that forage widely, eating many plant species, are more likely to use long-range senses like color and smell as primary sensory cues to draw them towards potential foods, or else they would waste much time in moving around to sample food texture with their forelimbs or mouth. Nevertheless, there is considerable evidence from field studies that the selection of foods, particularly leaves, by primates is heavily influenced by the avoidance of high dietary fiber (Waterman *et al.*, 1988; Rogers *et al.*, 1990; Ganzhorn, 1992; Janson and Chapman, 1999).

Fiber is a nutritional synonym for the plant cell wall, the structure that provides the basis for the mechanical properties of plant foods and which forms

Shaping Primate Evolution, ed. F. Anapol, R. Z. German, and N. G. Jablonski. Published by Cambridge University Press.

the main obstacle to their digestion. Since fiber is essentially colorless, tasteless, and odorless, it seems probable that mechanical texture is a major sensory criterion for fiber acceptance or rejection (Choong *et al.*, 1992). Whether feeding decisions that primates make on plant items are actually the result of a correlation between food attributes perceived at long range, such as color, and short-range cues like texture (Dominy and Lucas, 2001) or, instead, are the result of specific acceptance/rejection decisions based on manipulation with the hands or mouthfeel, the influence of texture on feeding patterns is probable.

Fiber and texture are often treated as equivalents, the latter being the physical manifestation of the amount of cell wall present in a tissue. To examine whether this assertion is justified, we have to decide how to analyze the mechanical properties that cell walls exhibit.

Mechanical properties of plant cells

The mechanical behavior of a plant part can be subdivided into two domains. As a reaction to load, there is initially a *deformation domain* whereby the disturbance to equilibrium produced by the stress (force divided by the unit area of the solid that it acts on) is adapted to by displacements that produce strains in the object (strains being linear or angular distortions). There are two types of deformation. The initial phase is largely elastic, when strains are reversible if the load is removed. Dependent on the loading regime, this can be followed by a plastic phase in which strains set permanently. Later, there may be a *fracture domain* characterized by a crack that soaks up much of the energy that has previously been causing the solid to distort, and which grows further as additional available energy is fed into it.

To understand the mechanical behavior of plant cells in the fracture domain, we need to know how mechanical properties control these domains or their boundaries. Definitions of the most relevant properties, Young's modulus, yield stress, and toughness, can be found in basic accounts written variously for materials scientists (e.g., Ashby and Jones, 1980, 1986), biologists (e.g., Vincent, 1990, 1992), or ecologists (e.g., Lucas *et al.*, 2000) and only a bare outline is given here. The Young's modulus is a measure of the elastic deformability of a solid under load and is defined as the stress–strain gradient in the elastic phase. Its units are megapascals (1 MPa = 10^6 Pa) or gigapascals (1 GPa = 10^9 Pa), where a Pascal (Pa) is a load of 1 newton that acts over a square meter. The yield stress defines the transition between deformational phases – from elastic (recoverable) to plastic (permanent) deformations. Its unit is also the megapascal (MPa), although the yield stress is a completely different type of quantity than the Young's modulus, the latter being calculated as the stress divided by a dimensionless strain. Lastly, toughness, the ability to resist crack growth, is defined as the energy consumed in growing a crack of given area. Its units are Joules per metre squared (J m^{-2}) or a unit one thousand times larger (kJ m^{-2}).

At the junction between the deformation and fracture domains lies the point of fracture initiation. For most of the history of mechanics, this has been characterized by the fracture stress. However, since the 1920s, it has been increasingly recognized that the fracture stress is object size-dependent (Atkins and Mai, 1985) and that other, more fundamental, parameters must be underpinning fracture onset. One such parameter is the critical stress intensity factor (K_{IC}), a concept from linear elastic fracture mechanics that is shorthand for a quantity roughly defined as the square root of the product of toughness and Young's modulus. The strength of a material, in its lay meaning, can be interpreted in two ways, either as the point of fracture initiation or as the onset of permanent strain. While the former is characterized by K_{IC}, the latter can be defined by the yield stress, often measured indirectly by an indentation (or hardness) test.

The cell wall of plants is an organized mixture, a mechanical composite. Cellulose, in the form of very fine and stiff crystalline strands, is set in a glue-like matrix. This matrix often consists just of low-stiffness hemicelluloses of various non-crystalline forms, but these can also be combined with lignin, an amorphous, highly cross-linked polymer. Without going into actual structural details, which are exceedingly complex, we can indicate in general terms what composite construction like this achieves. A mixture places bounds on the values of Young's modulus and yield stress, such that a composite cannot be stiffer or stronger than its stiffest or strongest constituent. There are no gains in these properties, only losses, by mixing components to form a composite. Toughness, however, is exempt from this "rule of mixtures," typically being hundreds or even thousands of times above that of any of its components tested separately, provided, that is, the composite is correctly organized (Atkins and Mai, 1985). A simple conclusion is that a composite like plant cell wall is likely to be designed to be optimally tough rather than stiff or strong, the latter both being maximized by single-component construction. How tough is the cell wall?

The amount of cell wall in a tissue can be calculated by finding the volume fraction of the cell (V_c) which the wall occupies. Lucas *et al.* (1995, 1997a, 1997b) conducted a survey of 82 plant tissues and plant-derived materials, each of which was sufficiently homogeneous for the cell-wall volume fraction of the tissue to be estimated. These tissues varied from watermelon flesh, with very large thin-walled cells (Lucas *et al.*, 1995) to seed shells with such thick walls that they occupy 95% of the tissue (i.e., $V_c = 0.95$). The toughness of each of these tissues was tested by a cutting technique described in Lucas *et al.* (1997a, 1997b, 2000). This method can isolate two contributing causes of toughening: (a) fracture events in the cell wall itself, and (b) the contribution of the cellular

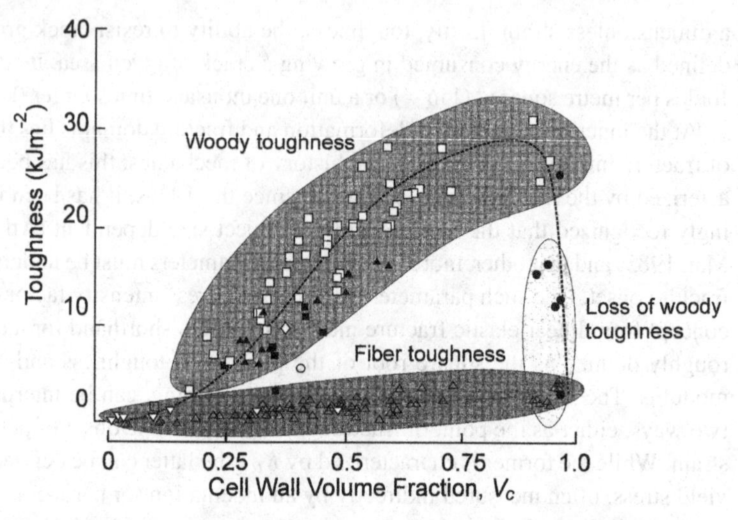

Figure 9.1. Results of a survey of the toughness of 82 plant tissues or plant-based materials plotted against the volume fraction of the tissue occupied by cell wall (V_c), roughly synonymous with the term "fiber content." The toughness attributable to the presence of cell wall can thus be called "fiber toughness"; values are indicated by the grayed-out upright triangles (▲). The regression line through these values shows that fiber toughness is proportional to V_c. Extrapolating to $V_c = 1.0$ predicts a toughness for the cell of 3.45 kJ m^{-2}. Overall toughness, due both to the plastic buckling mechanism described in the text and to fiber toughness, is shown by the other symbols: ▽ tissues with primary cell walls only; ■ pods/seed coats; ◇ woody gall; □ tropical woods; ▲ temperate woods; ○ climber; • seed shells. The dotted line follows the general trend. Elongated woody cells have a peak toughness about ten times that due to cell wall, exhibiting "woody toughness." At $V_c > 0.9$, the wall becomes too thick, and the cellular lumen too small, for the plastic buckling mechanism to operate, so there is a "loss of woody toughness."

framework. The key to partitioning the fractional contributions of these causes lies in the dependence of toughness on tissue thickness. When specimen thickness is < 0.5 mm, toughness is proportional to thickness. Above this thickness, toughness tends to plateau out. To obtain information on energy absorption due to fracture purely within the cell wall, toughness was regressed on thickness for each tissue and the relationship extrapolated to zero thickness so as to find the "intercept toughness" (Lucas *et al.*, 2000). For all 82 tissues, intercept toughness was found to be proportional to V_c (Fig. 9.1). By extrapolating to $V_c = 1.0$, an indirect estimate of the toughness of the cell wall could be obtained. This was 3.45 kJ m^{-2}, a value that can be interpreted, perhaps controversially, as "fiber toughness" – the toughness of the cell wall.

The gray upright triangles in Figure 9.1 represent fiber toughness for all tissues. There is a lot of scatter around the regression line, but it is difficult to attribute this variation to the structure of the cell wall (e.g., presence/absence of a secondary cell wall or wall organization) and the standard error of estimation of the intercepts for any tissues is an important factor. The real surprise is that 3.45 kJ m^{-2} is not very high for a composite. Although well above the rock-bottom toughness of single-component structures like gels (~1 J m^{-2}) or common ceramics (~10 J m^{-2}), and probably ten times that of any of the components of a cell wall if they could be tested in isolation, 3.45 kJ m^{-2} should be compared to 50 kJ m^{-2} or more that is achievable in artificial carbon, boron, or glass fiber composites (Atkins and Mai, 1985). In biological tissues, a comparison with the toughness of animal structures designed for attack or defense, like nails (Pereira *et al.*, 1997) and skin (Purslow, 1983), shows that >15 kJ m^{-2} should be anticipated for toughness to provide a reasonable defensive shield. While plant cell wall *per se* seems not to provide this defense, could there be any non-fiber mechanism in plant tissues available to elevate toughness?

The slope of the toughness–tissue thickness plot provides information on plastic toughening mechanisms in tissues and, as shown in Figure 9.1, seems to provide up to ten times the toughness that fiber alone does. The cause seems to lie in a truly cellular mechanism. Elongated "woody" plant cells, some of which are, confusingly, called fibers (the plural here always refers to cell type while singular denotes nutritional fiber), have parallel cellulose microfibrils in the S2 layer of their secondary cell walls that spiral at an angle of about 15°–30° to the cellular axis. It is known that, at critical stresses, this type of wall structure buckles plastically into the cellular lumen (Page *et al.*, 1971). In so doing, large amounts of energy are absorbed (Gordon and Jeronimidis, 1980). It appears that this buckling mechanism elevates toughness well above fiber toughness levels, but only in woody tissues. The mechanism acts non-linearly, peaking at about $V_c = 0.8$. Above this, toughness starts to plateau with increasing V_c and falls precipitously at $V_c > 0.9$ or so. This is because, at such high fiber levels, the buckling mechanism is frustrated by the fact that the cell wall is so thick that there is insufficient lumen for the wall to buckle into. Close to $V_c = 0.95$, the mechanism hardly operates at all and toughness appears to revert to levels due to nutritional fiber.

I conclude from this discussion that toughness and fiber are not physico-chemical equivalents: the toughness of plant tissues depends both on wall structure and on cellularity, and not just on the volume fraction that the wall occupies. Given this, feeding decisions based on food texture will not directly reflect fiber content. A nutritionist might, for example, know from objective tests that a piece of light balsa wood, a very young leaf, some types of fruit flesh, and root storage organs have similar fiber contents. Yet the subjective (sensory) mouthfeel of the balsa will be very different to the other tissues because it is several times tougher and much more difficult to chew. To human consumers, the acceptability of fruit pods (like peas or beans) often depends on the presence/absence of sclerenchyma fibers even though these constitute only a small fraction of their volume. Similarly, the inclusion of such fibers in mature leaves makes little difference to nutritional fiber content, but would provide a powerful motive for a primate to accept or reject these foods due to the substantial reduction in the breakdown rate that these woody cells impart.

There is, however, another reason why the toughness of plant foods does not reflect fiber content: cracks in dense tissues tend to pass between cells. Such intercellular fracture can leave the cell wall completely intact, particularly if the tissue is dry (Williamson and Lucas, 1995). According to Ashby *et al.* (1985), in woody tissues where $V_c > 0.2$, cracks switch from a preferential course through the cell wall to a path that tracks between cells whenever possible. The cost of intercellular fracture has been measured only in woody tissues but depending on the straightness of the crack path, lies between 100 and 1000 J m^{-2}, being independent of V_c.

The nature of mechanical defenses

Virtually all plant parts that are sufficiently exposed to herbivores are going to be defended against their attack. As an evolutionary response, herbivores should show adaptations to cope with these defenses, which if they are mechanical will include the dentition. There are two general strategies for mechanical defense seen in biological tissues, both of which refer to fracture resistance (Lucas *et al.*, 2000). The first type of defense aims to stop cracks from starting. This can be called *stress-limited defense* because the aim is to make a crack start at stresses higher than herbivores can generate. As shown more precisely elsewhere (Ashby, 1999), stress-limited defenses will generally involve either a high yield stress or high K_{IC}. The second type of defense involves trying to stop cracks that have already started from growing. This is called *displacement-limited defense* because it is usually associated with large displacements. As shown by Ashby (1999) in a general context or Agrawal *et al.* (1998) for mastication, these defenses are typified not just by high toughness, but instead by a high value for the square root of (toughness/Young's modulus). If, tissues are very thin (say, 10–20 cell layers in thickness or less) and loading is very localized (which is a probable feature of an optimized masticatory apparatus), then toughness alone should dominate defensive potential (Agrawal *et al.*, 1998).

The structures of plants designed in these two ways look very different. Plants with prominent stress-limited defenses, trying to prevent crack initiation, tend to have hard outer surfaces. A dense outer layer raises Young's modulus, particularly in bending, and elevates K_{IC}. There may also be hard sharp features, such as spines, thorns, and stiff hairs, which act to deter herbivores from even contacting the plant. These deterrents often have amorphous silica incorporated in them because this single-component construction is harder than plant cell wall material. Some hard tissues are, however, defended by their very thick cell walls – many seed shells are like this. In contrast, displacement-limited defenses, those that try to prevent cracks from growing, are typified by a woody structure, not particularly disposed on the exterior, but throughout the interior wherever possible.

Most ecological accounts emphasize chemical defenses in plants, regarding mechanical defenses either as exceptions or as phylogenetically ancient, predating the development of an extensive secondary chemistry. This is partly because stress-limited defenses are usually the only type of mechanical defense to be recognized. Plants, or plant parts, that express stress-limited defenses are usually low in secondary chemicals simply because their strategy is vested in the avoidance of cell opening by cracks. Most secondary plant compounds are stored inside cells, so parsimonious defense predicts an absence of these chemicals. However, stress-limited defenses are not that common because in order to have a high yield stress or Young's modulus, they need to be dense; ideally, being made from a single component (Lucas *et al.*, 2000). The optimal tissue organized this way is, therefore, dead. In contrast, displacement-limited defenses constrain crack growth. Cells are inevitably opened in these plant parts, and thus chemicals have a defined defensive role. As explained earlier, the composite nature of the plant cell wall means that it will have a high value for (toughness/Young's modulus). Accordingly, displacement-limited defenses are likely to predominate in plants.

Ingestion vs. mastication

Displacements are not limiting during ingestion because the animal can maneuver itself against the plant to try to detach part of it. Much more likely, the resistance of any plant part to ingestion, whether specifically developed as a defense by the plant or not, is going to be stress-limited, which thus puts a premium on preventing crack initiation. In contrast, during mastication, a process involving repeated fragmentation, displacements are restricted by the gape that can be generated between the jaws, which is always small and determined actually by the largest-sized particles currently being chewed. The latter has gained some support from masticatory experiments in humans (Agrawal *et al.*, 1997; Lucas *et al.*, 2002). J. F. V. Vincent (pers. comm.) has shown that K_{IC} is an important component of sensory perception by humans of the texture of fruits during ingestion, which could relate to textural evaluations made by primates.

Anti-ingestion defenses are likely to predominate in plants whenever major herbivores are large enough to threaten the survival of a plant by a single bite, i.e., when one fracture could be life-threatening for the plant. Savanna grasses are a familiar contemporary example. These small plants are grazed predominantly by large herbivores. Their long thin parallel-veined leaves have reinforcing thick-walled sclerenchyma that is very superficially positioned (to resemble the sandwich construction referred to by Gibson *et al.*, 1988). Resistance to the tensile element of the bite of large ungulates is maximized by these sclerenchyma fibers being aligned parallel to the direction of the bite force. This stress-limited design appears to explain the parallel venation of the leaves of many plants, including most monocotyledonous angiosperms. Such plants often also contain large amounts of superficially located silica or calcium oxalate, which is consistent with the same defensive strategy, as also is the general absence of secondary chemicals such as tannins or alkaloids. Such grasses must have been subject to predation by relatively large herbivores for much of their evolutionary history. There could be many other examples, but recent extinctions can confuse the contemporary picture because plants evolve more slowly than animals. The survival of many stress-limited defenses in the neotropics has been associated with megafauna such as gomphotheres, which died out about 10 000 years ago (Janzen and Martin, 1982).

In contrast to the above, anti-mastication defenses will typify plants that are attacked by very small "chewing" herbivores. It is extremely difficult to stop all damage at small scale. In particular, leaves are very difficult to protect because photosynthesis is compromised by any structures positioned in the light path leading to chloroplasts. The major predators of most angiosperms are "chewing" invertebrates (Leigh, 1999). This could be why displacement-limited defenses seem to predominate in most angiosperms.

Postcanine tooth form

Against this background of plant defenses, we can look at the design of features on molars in relation to plant tissue toughness. It seems reasonable that, without considering food type, the most efficient contact between tooth and food is a point contact, such as can be produced by a cusp (Lucas, 1979). When plant toughness is low, as it is when $V_c < 0.2$ or > 0.95 (Fig. 9.2), cusps can propagate cracks easily. Tissues with very low V_c include fruit flesh (Lucas *et al.*, 1995) and the mesophyll of leaves (Choong, 1996). Note that a frugivorous primate with cusped molars can chew very young leaves, which have relatively undifferentiated cells. For most such tissues, fragmentation of the tissue seems less important than opening the maximum amount of cells per chew. Blunt

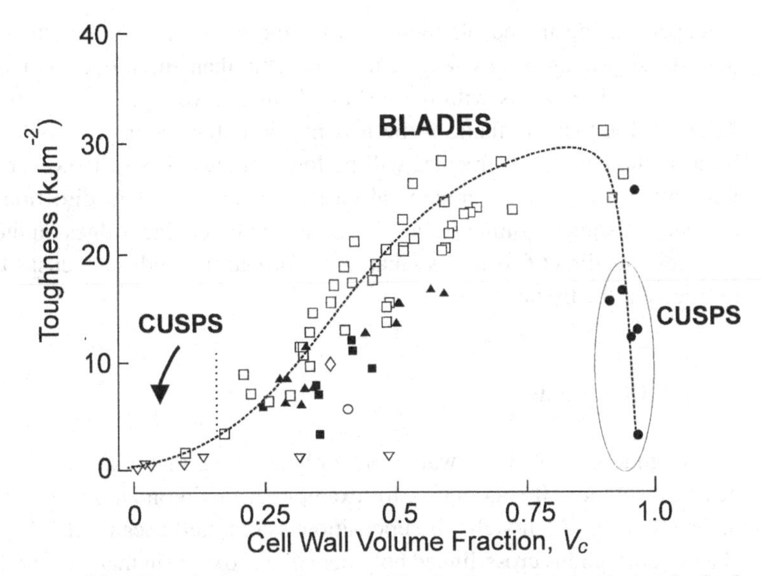

Figure 9.2. Cusped primate molars are predicted to be associated with plant foods of low toughness. These foods can be either low or high in fiber. In contrast, sharp blades on molars are required when toughness is high. Symbols as in Figure 9.1.

bulbous cusps have a distinct advantage over sharp cusps here in that their "zone of action," the region around the cusp tip that is capable of fracturing cells, is enlarged (Lucas and Luke, 1984).

When toughness is higher (i.e., V_c lying between 0.2 and 0.8), crack growth starts to be impeded. Fruit pods and some seed coats have overall cell-wall volume fractions in this range. Bladed features, i.e., extensions of a point contact on molars, are the optimal solution to driving these cracks, the proximity of the blade edge to the crack tip acting to maintain crack direction and so keep the crack moving. Additionally, most leaves, except the youngest, are toughened by sclerenchyma fibers that are usually arranged around veins in the form of a sheath. Whereas leaf mesophyll has a V_c of 0.03–0.05, these fibers have a V_c variously estimated between 0.8 and 0.94 (Gibson *et al.*, 1988; Choong, 1996), which is approximately in the optimal range for toughness. Even a few such fibers resist leaf fragmentation very effectively and, being elongated cells, they frustrate the propagation of free-running cracks in all but one direction – along the side of cells. Since venation in most dicotyledonous angiosperm leaves is in the form of a complex net, primate herbivores need to direct fractures across veins and so need bladed molars.

At very high V_c, toughness suddenly decreases (Fig. 9.2) because the plastic buckling mechanism cannot operate without a cellular lumen. Some seed shells lie in this range (Lucas and Peters, 2000). These seeds show stress-limited defenses, having traded off toughness for the high yield stress produced by high density. Blunt cusps will be more durable than sharp ones and yet still produce small contacts with these shells (which have high Young's moduli). Additionally, there is liable to be less plastic deformation with blunt cusps because the stresses at fracture will be lower (Lawn, 1993). However, radial fractures arising from symmetrical cusps will have random directions. The presence of some asymmetry, such as produced by distinct ridges on the sides of cusps, can direct fractures so that those formed from adjacent cusps link up to fragment the tissue.

Discussion

The toughness of plant cell wall is probably ten times that of any of its components in isolation. It is doubtful, for example, if lignin on its own would have much toughness (although this supposition is common) because it belongs to a class of amorphous cross-linked polymers like epoxy resin that are very brittle, i.e., the opposite of tough (Atkins and Mai, 1985). An extremely low toughness for pure crystalline cellulose is already known (Hiller *et al.*, 1996). Plant cell wall may provide some obstruction to herbivory, but nothing like that which follows from the cellular toughening mechanism found in woody cells. Thus, woody tissues peak in toughness ten times above that of the cell wall and probably one hundred times over that of their components. It is these tissues that really obstruct crack growth and the need for bladed molars in herbivorous primates appears very strongly linked to consumption of foods containing these cells.

When the volume fraction of cell wall in a tissue is both very low and very high, intracellular toughness is quite low and cracks can propagate quite freely. If the volume fraction of the cell wall is very high, then propagation between cells (intercellular cracks) can result in the toughness of a seed shell being just as low as that of fruit flesh (Williamson and Lucas, 1995). Thus, even though a woody shell might protect a seed, it may not be tough like wood itself. Many seed-eating primates appear to have cusped molars, superficially resembling those of frugivorous primates. Similarity in the fracture resistance of their typical foods may explain this. A frugivorous primate could eat very young leaves without the need for blades. For all but the youngest leaves, however, bladed molars would be needed because woody fibers would be differentiating and starting to lay down their secondary cell walls. Many leaf-eating primates also eat unripe fruits containing immature seeds. These seed coats may possess woody toughness, even if in some species this is lost in the covering of the mature seed as cell walls thicken above 95% of cell volume. Bladed molars could serve equally well to comminute both leaves and immature seeds.

Conventional wisdom has plants fighting off the physical attack of animals with toxic chemicals. Yet the cell wall of plants is clearly an attempt to protect all plant tissues mechanically, being seriously compromised only in tissues where the wall would jeopardize cell function, as in light reception for photosynthesis, or where no defense is intended at all. A key example of the latter is the flesh of fruits because this tissue is provided as a reward for a (vertebrate) consumer that will later disperse seed(s) embedded in the flesh either by spitting out or defecating them. However, even then, the need to ward off fruit thieves often means that the flesh is mechanically protected by a covering in the form of a fruit peel (Janson, 1983), which the target disperser alone might ideally be capable of removing. We can generalize and suggest that fruits generally have a stress-limited, anti-ingestion, design. As has been recognized by countless authors, incisal design is critical for most primate frugivores, the flesh generally having little or no resistance to mastication. The limit to the defense lies in the capacity of the target disperser to remove the peel. In contrast, leaves have anti-mastication defenses that because of their thinness is controlled essentially by their toughness. Accordingly, it is molar design that is critical in adaptations to folivory on all but the youngest leaves.

There is a large body of evidence showing that humans can make very sensitive texture judgements in the mouth. It seems unlikely that texture could be responsible for the degree of aversion that some secondary chemicals can provoke: errors of judgement over texture slow feeding, but do not kill. However, cell walls are so important to digestion by herbivores that it seems reasonable to suppose that other primates share this sensory ability. It surely must have evolved so that they and our ancestors could make many feeding decisions on textural grounds.

Acknowledgments

This book is dedicated to the life and work of Charles Oxnard and no chapter would be complete without an indication of Charles's influence. In my case, this is very personal: his courage, immense mental stamina, and clarity of thought have been inspirational. In the universities of the seaports of Southeast Asia, there is not always that much respect for research that does not have an anticipated application lurking in its immediate future. Back in 1988, I really needed someone to lift me out of the intellectual stagnation that this type of management demand can produce. Charles was that person. This work was supported by the Research Grants Council of Hong Kong.

References

Agrawal, K. R., Lucas, P. W., Prinz, J. F., and Bruce, I. C. (1997). Mechanical properties of foods responsible for resisting food breakdown in the human mouth. *Arch. Oral Biol.*, **42**, 1–9.

Agrawal, K. R., Lucas, P. W., Bruce, I. C., and Prinz, J. F. (1998). Food properties that influence neuromuscular activity during human mastication. *J. Dent. Res.*, **77**, 1931–1938.

Agrawal, K. R., Lucas, P. W., and Bruce, I. C. (2000). The effect of food fragmentation index on mandibular closing angle in human mastication. *Arch. Oral Biol.*, **45**, 577–584.

Ashby, M. F. (1999). *Materials Selection in Mechanical Design*. 2nd Edn. Oxford: Butterworth-Heinemann.

Ashby, M. F. and Jones, D. R. H. (1980). *Engineering Materials 1*. Oxford: Pergamon Press.

(1986). *Engineering Materials 2*. Oxford: Pergamon Press.

Ashby, M. F., Easterling, K. E., Harryson, R., and Maiti, S. K. (1985). The fracture and toughness of woods. *Proc. Roy. Soc. Lond. A*, **398**, 261–280.

Atkins, A. G. and Mai, Y.-W. (1985). *Elastic and Plastic Fracture*. Chichester: Ellis Horwood.

Choong, M. F. (1996). What makes a leaf tough and how this affects the pattern of *Castanopsis fissa* leaf consumption by caterpillars. *Funct. Ecol.*, **10**, 668–674.

Choong, M. F., Lucas, P. W., Ong, J. Y. S., Pereira, B. P., Tan, H. T. W., and Turner, I. M. (1992). Leaf fracture toughness and sclerophylly: their correlations and ecological implications. *New Phytol.*, **121**, 597–610.

Dominy, N. J. and Lucas, P. W. (2001). The ecological value of trichromatic vision to primates. *Nature*, **410**, 383–386.

Ganzhorn, J. U. (1992). Leaf chemistry and the biomass of folivorous primates in tropical forests. *Oecol.*, **91**, 540–547.

Gibson, L. J., Ashby, M. F., and Easterling, K. E. (1988). Structure and mechanics of the iris leaf. *J. Mater. Sci.*, **23**, 3041–3048.

Gordon, J. E. and Jeronimidis, G. (1980). Composites with high work of fracture. *Philos. Trans. Roy. Soc. Lond. A*, **294**, 545–550.

Hiller S., Bruce D. M., and Jeronimidis G. (1996). A micropenetration technique for mechanical testing of plant cell walls. *J. Text. Stud.*, **27**, 559–587.

Janson, C. H. (1983). Adaptation of fruit morphology to dispersal agents in a neotropical rain forest. *Sci.*, **299**, 187–189.

Janson, C. H. and Chapman, C. A. (1999). Resources and primate community structure. In: *Primate Communities*, ed. J. G. Fleagle, C. H. Janson, and K. E. Reed. Cambridge: Cambridge University Press. pp. 237–267.

Janzen, D. H. and Martin, P. S. (1982). Neotropical anachronisms: the fruits the gomphotheres ate. *Sci.*, **215**, 19–27.

Lawn, B. (1993). *Fracture of Brittle Solids*. 2nd Edn. Cambridge: Cambridge University Press.

Leigh, E. G., Jr. (1999). *Tropical Forest Ecology: a View from Barro Colorado Island*. New York, NY: Oxford University Press.

Lucas, P. W. (1979). The dental–dietary adaptations of mammals. *Neues Jahrb. Geol. Palaeontol.*, **8**, 486–512.

(in press). *Dental Function Morphology: How Teeth Work.* Cambridge: Cambridge University Press.

Lucas, P. W. and Luke, D. A. (1984). Chewing it over – basic principles of food breakdown. In: *Food Acquisition and Processing in Primates*, ed. D. J. Chivers, B. A. Wood, and A. Bilsborough. New York, NY: Plenum Press. pp. 283–302.

Lucas, P. W. and Peters, C. R. (2000). Function of postcanine tooth shape in mammals. In: *Development, Function and Evolution of Teeth*, ed. M. F. Teaford, M. M. Smith, and M. Ferguson. Cambridge: Cambridge University Press. pp. 482–289.

Lucas, P. W., Darvell, B. W., Lee, P. K. D., Yuen, T. D. B., and Choong, M. F. (1995). The toughness of plant cell walls. *Philos. Trans. Roy. Soc. Lond. B*, **348**, 363–372.

Lucas, P. W., Tan, H. T. W., and Cheng, P. Y. (1997a). The toughness of secondary cell wall and woody tissue. *Philos. Trans. Roy. Soc. Lond. B*, **352**, 341–352.

Lucas, P. W., Cheng, P. Y., Choong, M. F., *et al.* (1997b). The toughness of plant tissues. In: *Proceedings of Plant Biomechanics 9*, ed. G. Jeronimidis and J.F.V. Vincent. Reading: Centre for Biomimetics, University of Reading. pp. 109–114.

Lucas, P. W., Turner, I. M., Dominy, N. J., and Yamashita, N. (2000). Mechanical defenses to herbivory. *Ann. Bot.*, **86**, 913–920.

Lucas, P. W., Prinz, J. F. Agrawal, K. R., and Bruce, I. M. (2002). Food physics and oral physiology. *Food Qual. Pref.*, **13**, 203–213.

Page, D. H., El-Hosseiny, F., and Winkler, K. (1971). Behaviour of single wood fibres under axial tensile strain. *Nature*, **229**, 252–253.

Pereira, B. P., Lucas, P. W., and Toh, S. H. (1997). Ranking the fracture toughness of mammalian soft tissues using the scissors cutting test. *J. Biomech.*, **30**, 911–94.

Purslow, P. P. (1983). Measurement of the fracture toughness of extensible connective tissue. *J. Mater. Sci.*, **18**, 3591–3598.

Rogers, M. E., Maisels, F., Williamson, E. A., Fernandez, M., and Tutin, C. E. G. (1990). Gorilla diet in the Lopé Reserve, Gabon: a nutritional analysis. *Oecol.*, **84**, 326–339.

Vincent, J. F. V. (1990). Fracture in plants. *Adv. Bot. Res.*, **17**, 235–287.

Vincent, J. F. V. (1992). *Biomaterials: a practical approach.* Oxford: IRL Press.

Waterman, P. G., Ross, J. A .M., Bennett, E. L., and Davies, A. G. (1988). A comparison of the floristics and leaf chemistry of the tree flora in two Malaysian rain forests and the influence of leaf chemistry in populations of colobine monkeys in the Old World. *Biol. J. Linn. Soc.*, **34**, 1–32.

Williamson, L., and Lucas, P. W. (1995). The effect of moisture content on the mechanical properties of a seed shell. *J. Mater. Sci.*, **30**, 162–166.

10 *Convergent evolution in brain "shape" and locomotion in primates*

WILLEM DE WINTER
Leiden Experts on Advanced Pharmacokinetics and Pharmacodynamics

Introduction

For many years, comparative studies have indicated that the sizes of mammalian brain components are mostly related to the size of the brain as a whole, suggesting that all mammals share smaller or larger versions of essentially the "same" brain (Preuss 1995). This prompted the view that mammalian brain evolution can be represented along a one-dimensional scale of encephalization or "general intelligence" (e.g., Jerison, 1973; Eisenberg, 1981; Harvey and Krebs 1990). Recently, such allometric regularities in brain morphology have been attributed to a highly conservative, uniformly mammalian "Bauplan," which is thought to constrain the possible evolutionary responses to selection for specific neural reorganizations to those stemming from an orderly expansion of the brain as a whole (Finlay and Darlington, 1995; Finlay *et al.*, 1998). From 1993 to 2001, I had both the privilege and the pleasure to work closely with Charles Oxnard, adapting his hypothesis-free approach to multivariate morphometrics (e.g., Oxnard, 1984), combined with modern data visualization techniques, to the comparative study of the brain in mammals. This allowed us to analyze a large number of brain variables across a broad range of mammals simultaneously (de Winter 1997).

We were able to show that the relative proportions of different systems of functionally integrated brain structures vary independently between different mammalian orders, demonstrating separate evolutionary radiations in mammalian brain organization (de Winter and Oxnard, 1996, 1997, 2001; de Winter, 1997). Moreover, in a recent publication we identified, within each major order, clusters of unrelated species that occupy similar behavioral niches and have convergently evolved similar brain proportions (de Winter and Oxnard, 2001). Taken together, these findings provide compelling evidence of selective brain

Shaping Primate Evolution, ed. F. Anapol, R. Z. German, and N. G. Jablonski. Published by Cambridge University Press.

reorganizations in response to changes in lifestyle and niche, and hence contradict the once dominant view that mammalian brain evolution is constrained to a unitary increase in overall brain size alone. In contrast, they point to a complex interplay between developmental constraints and adaptive plasticity, and suggest new insights in the evolutionary processes that have shaped brain and intelligence in mammals (de Winter and Oxnard, 2001).

However, in our previous publications we focused on the implications of my findings for mammalian brain evolution in general, and hence were not able to provide detailed information on the various mammalian orders separately. In the following, therefore, I present the detailed findings for the primates.

The first part of this paper provides a detailed description of the distribution of species within the primate dispersion identified previously in de Winter and Oxnard (2001). This distribution coincides with a spectrum of locomotion that is also apparent in Charles Oxnard's biomathematical studies of primate limbs (Oxnard, 1984), ranging from hind-limb-dominated leaping and squirrel-like scurrying, through quadrupedal running and jumping, to forelimb-dominated suspensory climbing and armswinging.

The second part provides a detailed description of the associations of brain proportion variables that account for the separations of species described in the first part, together with discussion of their behavioral correlates. The spectrum of lifestyles implied by the locomotor descriptions in the first part corresponds to the expansion of brain structures in the second that are involved in the highest levels of the motor hierarchy relative to those involved in lower levels. This association may, therefore, reflect an evolutionary increase in voluntary over relatively more stereotyped control of behavior.

Materials and methods

This study applies a descriptive multivariate morphometric approach (Oxnard, 1984; de Winter, 1997; de Winter and Oxnard, 2001) to the comparative study of quantitative data on the mammalian brain. Rather than using classical statistics to test a priori hypotheses about the data, this approach aims to explore the multidimensional structure of the data in all its richness and intricate detail. It employs unsupervised, descriptive multivariate statistics and data visualization. These methods are robust to (nested) interdependencies between observations, and are widely used in many different fields to identify clusters of interrelated data points (e.g., Cleveland, 1993; Kzranowski and Marriott, 1994; Hastie *et al.*, 2001). Here labeling of data points is used to distinguish spatial relationships between data points associated with phylogeny from those associated with, e.g., ecology and lifestyle.

The data on which this study is based consist of the volumes of 11 non-overlapping brain parts, measured on 921 individuals belonging to 363 different species across 5 different mammalian orders. These include 264 different species of bats, 50 species of insectivores, 44 species of primates, 3 species of tree shrews, and 2 species of elephant-shrews. The brain parts measured consist of the four fundamental divisions of the brainstem: medulla (+ reticular formation), cerebellum (+ brachium, nuclei pontis), midbrain (– reticular nucleus), and diencephalon; and the seven main subdivisions of the forebrain: striatum, olfactory bulb, paleocortex (+ amygdala), septum, hippocampus, schizocortex (entorhinal, perirhinal and presubicular cortices), and neocortex (isocortical gray + underlying white matter).

Summaries of this dataset for primates and insectivores were published by Stephan *et al.* (1981, 1991). These summaries have in common that for species in which more than one brain was measured only the average volumes are given. Furthermore, they present the data after a conversion that takes into account the difference between individual brain weight and the species average brain weight (Stephan *et al.*, 1981). Here, in contrast, the original raw, non-standardized individual measurements are used, kindly provided to me in digital format by Dr. Heinz Stephan of the Max Planck Institut für Hirnforschung in Frankfurt, Germany. The procedures used in obtaining these data are described in detail by Stephan *et al.* (1981, 1991).

In order to obtain measures that are more sensitive to variations in the functional, systemic interdependence of these brain parts, the data were rescaled to reflect the functional proportions between their major inputs and outputs (de Winter, 1997; see also Deacon, 1990). Two major sources of input–output projections were identified in this context: peripheral projections associated with varying body size, and hence, by proxy, with the size of the medulla (Passingham, 1975; Dunbar, 1992); and internal projections associated with variations in the size of the neocortex (de Winter, 1997; see also Barton and Harvey, 2000). The size of each brain part was therefore expressed in proportion to both medulla and neocortex within the same specimen. This resulted in 19 different brain structure proportions across different developmental growth fields (Deacon, 1990). Although derived mathematically, these proportions represent empirically measurable properties that are intrinsic to each specimen (de Winter, 1997).

Following univariate and bivariate examination of both the raw and the transformed datasets, a principal components analysis was applied to the correlation matrix of the functionally rearranged brain data. The first three components reduced the dimensionality of the data space from 19 to 3 while preserving 85% of its total variance, and hence were retained for further analysis. Phylogenetic and lifestyle associations among species, within the data subspace spanned by these first three principal components, were investigated interactively by dynamic three-dimensional computer representations with data points rendered as color-coded spheres with colored highlights (de Winter, 1997; de Winter and Oxnard, 2001). These components were rotated to align with the directions of major dispersion for the three largest mammalian orders. This allowed identification through biplots (Fig. 10.1) of the associations between the variables that contributed most to those dispersions. Biplots combine graphical displays of the relationships between species and variables into single plots, thus showing the interrelationships between the two.

Clusters of species

Figure 10.1 shows the distribution of species along principal axis of the primate dispersion in the proportional brain data space, and combines a scatterplot of this axis against the insectivore axis (de Winter and Oxnard, 2001) with boxplots for the major primate groups. This figure shows that the distribution of the species along the primate dispersion is broken up into three distinct clusters and one spectacular outlier:

- Hominoids and atelines with the human specimen as outlier
- Old World monkeys plus New World howlers and sakis
- Marmosets and tamarins, owl monkeys and titis, tarsiers, and all strepsirhines

Each of these clusters comprises representatives of different lineages of both Old and New World primates that have evolved independently for at least the past 30 million years. The following section gives for each cluster of species a detailed description of its composition, immediately followed by a brief summary of findings from the literature on the eco-ethological commonalities of these species. This section is followed by a more general discussion of these clusters in relation to the evolution of primate locomotion.

Species composition of the main clusters

Hominoids and atelines

In Figure 10.1 the outlying position of the single human specimen stands out at the far end of the primate dispersion. The gap between *Homo* and its closest neighbors, the two great apes *Pan* and *Gorilla*, was found to span almost half of the total range of the primates along this axis (46%). Hence, although the human specimen extrapolates a trend that was already present in the nonhuman primates, it is clearly an outlier to an extent similar to the entire spread of the other

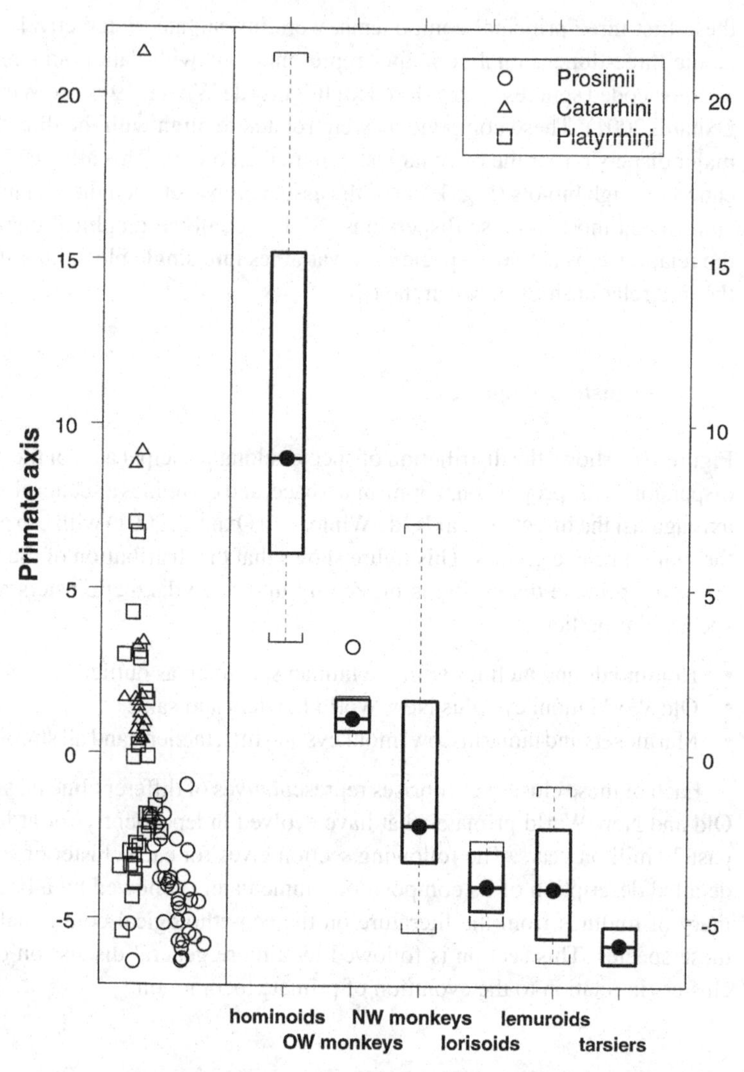

Figure 10.1. Scatterplot of the primate radiation in the brain data space, combined with boxplots for the main primate groups on the same axis. In the scatterplot the *x*-axis corresponds to the insectivore axis in Figure 10.5, and different symbols identify various subgroups of these main primate groupings. The boxplots are ordered according to decreasing median scores (black dots) along the x-axis. Separate boxplots for the Old and New World monkeys highlight the much wider range of variation in the latter. In fact, the New World monkeys (ceboids) can be seen to span nearly the entire range for the nonhuman primates: if they were removed from the plot, marked discontinuities between the remaining groups would be apparent.

primates. Further down this axis, the distributions of the remaining hominoids (represented by great apes *Pan* and *Gorilla*, and the lesser ape *Hylobates*) and the New World atelines (represented by the spider and woolly monkeys *Ateles* and *Lagothrix*) overlap almost completely, even though the hominoids are phylogenetically more closely related to the Old World monkeys in the second cluster of Figure 10.1. By comparison, in the analyses of the raw measurement data the atelines are separated from the apes by the large terrestrial baboons and mangabeys. Hence, even though the atelines have smaller brains than these large terrestrial Old World monkeys, they nevertheless score higher on the primate axis of the proportional brain data, and therefore have a relatively larger neocortex, striatum, and cerebellum in proportion to the size of their medullae.

A common distinctive feature of hominoids and atelines alike, not shared by any of the Old World cercopithecines and only marginally by some colobines, is their ability to routinely suspend themselves below branches by their arms in locomotion and feeding (e.g., Tuttle, 1975; Fleagle, 1976; Mittermeier, 1978; Oxnard, 1984; Cant, 1986; Martin, 1990). This enhances their ability to reach supports in many directions and provides them with a wide variety of locomotor patterns and feeding postures, thus allowing them to exploit arboreal niches not otherwise accessible to primates of their size. The observed convergence of apes and atelines along the primate axis is therefore consistent with the notion that, having independently evolved the same kind of forelimb-dominated climbing–feeding complex that is considered the primary hominoid adaptation, the atelines are the New World locomotor analog of the apes (Fleagle, 1976; Mittermeier, 1978; Cant, 1986).

Old World monkeys, howlers and sakis

Figure 10.1 also demonstrates a clear discontinuity between the overlapping distributions of the suspensory apes and atelines and a neighboring cluster in which the distributions of the Old World monkeys and the New World howlers and sakis coincide. The core of this cluster consists of all of the Old World monkeys except the macaque, including the mangabey (*Cercocebus albigena*), the patas monkey (*Erythrocebus patas*), the baboon (*Papio anubis*), the talapoin (*Miopithecus talapoin*), the guenons (*Cercopithecus ascanius* and *C. mitis*), the red colobus (*Colobus badius*), the douc langur (*Pygathrix nemaeus*), and the proboscis monkey (*Nasalis larvatus*). The Old World monkeys overlap with representatives of two of the remaining three New World cebid subfamilies, the howlers (*Alouatta seniculus*) and sakis (*Pithecia monachus*) at either end, while the cluster is bounded at its lower end along the primate axis by the common squirrel monkey (*Saimiri sciureus*).

Because it includes not only terrestrial and arboreal, but also frugivorous, folivorous (howlers, colobines), and partly insectivorous (*Saimiri*, *Miopithecus*) and even faunivorous (baboons) forms, this cluster appears to lump together both New and Old World monkeys from all walks of life, except perhaps for a broad similarity in their generally quadrupedal locomotor habits. Fleagle (1999) describes all Old World monkeys as primarily quadrupedal walkers and runners, with some good leapers particularly among the colobines. Although howlers do engage frequently in suspensory postures during feeding, they almost always hang from their legs and/or tail and engage only rarely in the arm-hanging activities typical of both atelines and hominoids (Cant, 1986; Bergeson, 1998). Tail-assisted hind-limb suspension is also an important positional behavior in sakis (Meldrum, 1998), and both these groups of true New World monkeys are now considered to be primarily arboreal quadrupeds (Ankel-Simons, 1983; Fleagle, 1999; Martin, 1990). Hence, the species in this cluster can be characterized as essentially quadrupedal walkers, climbers, or leapers.

Aotines, callitrichids, and tarsiers

Continuing further down the primate axis, Figure 10.1 shows that the above cluster of Old and New World monkeys is separated by another distinct gap from the remaining haplorhines (all aotines, marmosets, and tamarins, plus the tarsiers) and all of the strepsirhines (all lemuroids, lorisoids, and cheirogaleids). Beyond this gap, the aotines (*Aotus trivirgatus* and *Callicebus moloch*), the tamarins (*Saguinus tamarin* and *S. oedipus*), Goeldi's monkey (*Callimico goeldii*) and the marmosets (*Callithrix jacchus* and *Cebuella pygmea*) further project in the haplorhine plane. The strepsirhines then gradually branch off along the insectivore axis to link up with the megabats (see Figure 10.5).

The cebid owl monkeys are the only nocturnal anthropoids and, like the titis, have a predominantly frugivorous diet that to varying degrees is supplemented with arthropods (Fleagle, 1999). The marmosets, tamarins, and Goeldi's monkey are the only clawed anthropoids. As a group, they are more or less omnivorous and may include varying proportions of insects, fruit, gums, saps, and nectar in their diets (e.g., Fleagle, 1999; Martin, 1990). Based on their limb proportions, Erikson (1963) grouped all anthropoid species in this cluster into a single locomotor type as springers, reflecting their hind-limb-dominated springing and squirrel-like scurrying locomotor habits. As typical vertical clingers-and-leapers, the tarsiers fit this category rather well (Hershkovitz, 1973; Fleagle, 1999; Martin, 1990). Grading towards the tarsiers, the squirrel-like pygmy marmosets (*Cebuella*) are the lowest-scoring haplorhines on the primate axis. Their locomotor skeleton is described by Hershkovitz (1973, p. 45) as "more nearly generalized than that of any other living primate."

Lemuroids, lorisoids, and cheirogaleids

At the junction between the haplorhine and strepsirhine branches in Figure 10.1, the nocturnal owl monkeys merge with the diurnal brown lemurs (*Lemur fulvus*), indris (*Indri indri*), and sifakas (*Propithecus verreauxi*). The single specimen of the nocturnal aye-aye (*Daubentonia madagascariensis*) is located to the right of this cluster, well outside the plane of the haplorhine distribution.

Figure 10.3 provides a more detailed overview of the strepsirhine distribution along the major primate axis. It clearly shows the strong overlap of the lemurs and indriids along this axis. Due to the relatively outlying position of the diurnal ruffed lemur (*Varecia variegata*), the true lemurs also show some overlap with the nocturnal lorises, represented by the slender loris (*Loris tardigradus*), slow loris (*Nycticebus coucang*), and potto (*Perodicticus potto*). Further down the major primate axis these are followed by the nocturnal, vertical clinging-and-leaping woolly lemur (*Avahi laniger*) and sportive lemur (*Lepilemur ruficaudatus*). The sportive lemurs connect with the nocturnal bushbabies (*Galago senegalensis*, *Otolemur crassicaudatus*, and *Galagoides demidoff*), which in their turn overlap with the cheirogaleids. The latter are represented by the greater and fat-tailed dwarf lemurs (*Cheirogaleus major* and *C. medius*) and the mouse lemurs (*Microcebus murinus*). They connect the strepsirhine branch with the flying foxes of the genus *Pteropus* in the megabat cluster (see de Winter, 1997, Figure 5; de Winter and Oxnard, 2001).

The strepsirhines, and in particular the Malagasy lemurs, have radiated into an extremely diverse array of species with dietary, activity-timing, and locomotor adaptations for exploiting a wide range of ecological conditions. Their locomotor habits are usually differentiated into three broad types, although many subcategories are distinguished (e.g., Hershkovitz, 1973; Martin, 1990; Oxnard *et al.*, 1990; Fleagle, 1999). The first type is a generalized form of hind-limb-dominated quadrupedal and relatively horizontal branch-walking, squirrel-like scurrying, and springing. This locomotor type is often considered representative of the ancestral primate condition (e.g., Charles-Dominique, 1977; Martin, 1990), and finds its typical expression in the cheirogaleids and some bushbabies (here *Galagoides demidoff*). The locomotor habits of the much larger true lemurs (Lemurinae) and aye-aye have been associated with this type (Oxnard *et al.*, 1990). The second type overlaps with the first and includes highly specialized, hind-limb-dominated forms of vertical springing or vertical clinging-and-leaping between vertical supports. Two subgroups are generally identified: one includes the sportive lemurs (*Lepilemur*), indris, and woolly lemurs (*Avahi*), while the other includes some bushbabies and the tarsiers (here *Galago senegalensis*). The third type is a quadrupedal (or "quadrumanous") form of horizontal, sloth-like slow climbing that is exclusive to the lorises (Lorisinae).

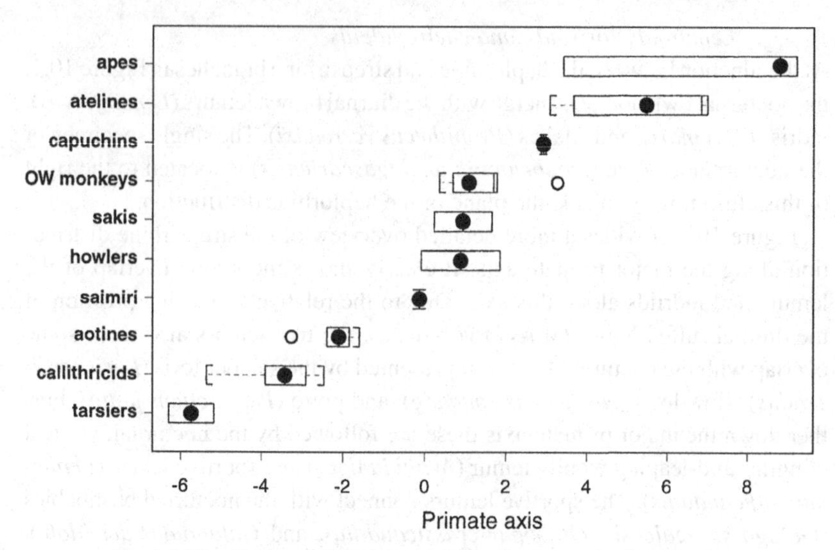

Figure 10.2. Differentiations in brain proportions among haplorhine primates. Boxplots showing the distribution of the haplorhines along the primate major axis. Three discontinuous groupings are apparent. The ranges for the apes (gibbon, gorilla, and chimpanzee) and the New World atelines completely overlap but are clearly separate from the next grouping. In this second cluster the Old World cercopithecids and colobids overlap completely with the New World sakis (Pithecinae), howlers (Alouattinae), and squirrel monkey (*Saimiri*). The capuchins and the outlying cercopithecid macaque are intermediary between this cluster and the ape/ateline grouping. The third cluster contains a range of partly overlapping groups consisting of the nocturnal owl monkeys (*Aotus*), followed by the clawed tamarins and marmosets, and finally the tarsiers. These three clusters closely match the three locomotor groupings proposed by Erikson (1963).

Three stages in the evolution of primate locomotor patterns

The three clusters in the anthropoid distribution along the primate axis in Figure 10.1 very closely match three locomotor categories proposed in a seminal paper on primate locomotion by Erikson (1963). Erikson noted that the New World primates are much more diverse in locomotor types than the Old World monkeys, on the one hand preserving the primitive proportions of the generalized arboreal quadrupeds found among the strepsirhines, and on the other hand presenting varying degrees of adaptations to forelimb suspensory climbing found elsewhere only among the anthropoid apes (Erikson, 1963). This observation is clearly mirrored in Figures 10.2 and 10.3, where the New World monkeys are shown to span nearly the entire non-human range along the primate axis.

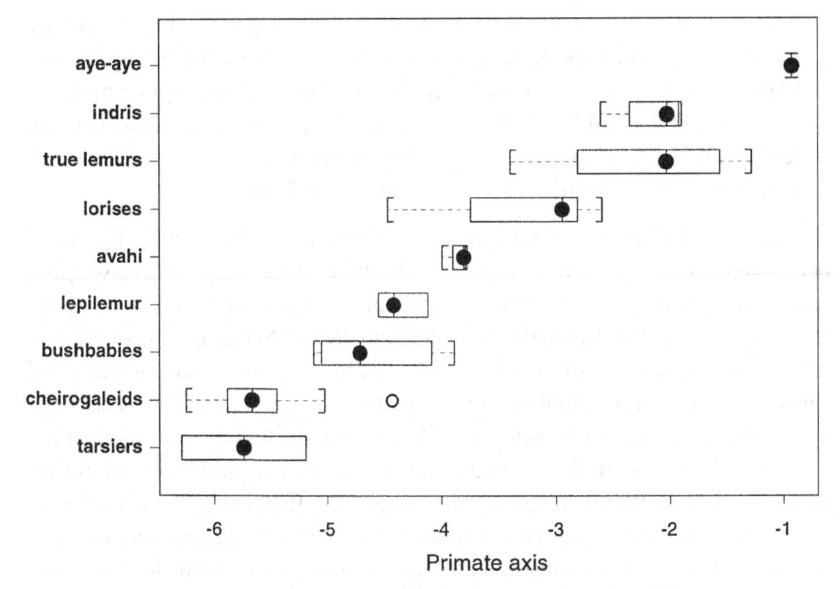

Figure 10.3. Differentiations in brain proportions among strepsirhine primates, with tarsiers for comparison. Boxplots showing the distribution of different strepsirhine groups along the primate major axis. It shows a gradient which ranges from the larger diurnal forms such as the elusive aye-aye (*Daubentonia*), the spectacular vertical clinging-and-leaping indriids, and the quadrupedal true lemurs, via the deliberate slow-climbing lorisids, through the smaller vertical clinging-and-leaping types (avahis, sportive lemurs, and some galagos), to the small, horizontal branch-scurrying cheirogaleids.

Based on their limb proportions, Erikson divided the New World monkeys into three general locomotor types (see also Hershkovitz 1973): springers, climbers, and brachiators.

Springers are small, generalized, hind-limb-dominated quadrupedal branch runners and springers with long slender trunks and relatively short limbs, but with hind limbs markedly longer than forelimbs. New World representatives are aotines, marmosets, and tamarins, paralleled in the Old World by most or all strepsirhines.

Climbers are medium-sized quadrupedal walkers, climbers, and leapers with moderately long limbs, with forelimbs beginning to approach the length of hind limbs. Arms become more important as a hoisting and climbing apparatus (Hershkovitz 1973). New World representatives are sakis, uakaris, capuchin, and squirrel monkeys, paralleled by all Old World monkeys.

Brachiators are large, forelimb-dominated suspensory climbers and arm-swingers, with forelimbs as long as to markedly longer than hind limbs, and accompanied by a prehensile tail. Arms become a principal and virtually independent locomotor organ (Hershkovitz, 1973). New World representatives are the atelines, in this study represented by woolly and spider monkey, paralleled in the Old World by the hominoid lesser and great apes. Erikson tentatively included the New World howlers in this group, but these are now regarded as primarily quadrupedal climbers: see below.

Some comments are in order here. First, the term "brachiator" has since been dropped as a generic term for all forelimb suspensory or armswinging species, and is now reserved for the specialized ricochetal brachiation of the lesser apes (Hylobatidae) only (e.g., Martin, 1990). Second, Erikson (1963) identified several taxa as transitional between his various groups: howlers and capuchins between his climber and brachiator, and squirrel monkeys between his climber and springer groups. In the brain data, the transitional status of the cebines is borne out in their marginal positions relative to the central cluster of Old and New World monkeys in Figure 10.2. Especially the capuchin is clearly intermediate between the "climber" and "hanger" clusters. Field studies of howlers have since demonstrated that they rarely engage in true forelimb suspensory behaviors and therefore are more appropriately classed as quadrupedal climbers (Cant, 1986). Third, Erikson (1963) noted significant parallels in Old and New World locomotor types. In terms of their limb proportions, he found all Old World monkeys to fall within his climber type, even those with secondarily evolved terrestrial habits. Similarly, he reported that in the Old World his New World springer group was paralleled by the prosimians, and his brachiator group by the anthropoid apes. Finally, although Erikson's groupings are defined in terms of locomotion, many other traits tend to be associated with these locomotor traits (e.g., body size, diet, habitat, etc.) in a particular lifestyle (Oxnard *et al.*, 1990). Therefore it is probably best to think of these groupings in terms of lifestyle groupings that happen to have locomotor differences as their most salient feature.

Figure 10.4 shows the distribution of Erikson's three locomotor groupings in the brain data, although with labels which try to avoid the discussions on Erikson's original terms. It is striking to see how clearly Erikson's three general locomotor types are separated along the primate axis in Figure 10.4.

Martin (1990) identified three stages in the evolution of primate locomotor patterns: an ancestral primate stage, an ancestral simian stage, and an ancestral hominoid stage. Although not discussed by Martin, these stages show a marked affinity with Erikson's scheme, and coincide with the major groupings of primates found in Figure 10.4.

Martin's ancestral primate is a small-bodied, hind-limb-dominated arboreal form with grasping feet by virtue of a widely divergent hallux, a diagonal-sequence gait, and special adaptation of the tarsus. It is thought to have

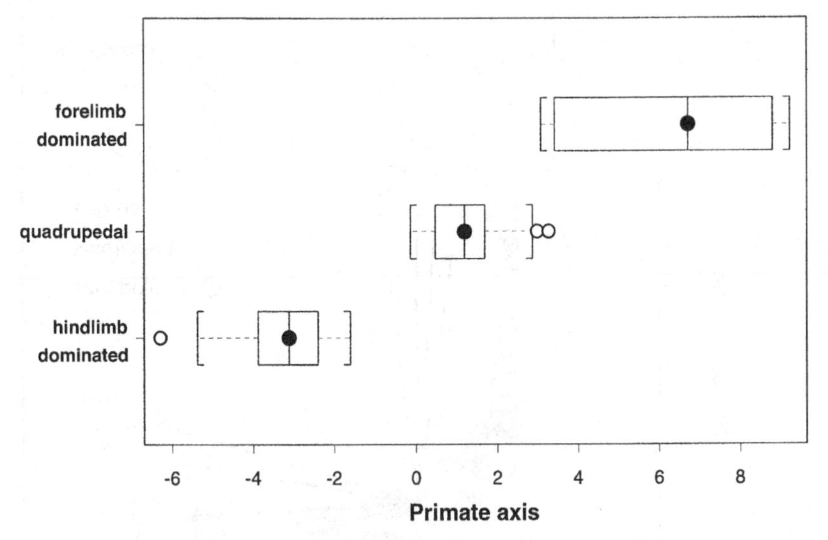

Figure 10.4. Locomotor differentiations in brain proportions among haplorhine primates. Boxplots showing the clear separations, along the primate axis, of Erikson's three locomotor groupings of haplorhine primates. These broadly defined locomotor groups each comprise representatives of separate Old and New World lineages that have independently evolved similar brain proportions. The apes and their New World ateline cousins (Erikson's "brachiators") are considered to share a forelimb-dominated climbing–feeding complex. As a group, they are differentiated from the essentially quadrupedal Old World monkeys and the non-ateline cebids, except the owl monkey ("climbers"). This latter group, in turn, is separated from the hind-limb-dominated owl monkeys, marmosets, and tamarins, and the tarsiers ("springers").

inhabited the nocturnal fine branch niche, where its locomotion involved both "quadrumanous" climbing and leaping (Martin 1990). Many of these ancestral primate features appear to have been preserved in the various smaller representatives of Erikson's springer group, and especially in the dwarf lemurs (Cheirogaleidae) and bushbabies (Galagidae) at the base of the primate axis in Figures 10.2 and 10.4 (Charles-Dominique and Martin, 1970; Cartmill, 1974; Charles-Dominique, 1977).

Because of a marked increase in size, perhaps associated with a shift to diurnal habits, representatives of the second, ancestral simian stage in Martin's scheme would have been better suited for travel along trunks and relatively broad branches. As a result, in spite of retention of certain fundamental features of hind-limb domination, such as a diagonal gait pattern and a posterior location of the center of gravity, Martin postulates that the ancestral simians would have evolved a pronounced quadrupedal tendency that has persisted in the modern monkeys of the Old and New Worlds (Martin, 1990). Hence, Martin's second stage in the evolution of primate locomotion shares important characteristics

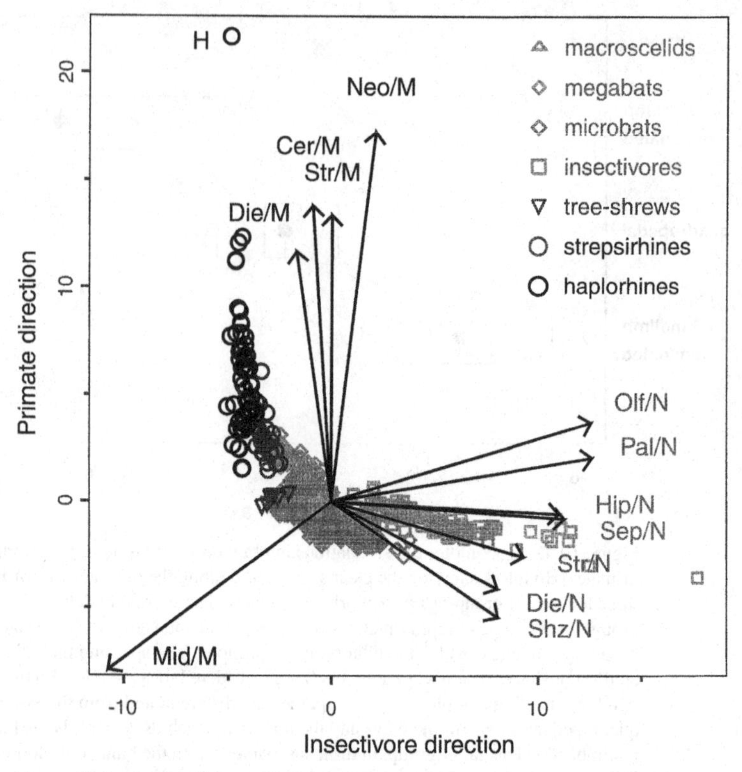

Figure 10.5. The primate radiation in its mammalian context. Biplot of the rotated principal components of the proportionally transformed brain data corresponding to the primate and insectivore directions in the overall mammalian brain data space (de Winter and Oxnard, 2001). Different symbols are used to identify the different mammalian groups, with the nonprimate groups dimmed. The primates are dispersed along a single axis which is at right angles to the insectivore dispersion, with the exception of a medial shift of the strepsirhines along the insectivore axis. Note the position of the tree-shrews at the base of the primate/strepsirhine dispersion. The arrows indicate the relative contributions of the various brain proportions to the species dispersions in the plot. Two clusters of variables are apparent, one associated with the primate dispersion and the other with the insectivore trend. The primate dispersion coincides primarily with the medullary proportions ("/M") of **cer**ebellum, **str**iatum, **die**ncephalon, and **neo**cortex; the insectivore direction with the neocortical proportions ("/N") of **sep**tum, **hip**pocampus, **pal**eocortex, and **str**iatum.

with Erikson's climber type, in particular a more equal involvement of fore and hind limbs in locomotion. In Figures 10.2, 10.3, and 10.5 the modern exponents of this stage can be seen to form a distinct, separate cluster halfway along the primate axis.

Martin proposed that a further increase in size in the ancestral hominoids, his third stage in primate locomotor evolution, would have coincided with a shift from above-branch quadrupedal locomotion to below-branch forelimb suspensory locomotion as the predominant mode of arboreal activity, accompanied by a further lengthening of the forelimbs relative to the hind limbs. Clearly, although not mentioned as such by Martin, this hominoid development would appear to have been paralleled by the atelines in the New World (Fleagle, 1976; Mittermeier, 1978; Cant, 1986). Hence, this stage closely resembles Erikson's brachiator type, and coincides with the convergence of the atelines with the apes at the positive end of the primate axis in Figures 10.1, 10.3, and 10.5.

Clusters of variables

In the following sections, the contributions of the various brain-structure proportions to the dispersions documented above are discussed. As already noted, Figure 10.5 displays a biplot of the first two varimax rotated principal components of the proportional brain data. In addition to showing species dispersions, it also shows the clusters of brain-proportion variables that are responsible for the separation of species in the brain data space discussed above (see also de Winter and Oxnard, 2001).

Hence, Figure 10.5 shows that the dispersion of species in the primate direction is produced by a concerted increase in the proportions, relative to the medulla, of cerebellum, striatum, neocortex, and diencephalon. All of these structures are heavily interconnected. That is, almost all regions of the neocortex project to the striatum and cerebellum, each of which, via various thalamic nuclei in the diencephalon, projects back to most areas of the frontal lobe of the cerebral cortex.

Parallel loops involving neocortex and striatum

In all mammals, the dorsal striatum (caudate nucleus and putamen) receives convergent inputs from the entire cerebral cortex with the exception of the primary visual and auditory areas (Houk, 1995; Marin *et al.*, 1998; Mink, 1999). Because these corticostriatal projections retain a roughly topographical organization, they give rise to complex overlapping representations of the functional cortical map at the striatal level (Marin *et al.*, 1998), providing an anatomical framework for the integration and transformation of information from several areas of the cerebral cortex (Mink, 1999). This information is then funneled back to the frontal areas of cortex via the pallidum and the ventral anterior and lateral nuclei of the thalamus (e.g., Houk, 1995; Mink, 1999). Hence, in mammals the striatum appears to integrate comprehensive information about the behavioral situation and its environmental context that may be used for predicting future events, planning the appropriate actions and predicting their outcome (Schultz *et al.*, 1995).

In primates and humans, the caudate nucleus of the dorsal striatum is directly linked in several closed loops via pallidum and ventral anterior thalamus with association areas in the prefrontal cortex (Middleton and Strick, 1994; Strick *et al.*, 1995). Through these loops the striatum is implicated in the motor component of working memory (Goldman-Rakic, 1995; Houk, 1995), and hence in the representationally guided, cognitive planning of the order and timing of future behavior (Fuster, 1989). It is thought, therefore, that the striatum is closely tied in with the highest levels of the motor control hierarchy, where it appears to be involved in both the cognitive and motor processing underlying the "purposiveness" of voluntary behavior, and the strategic planning of action (Brooks, 1986; Fuster, 1989; Middleton and Strick, 1994; Houk, 1995).

Loops involving neocortex and cerebellum

The cerebellum receives sensory, motor, perceptual, and cognitive information from all parts of the nervous system that it projects back to all parts of the motor system except the basal ganglia (e.g., Butler and Hodos, 1996; Voogd and Glickstein, 1998). Accordingly, the basic role of the cerebellum in mammals is generally thought to involve the coordination and fine control of movement and posture (e.g., Bastian *et al.*, 1999).

In anthropoid, and especially hominoid, primates the lateral cerebellum is strongly expanded (Voogd and Glickstein, 1998). Via the pons, it receives extensive projections from the frontal, parietal, and visual cortices. These corticocerebellar projections are funneled back, via the dentate nucleus and the ventral lateral nucleus of the thalamus, to the motor, premotor, and prefrontal association areas of the frontal cortex. Through these projections, the lateral cerebellum is thought to play a crucial role in the planning and control of voluntary movement of the extremities (e.g., Bastian *et al.*, 1999). Evidence for closed loops linking the lateral cerebellum with the prefrontal cortex suggests that it also participates in certain higher cognitive functions (Middleton and Strick, 1994; Bastian *et al.*, 1999).

In anthropoid primates, therefore, both the dorsal striatum and the lateral cerebellum are closely linked through successive recursive loops with the motor, premotor, and prefrontal areas of the frontal lobe of the cerebral cortex (Middleton and Strick, 1994). These loops are relayed by the ventral lateral and ventral posterior thalamic nuclei in the diencephalon. Hence, although anatomically these various brain parts are located at very different levels of the neuraxis, functionally they are closely integrated into a distributed system that is thought to mediate both the cognitive and motor aspects of the planning and control of purposive, voluntary behavior (e.g., Brooks, 1986; Shepherd, 1994; Middleton and Strick, 1994; Schieber, 1999).

The distributed motor hierarchy

It follows that the cluster of variables responsible for the main primate dispersion in Figure 10.5 reflects a proportional expansion of all the highest levels of the distributed motor hierarchy relative to the medulla (e.g., Brooks, 1986; Schieber, 1999). The size of the medulla has been put forward as a useful measure for the total amount of sensory–motor projections between brain and spinal cord (Passingham, 1975). However, in Stephan's data the medulla is measured as including the entire reticular formation of the brainstem (e.g., Stephan *et al.*, 1991: see Materials and methods), and thus also includes the primary centers and pathways for the automatic control of posture and locomotion (e.g., Brooks, 1986; Shepherd, 1994; Butler and Hodos, 1996; Baker, 1999).

This suggests that the primate direction may be thought of as representing an expansion of the highest, relative to the lowest, levels of the motor hierarchy. Hence, it may reflect a gradual evolutionary increase in voluntary or purposive, over more stereotyped or instinctive, control of behavior. Such an interpretation would match the previously discussed spectrum of complexity and flexibility of locomotion and behavior in general reflected by the ordination of species along this trend (de Winter and Oxnard, 2001).

Summary and conclusions

In conclusion, the distribution of the primates in the data space of the 19 internal brain-structure proportions was found to reflect a spectrum of locomotion that is also apparent in biomathematical studies of primate limbs (Oxnard, 1984), ranging from squirrel-like scurrying and hind-limb-dominated springing, through quadrupedal climbing, walking, and jumping, to forelimb-dominated suspensory climbing and armswinging. This spectrum was broken up into three groupings, each comprising representatives of different primate radiations that have independent evolutionary histories. These groupings were found to correspond closely to lifestyle groupings characterized by Erikson's (1963) three primate locomotor types, and appear to reflect Martin's (1990) three stages in the evolution of primate locomotor patterns. Taken together, these findings suggest that convergent evolution in relation to lifestyle in general and locomotor habits in particular may have played an important role in the evolution of the primate brain.

The ordination of species reflecting the above locomotion spectrum was found to be associated with the proportional expansion of neocortex, striatum, cerebellum, and diencephalon relative to the size of the medulla. These brain divisions, or parts thereof, are closely interconnected through parallel loops involved in the highest level of the distributed motor hierarchy that is thought to mediate both the cognitive and motor aspects of the planning and control of purposive, voluntary behavior (e.g., Brooks, 1986; Shepherd, 1994; Middleton and Strick, 1994; Schieber, 1999). Hence, these findings add a new dimension to the hypothesis that, because the primate postcranial skeletal framework is essentially primitive, one of the hallmarks of primate evolution has been the development of relatively sophisticated central nervous control of locomotion (Martin, 1990).

Furthermore, new data are accumulating which indicate that important structures and nuclei in cerebellum, striatum, diencephalon, and neocortex act in closed loops in the cognitive planning of action in advance of its execution. A concerted expansion of these structures across primates in association with lifestyle and locomotor differentiations, as proposed here, suggests that primate brain evolution may have been driven by a trend towards increasing behavioral flexibility which culminates in our own species.

Additional research is needed to verify and substantiate the hypotheses generated by this work. Currently, canonical variates analyses on the present data are being carried out to verify and further investigate the separations and groupings of species identified in this study. But present findings should also stimulate further comparative neuroanatomical and neuroimaging studies that can elucidate the detailed neural and neurobehavioral correlates of the species associations identified here, as well as comparative field work on locomotor habits and lifestyle in primates.

Acknowledgments

Thanks are due to Heinz Stephan for his provision of much unpublished original data, and to Lutz Slomianka, Alan Harvey, and Michael Arbib, who provided important neurobiological discussion. None of this work would have been possible, however, without the generous support and inspirational mentorship of Charles Oxnard, who on several occasions stepped in when times were at their most difficult. Much of the work was carried out while I was his doctoral student in Anatomy and Human Biology and then Research Fellow in the Centre for Human Biology at the University of Western Australia. It was supported by ARC Large Grants to Charles Oxnard.

References

Ankel-Simons, F. (1983). *A Survey of Living Primates and Their Anatomy*. New York, NY: MacMillan.

Baker, J. (1999). Supraspinal postural control: the medial "postural" system. In: *Fundamental Neuroscience*, ed. M. J. Zigmond, F. E. Bloom, S. C. Landis, J. L. Roberts, and L. R. Squire. San Diego, CA: Academic Press. pp. 913–930.

Barton, R. A. and Harvey, P. H. (2000). Mosaic evolution of brain structure in mammals. *Nature*, **405**, 1055–1058.

Bastian, A. J., Mugnaini, E., and Thach. W. T. (1999). Cerebellum. In: *Fundamental Neuroscience*, ed. M. J. Zigmond, F. E. Bloom, S. C. Landis, J. L. Roberts, and L. R. Squire. San Diego, CA: Academic Press. pp. 973–992.

Bergeson, D. (1998). Patterns of suspensory feeding in *Alouatta palliata*, *Ateles geoffroyi*, and *Cebus capucinus*. In: *Primate Locomotion: Recent Advances*, ed. E. Strasser, J. G. Fleagle, A. Rosenberger, and H. McHenry. New York, NY: Plenum Press.

Brooks, V. B. (1986). *The Neural Basis of Motor Control*. New York, NY: Oxford University Press.

Butler, A. B. and Hodos, W. (1996). *Comparative Vertebrate Neuroanatomy*. New York, NY: Wiley-Liss.

Cant, J. G. H. (1986). Locomotor and feeding postures of spider and howling monkeys. *Folia Primatol.*, **46**, 1–14.

Cartmill, M. (1974). Rethinking primate origins. *Science*, **184**, 436–443.

Charles-Dominique, P. (1977). *Ecology and Behaviour of Nocturnal Primates*. London: Duckworth.

Charles-Dominique, P. and Martin, R. D. (1970). Evolution of lorises and lemurs. *Nature*, **227**, 257–260.

Cleveland, W. S. (1993). *Visualizing Data*. Summit, NJ: Hobart Press.

de Winter, W. (1997). *Perspectives on Mammalian Brain Evolution*. Ph.D. thesis, University of Western Australia.

de Winter, W. and Oxnard, C. E. (1996). Multivariate morphometrics of the primate brain and its mammalian context. *Proc. Int. Primat. Soc.*, **16**, 478.

(1997). The primate brain and its mammalian context: a morphometric study of volumetric measures of individual brain components. *Amer. J. Phys. Anthropol.*, Suppl. **24**, 243.

(2001). Evolutionary radiations and convergences in the structural organization of mammalian brains. *Nature*, **409**, 710–714.

Deacon, T. W. (1990). Rethinking mammalian brain evolution. *Amer. Zool.*, **30**, 629–705.

Dunbar, R. I. M. (1992). Neocortex size as a constraint on group size in primates. *J. Hum. Evol.*, **20**, 469–493.

Eisenberg, J. F. (1981). *The Mammalian Radiations*. London: Athlone.

Erikson, G. E. (1963). Brachiation in New World monkeys and in anthropoid apes. *Symp. Zool. Soc. Lond.*, **10**, 134–164.

Finlay, B. L. and Darlington, R. B. (1995). Linked regularities in the development and evolution of mammalian brains. *Science*, **268**, 1578–1584.

Finlay, B. L., Hersman, M. N., and Darlington, R. B. (1998). Patterns of vertebrate neurogenesis and the paths of vertebrate evolution. *Brain Behav. Evol.*, **52**, 232–242.

Fleagle, J. G. (1976). Locomotion and posture of the Malayan siamang and implications for hominoid evolution. *Folia Primatol.*, **26**, 245–269.

(1999). *Primate Adaptation and Evolution*, 2nd edn. San Diego, CA: Academic Press.

Fuster, J. M. (1989). *The Prefrontal Cortex*. 2nd edn. New York, NY: Raven Press.

Goldman-Rakic, P. S. (1995). Anatomical and functional circuits in prefrontal cortex of nonhuman primates. *Adv. Neurol.*, **66**, 51–63.

Harvey, P. H. and Krebs, J. R. (1990). Comparing brains. *Science*, **249**, 140–146.

Hastie, T., Tibshirani, R., and Friedman, J. (2001). *The Elements of Statistical Learning*. New York, NY: Springer.

Hershkovitz, P. (1973). *Living New World Monkeys* (Platyrrhini). Chicago, IL: University of Chicago Press.

Houk, J. C. (1995). Information processing in modular circuits. In: *Models of Information Processing in the Basal Ganglia*, ed. J. C. Houk, J. L. Davies, and D. G. Beiser. Cambridge, MA: MIT Press. pp. 3–9.

Jerison, H. J. (1973). *Evolution of the Brain and Intelligence*. New York, NY: Academic Press.

Krzanowski, W. J., and Marriott, F. H. C. (1994). *Multivariate Statistics*. London: Edward Arnold.

Marin, O., Smeets, W. J. A. J., and Gonzalez, A. (1998). Evolution of basal ganglia in tetrapods. *Trends Neurosci.*, **21**, 487–494.

Martin, R. D. (1990). *Primate Origins and Evolution*. Princeton, NJ: Princeton University Press.

Meldrum, D. J. (1998). Tail-assisted hind limb suspension as a transitional behavior in the evolution of the platyrrhine prehensile tail. In: *Primate Locomotion: Recent Advances*, ed. E. Strasser, J. G. Fleagle, A. Rosenberger, and H. McHenry. New York, NY: Plenum Press. pp. 145–156.

Middleton, F. A. and Strick, P. L. (1994). Anatomical evidence for cerebellar and basal ganglia involvement in higher cognitive function. *Science*, **266**, 458–461.

Mink, J. W. (1999). Basal ganglia. In: *Fundamental Neuroscience*, ed. M. J. Zigmond, F. E. Bloom, S. C. Landis, J. L. Roberts, and L. R. Squire. San Diego, CA: Academic Press. pp. 951–972.

Mittermeier, R. A. (1978). Locomotion and posture in *Ateles geoffroyi* and *A. paniscus*. *Folia Primatol.*, **30**, 161–193.

Oxnard, C. E. (1984). *The Order of Man: a Biomathematical Anatomy of the Primates*. New Haven, CT: Yale University Press.

Oxnard, C. E., Crompton, R. H., and Lieberman, S. S. (1990). *Animal Lifestyles and Anatomies: the Case of the Prosimian Primates*. Seattle, WA: University of Washington Press.

Passingham, R. E. (1975). The brain and intelligence. *Brain Behav. Evol.*, **11**, 1–15.

Preuss, T. M. (1995). The role of the neurosciences in primate evolutionary biology. In: *Primates and Their Relatives in Phylogenetic Perspective*, ed. R. D. E. MacPhee. New York, NY: Plenum Press. pp. 333–362.

Schieber, M. H. (1999). Voluntary descending control. In: *Fundamental Neuroscience*, ed. M. J. Zigmond, F. E. Bloom, S. C. Landis, J. L. Roberts, and L. R. Squire. San Diego, CA: Academic Press. pp. 931–950.

Schultz, W., Apicella, P., Romo, R., and Scarnati, E. (1995). Context-dependent activity in primate striatum reflecting past and future behavioral events. In: *Models of Information Processing in the Basal Ganglia*, ed. J. C. Houk, J. L. Davies, and D. G. Beiser. Cambridge, MA: MIT Press. pp. 11–26.

Shepherd, G. M. (1994). *Neurobiology*. New York, NY: Oxford University Press.

Stephan, H., Frahm, H. D., and Baron, G. (1981). New and revised data on volumes of brain structures in insectivores and primates. *Folia Primatol.*, **35**, 1–29.

Stephan, H., Baron, G., and Frahm, H. D. (1991). *Insectivora*. London: Springer-Verlag.

Strick, P. L., Dum, R. P., and Picard, N. (1995). Macro-organization of the circuits connecting the basal ganglia with the cortical motor areas. In: *Models of Information Processing in the Basal Ganglia*, ed. J. C. Houk, J. L. Davies and D. G. Beiser. Cambridge, MA: MIT Press. pp. 117–129.

Tuttle, R. (1975). Parallelism, brachiation and hominid phylogeny. In: *Phylogeny of Primates*, ed. W. P. Luckett and F. S. Szalay. New York, NY: Plenum Press. pp. 447–480.

Voogd, J. and Glickstein, M. (1998). The anatomy of the cerebellum. *Trends Neurosci*, **21**, 370–375.

Part III
In vivo *organismal verification of functional models*

Though it would appear from my early work that I primarily used anatomical inference for the "functional" explanations for the morphological, especially morphometric, adaptations that I saw in my studies, I nevertheless was always interested in the "real" biomechanics, interested in testing, that is, the hypotheses resulting from anatomical inference. The best that we could do at the time for the shoulder was to rely on the pioneering electromyographic studies on the human shoulder carried out by Inman, Saunders, and Abbott during the Second World War. Doing electromyography on nonhuman primates seemed such a long shot in those days.

Yet in fact my first research grant (from the US Public Health Service as it was called then), awarded in 1962 while I was still in the UK, aimed in part to design an implantable telemetric device for recording electromyographic and strain-gauge information from freely moving primates. Stanley Salmons was the research fellow employed on that grant for that purpose. But of course, only a few weeks' reading was enough to demonstrate that we could not do it – it just was not possible to make the device small enough for that purpose in those days. Stanley Salmons, however, did go on to design the "buckle transducer'" a first device for measuring tension in tendons, so all was not lost. He is today Professor of Biomedical Engineering at the University of Liverpool and has been engaged in these latter years in studies making the latissimus dorsi muscle into an adjunct pump in cases of cardiac insufficiency.

Carrying out cine studies of what the animals were doing was a further aim that was never achieved. Cineradiography was an absolute pipedream. The technologies just were not far enough advanced in those days. Yet they came along so soon afterwards.

Such ideas, however, did exist in my mind even in those very early years. They were expressed in books and especially in postgraduate courses that I gave at the University of Chicago on bone/joint/muscle biomechanics. They have since been carried forward by many workers, only some of whom were my colleagues and students. Some of these colleagues may have been influenced by my expression of the ideas. More likely, however, they may have been stimulated by the deficits in the original work occasioned because I could not follow them up. The chapters below are a selection from among a very large body of work that currently exists in this area.

Charles Oxnard

11 *Jaw adductor force and symphyseal fusion*

WILLIAM L. HYLANDER, CHRISTOPHER J. VINYARD
Duke University

MATTHEW J. RAVOSA
Northwestern University

CALLUM F. ROSS
Stony Brook University

CHRISTINE E. WALL, KIRK R. JOHNSON
Duke University

Introduction

Research over the last 25–30 years has revealed a considerable amount about the basic mechanisms of mammalian mastication (e.g., van Eijden and Turkawski, 2001; Türker, 2002). This progress has been largely due to the development of new experimental procedures and techniques. On the other hand, there has been relatively little emphasis on employing these procedures and techniques so as to facilitate adaptive explanations for the evolution of the mammalian masticatory apparatus (Herring, 1993). It has been our intent over the last several years to do just that (Ross and Hylander, 1996; Hylander *et al.*, 1998, 2000, 2002, 2003; Ravosa *et al.*, 2000; Vinyard *et al.*, 2001, in press a; Wall *et al.*, 2002; Williams *et al.*, 2003). In recent years the functional morphology of the craniofacial region of primates and other mammals has attracted a significant amount of research interest (Weijs, 1994; Ross and Hylander, 1996, 2000; Spencer, 1998; Anapol and Herring, 2000; Daegling and Hylander, 2000; Dechow and Hylander, 2000; Herring and Teng, 2000; Hylander *et al.*, 2000; Lieberman and Crompton, 2000; Ravosa *et al.*, 2000). This is simply because there continue to be many unanswered research questions or problems. One persistent problem that has received a considerable amount of attention is related to the adaptive significance of symphyseal fusion in mammals. As noted by many, the ossification or fusion of the left and right sides of the lower jaw or dentaries has

Shaping Primate Evolution, ed. F. Anapol, R. Z. German, and N. G. Jablonski. Published by Cambridge University Press.

occurred independently in many different mammalian lineages (e.g., Beecher, 1977). In spite of extensive morphological and comparative studies, there is little agreement as to the ultimate reasons for symphyseal fusion (Beecher, 1979; Scapino, 1981) – that is, whether or not certain diets are normally associated with the evolution of symphyseal fusion, and also whether symphyseal fusion is an outcome of allometric factors (Beecher, 1977; Scapino, 1981; Ravosa and Hylander, 1994). This lack of consensus has caused us to refocus our attention on the more immediate or proximate factors that are related to symphyseal fusion. More specifically, we have been attempting to determine if certain types of biomechanical situations or loading patterns are plausibly associated with symphyseal fusion.

When most mammals chew food, they typically do so unilaterally but have bilateral recruitment of their jaw-adductor muscles, i.e., both the working-side (chewing side) and the balancing-side (non-chewing side) jaw muscles produce varying amounts of jaw muscle force. This fact has important consequences as to how various portions of the mandible are loaded, and is particularly important for loading of the mandibular symphysis. This is simply because during unilateral chewing or biting, muscle force from the balancing-side jaw muscles must be transferred through the symphysis to the working side, and this in turn results in various sorts of symphyseal loading regimes (Hylander, 1984).

Recently we tested two hypotheses about how balancing-side jaw-adductor muscles are recruited in primates (Hylander *et al.*, 1998). One of these, the *symphyseal fusion–muscle recruitment* hypothesis, states that increased balancing-side jaw-adductor muscle force is linked to the evolution of symphyseal fusion in anthropoid primates. That is, symphyseal fusion is an adaptation to strengthen the symphysis so as to prevent its structural failure due to increased stress associated with increased recruitment of balancing-side muscle force during forceful mastication. Similarly, the *allometric constraint-muscle recruitment* hypothesis states that increased balancing-side muscle force is simply linked to allometric constraints on jaw-adductor muscle force production associated with the evolution of increasing body size. That is, larger primates must recruit relatively higher levels of balancing-side muscle force so as to generate equivalent amounts of bite force during forceful mastication.

These two hypotheses, which are not mutually exclusive, were tested by analyzing mandibular corpus bone-strain data recorded from long-tailed macaques, thick-tailed galagos, and owl monkeys, as these subjects engaged in forceful mastication (Hylander *et al.*, 1998). The data indicate that for all species analyzed, peak working-side corpus strains are relatively large and similar in magnitude. Moreover, for these two anthropoid species the balancing-side strains are only slightly smaller than their working-side strains, whereas for galagos the balancing-side strains are much smaller than both their working-side strains and the balancing-side strains of anthropoids. As balancing-side corpus strains are predominately the result of balancing-side muscle force (Hylander, 1977), these data indicate that compared to anthropoids, galagos apparently recruit much less balancing-side muscle force during mastication.

Macaques and owl monkeys have a fully fused mandibular symphysis, which is the derived condition for crown anthropoids, whereas adult thick-tailed galagos have an unfused symphysis, which is a retention of the primitive mammalian condition. Furthermore, unlike the rigid symphysis of anthropoids, the symphysis of thick-tailed galagos is highly mobile and is not only structurally weaker than a fully fused symphysis, but is structurally weaker than the unfused symphyses of most other extant primates (Beecher, 1977, 1979). Thus, the bone strain data in combination with the above morphological observations support the symphyseal fusion–muscle recruitment hypothesis. Furthermore, the strain data refute the allometric constraint–muscle recruitment hypothesis (Hylander *et al*., 1998).

Muscle force direction, symphyseal stress, and symphyseal fusion

If increased recruitment of balancing-side adductor muscle force in anthropoids is indeed linked to the evolution of symphyseal fusion, it has been hypothesized that this increased force is either vertically (dorsoventrally) directed, transversely (mediolaterally) directed, or is some near-equal combination of these force components (Ravosa and Hylander, 1994; Ravosa and Simons, 1994; Ravosa, 1996; Hylander *et al*., 1998, 2000, 2002; Hogue and Ravosa, 2001). If vertically directed, then a likely functional correlate of this force is an increase in vertically directed bite force, i.e., an increase in crushing or vertical shearing forces along postcanine teeth. As an increase in vertically directed muscle force is likely to cause the symphysis to experience an increase in bending in the frontal plane (due to axial torsion of the mandibular corpora) and dorsoventral shear (Hylander, 1984), then perhaps symphyseal fusion is a structural adaptation to prevent mechanical failure by more effectively resisting increased symphyseal stresses due to these particular shearing and bending regimes.

If the increased balancing-side jaw-adductor muscle force in anthropoids is primarily transversely directed, then a likely functional correlate of this force is a relative increase in transversely directed bite force, i.e., a relative increase in grinding and/or transverse shearing forces along postcanine teeth. As an increase in transversely directed balancing-side muscle force causes an increase in lateral transverse bending or wishboning of the symphysis (Hylander, 1984), then perhaps symphyseal fusion is an adaptation to prevent its structural failure by more effectively resisting increased symphyseal stresses due to this particular bending regime.

Alternatively, perhaps the increased balancing-side muscle force is some near-equal combination of an increase in both vertically and transversely directed muscle forces, and therefore perhaps symphyseal fusion is an adaptation to more effectively resist increased stresses associated with an increase in some combination of all of the above loading regimes.

Purpose of this study

The main purpose of this study is to test various hypotheses that link symphyseal fusion to jaw muscle recruitment patterns during mastication. Towards this end, we summarize here the results of a detailed electromyographic (EMG) analysis of the superficial and deep masseters of baboons, macaques, owl monkeys, thick-tailed galagos, and ring-tailed lemurs. Unlike anthropoids and similar to thick-tailed galagos, ring-tailed lemurs have an unfused mandibular symphysis. Compared to galagos, however, ring-tailed lemurs have a stiffer and stronger symphysis (Beecher, 1977, 1979), and these differences in morphology may be reflected in corresponding differences in the motor patterns of their jaw muscles during mastication. This EMG analysis enables us to characterize activity patterns for one portion of the jaw-adductor muscle mass that has a relatively large vertical orientation, the superficial masseter, and one portion that has a relatively large transverse orientation, the deep masseter (Hylander and Johnson, 1994). Data such as these can provide insight as to whether there are major differences in muscle recruitment and firing patterns during mastication, and by inference in symphyseal loading patterns, between anthropoids on the one hand and strepsirhines with unfused symphyses on the other.

Hypothesis 1

In contrast to anthropoids, thick-tailed galagos and ring-tailed lemurs recruit relatively less overall balancing-side muscle force during the power stroke of mastication: i.e., bite force in galagos and lemurs may be generated almost entirely by the recruitment of force from their working-side muscles. If so, as EMG activity of the masseter is highly correlated with relative masseter force during the power stroke of mastication in primates (Hylander and Johnson, 1989, 1993), compared to anthropoids, galagos and lemurs are predicted to have relatively large working-side/balancing-side (W/B) ratios of peak EMG values for *both* the superficial and deep masseter muscles. If true, this outcome would support the hypothesis that increased vertically and transversely directed balancing-side muscle forces are linked to the evolution of symphyseal fusion in anthropoids. Furthermore, it would also support the hypothesis that symphyseal fusion is linked to symphyseal stress due to some combination of (1) dorsoventral shear, (2) frontal bending of the symphysis associated with twisting of the mandibular corpora, and (3) lateral transverse bending or wishboning.

Hypothesis 2

On the other hand, and similar to anthropoids, the galagos and lemurs may recruit a substantial amount of force from their balancing-side superficial masseter (Hylander *et al.*, 1992; Hylander and Johnson, 1997), but unlike anthropoids, they may recruit relatively little force from their balancing-side deep masseter. If so, the data will demonstrate that anthropoids and strepsirhines have similar or overlapping W/B EMG ratio values for the superficial masseter. For the deep masseter, however, this hypothesis predicts that anthropoids have a small W/B ratio whereas thick-tailed galagos and ring-tailed lemurs have a large W/B ratio. This outcome would support the hypothesis that the evolution of symphyseal fusion in anthropoid primates is linked to increased transversely directed balancing-side muscle forces associated with the deep masseter. Furthermore, it would also support the hypothesis that symphyseal fusion is linked to symphyseal stress due to wishboning.

Hypothesis 3

Another possibility is that, similar to anthropoids, thick-tailed galagos and ring-tailed lemurs may recruit a substantial amount of force from their balancing-side deep masseter, but, unlike anthropoids, they may recruit relatively little force from their balancing-side superficial masseter. If so, the data will demonstrate that anthropoids and strepsirhines have similar or overlapping W/B ratios for the deep masseter, but for the superficial masseter anthropoids will have a small W/B ratio, and thick-tailed galagos and ring-tailed lemurs will have a large W/B ratio. This outcome would support the hypothesis that the evolution of symphyseal fusion in anthropoid primates is linked to increased vertically directed balancing-side muscle forces. Furthermore, it would support the hypothesis that symphyseal fusion is linked to symphyseal stress due to a combination of dorsoventral shear and frontal bending.

Masseter firing patterns and wishboning

We also propose to determine whether thick-tailed galagos, ring-tailed lemurs, and owl monkeys exhibit the macaque and baboon pattern of late peak activity in the balancing-side deep masseter, because this firing pattern contributes to wishboning of the symphysis in macaques and presumably baboons (Hylander and Johnson, 1994). Most notably, this hypothesis predicts that the balancing-side deep masseter of anthropoids reaches peak activity late in the power stroke, at a time when vertically directed muscle force is rapidly declining during the most terminal portion of the power stroke (Hylander and Johnson, 1994). If it turns out that owl monkeys have this firing pattern whereas galagos and lemurs do not, this outcome would support the hypothesis that the evolution of symphyseal fusion in anthropoids is linked to transversely directed muscle forces.

In summary, the symphyseal fusion–muscle recruitment hypothesis (hypotheses 1, 2, and 3) predicts that anthropoids recruit relatively more balancing-side masseter force than do galagos and lemurs, and therefore that compared to these strepsirhines the W/B EMG ratios of the superficial and/or deep masseters are smaller for anthropoids. Hypotheses 1 and 2 are also compatible with the hypothesis that, unlike thick-tailed galagos and ring-tailed lemurs, anthropoids have a deep masseter firing pattern that causes the mandibular symphysis to experience wishboning during mastication.

Materials and methods

Subjects

Five baboons (*Papio anubis*), four macaques (three *Macaca fascicularis* and one *Macaca fuscata*), two owl monkeys (*Aotus trivirgatus*), three thick-tailed galagos (two *Otolemur crassicaudatus* and one *Otolemur garnettii*), and four ring-tailed lemurs (*Lemur catta*) served as subjects. All subjects, which were either young adults or subadults, are described in detail elsewhere (Hylander *et al.*, 2000, 2002). Prior to the experiments all subjects were trained to eat various foods in the laboratory. All subjects were used for at least two and as many as six separate experiments.

EMG electrodes and electrode placement procedure

As many as eight fine-wire indwelling bipolar electrodes were placed within the left and right middle portion of the superficial and deep masseter muscles according to procedures outlined elsewhere (Hylander *et al.*, 2000) (Fig. 11.1). All electrodes were inserted with the subjects under light sedation.

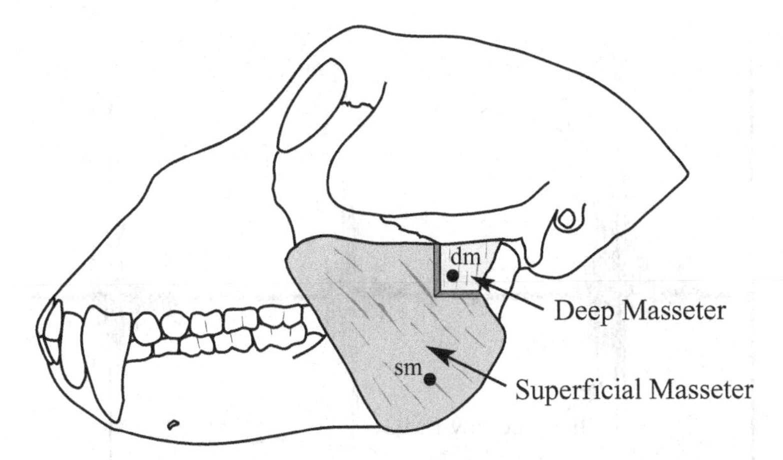

Figure 11.1. Lateral view of a macaque skull and the superficial and deep portions of the masseter muscle. The solid black dots indicate the location of the bipolar fine-wire electrodes in the deep (dm) and superficial (sm) portions of the masseter (from Hylander *et al.*, 2000).

Recording procedure

Prior to recovery from sedation, the subject was placed in a restraining chair (baboons, macaques, and lemurs) or restraining sling-suit (galagos and owl monkeys) especially designed to permit normal head, neck, and jaw movements during mastication. Once the animal fully recovered from sedation, it was fed pieces of apple skin and hard and/or tough foods, i.e., unpopped popcorn kernels, dried gelatin candy (dried gummy bears), and dried apricots and prunes. The EMG potentials were simultaneously amplified and filtered (bandpass 100–3000 Hz) and then recorded on an FM tape recorder at 15 inches (0.38 m) per second. Details of the recording procedure are the same as described previously (Hylander and Johnson, 1989, 1994). The data presented here are based on 55 separate experiments (16 baboon, 10 macaque, 12 owl monkey, 8 galago, and 9 lemur experiments). All EMG data were recorded as subjects vigorously chewed various food items. We continued to feed the various foods in an intermittent fashion and data were recorded until either we obtained a surplus of data or the animal refused to eat any additional food. At the conclusion of the recording session the animal was sedated, the electrodes were removed, and the animal was returned to its cage.

EMG quantification

For each chewing sequence analyzed the EMG data were played from the FM tape recorder into a 16-channel analog-to-digital converter (12-bit resolution),

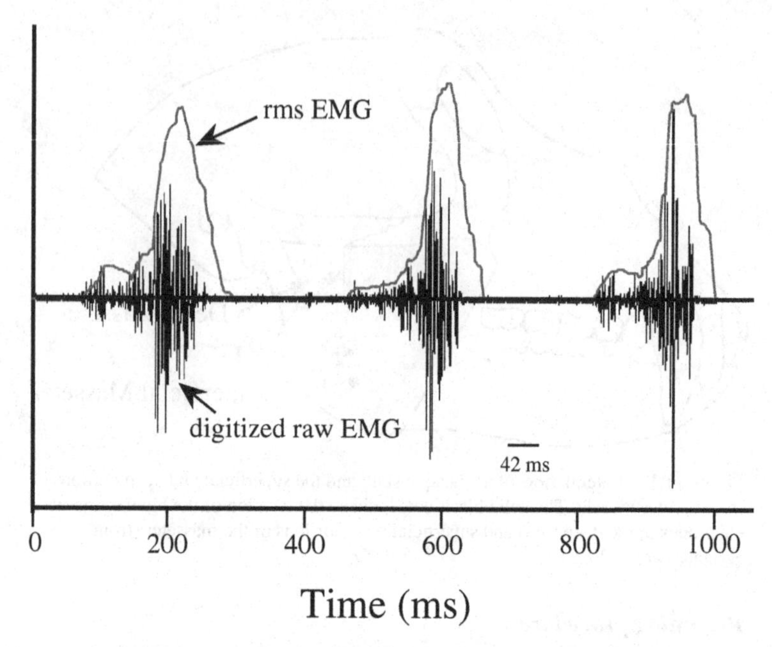

Figure 11.2. Digitized raw EMGs and the corresponding root-mean-square (rms) EMGs for the working-side superficial masseter of a male macaque during mastication of popcorn kernels. The rms values were calculated using a 42 ms time constant for reasons outlined in Hylander and Johnson (1993). The use of this time constant results in a EMG waveform that approximates the waveform of jaw muscle force. Each rms EMG value represents the root-mean-square of the raw digitized EMG values for the previous 42 ms. This results in a latency period between the apparent "peak" raw EMG and the actual peak rms EMG value. Moreover, the rms EMG value reaches zero 42 ms after the raw EMG appears to reach zero. There is also a latency period between peak rms EMG and peak muscle force. On average, peak muscle force follows peak rms EMG by about 30 ms (Hylander and Johnson, 1993).

and the digitized values were written to the hard disk of a microcomputer. Each channel was sampled and digitized at a rate of 10 000 Hz with a channel separation time of 6.25 microseconds. The digitized values were then read back into the microcomputer for subsequent processing and analysis.

Raw digitized EMG values (Fig. 11.2) were first filtered with a digital Butterworth band-pass filter (100–3000 Hz). The EMG was then rectified and smoothed by calculating the root-mean-square (rms) values from the raw digitized values for a 42 millisecond (ms) time constant (Fig. 11.2) (Hylander and Johnson, 1993). The rms values were calculated in 2 ms intervals for the entire chewing sequence. For each electrode and power stroke we then identified the maximum peak rms value for each experiment. The peak values for each experiment were then scaled by assigning a value of 1.0 to the largest value, and the remaining values were scaled in a linear fashion. Thus, for each power stroke and muscle there is a scaled value for peak EMG activity.

W/B EMG ratios

For each power stroke the scaled peak working-side EMG values for the superficial and deep masseter muscles were then divided by the scaled peak balancing-side EMG value. We refer to this value as the W/B EMG ratio. Means and standard deviations of this variable were determined for each experiment and subject. For each primate group grand means of the W/B EMG ratio were calculated based on experiment mean values. For reasons outlined elsewhere (Hylander and Johnson, 1994), a log transformation of the data was performed, and the transformed mean values of the W/B ratios for the left and right sides were combined and then transformed back to their original scale.

Due to inherent problems associated with the statistical testing of ratios, we intend to focus the analysis on the grand means and experiment means of these ratios. One of our main goals is to determine if the W/B ratios of the superficial and/or deep masseter in galagos and lemurs *consistently* differ from those of anthropoids.

Jaw muscle firing patterns

So as to establish a standard uniform procedure for the analysis of muscle firing patterns, the timing of peak EMG activity of the working-side and balancing-side deep masseters and the balancing-side superficial were compared to the peak EMG activity of our reference muscle, the working-side superficial masseter. Furthermore, the peak timing differences were also determined between (1) the working-side and balancing-side deep masseters, (2) the balancing-side superficial and deep masseters, and (3) the working-side deep and balancing-side superficial masseters.

Another main goal is to determine whether the masseter firing pattern for baboons, macaques, and owl monkeys differs from that for galagos and lemurs, and also whether only the anthropoids possess the masseter firing pattern thought to be associated with wishboning of the macaque symphysis (Hylander *et al.*, 1987). The wishboning firing pattern is characterized by peak activity of the working-side deep and balancing-side superficial masseters *preceding* peak activity of the working-side superficial masseter, and by peak activity of the balancing-side deep masseter *following* peak activity of the working-side superficial masseter (Hylander and Johnson, 1994) (Figs. 11.3 and 11.4). Thus, as

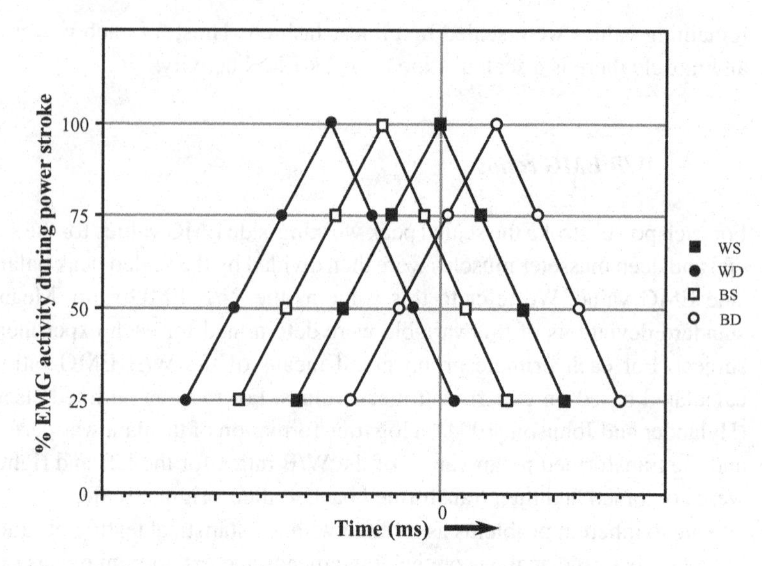

Figure 11.3. Predictions of masseter firing pattern associated with wishboning of the symphysis. The working hypothesis predicts the occurrence of this pattern only in those primates with a fully fused mandibular symphysis, i.e., in extant anthropoids, but not in those prosimians with a highly mobile mandibular symphysis, such as thick-tailed galagos. The working-side deep masseter peaks first and the balancing-side deep masseter peaks last. Furthermore, the balancing-side superficial masseter peaks after the working-side deep and before the working-side superficial masseter. The vertical line indicates peak activity in the reference muscle, the working-side superficial masseter. Abbreviations: WS, working-side superficial masseter; WD, working-side deep masseter; BS, balancing-side superficial masseter; BD, balancing-side deep masseter (from Hylander *et al.*, 2000).

noted earlier, this firing pattern results in the balancing-side deep masseter reaching peak force late in the power stroke, well after peak force of the balancing-side superficial masseter, at a time when overall vertically directed jaw-adductor muscle force is rapidly declining during the terminal portion of the power stroke.

Means and standard deviations were calculated for all timing values for each individual experiment and each subject. Grand means and mean standard deviations were then calculated from the mean values for each experiment. All tests of significance were of the grand means. The mean timing differences in peak EMGs between muscles were tested for significance at the 95% level by using a nonparametric test, the Wilcoxon signed-ranks test for paired comparisons. When peak EMG of a muscle was predicted to either precede or follow peak EMG of another muscle a one-tailed test of significance was utilized; otherwise, all tests of significance were two-tailed tests.

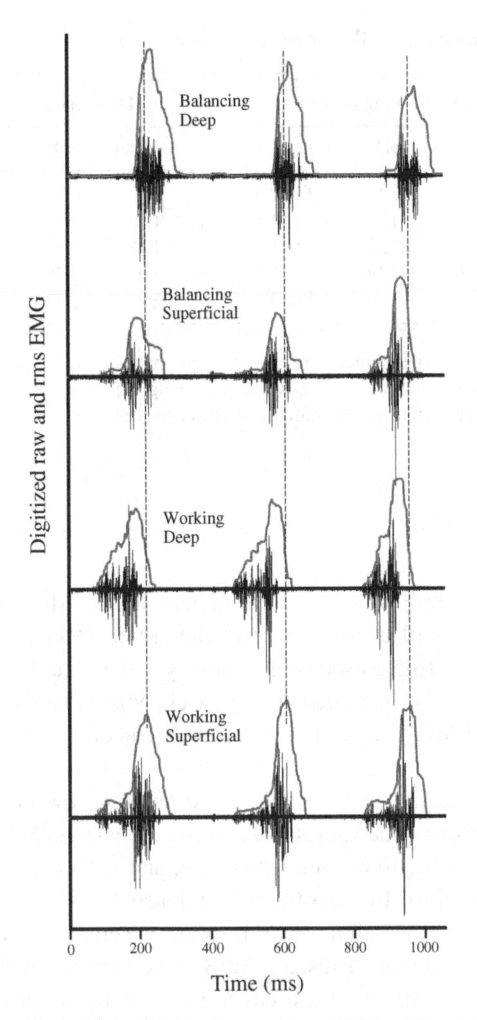

Figure 11.4. Digitized raw and rms EMGs of the working-side and balancing-side deep and superficial masseters in a male macaque during mastication of popcorn kernels. On average the working-side deep masseter is the first muscle to reach peak activity, and the balancing-side deep masseter is the last to reach peak activity. Moreover, peak activity in the balancing-side superficial masseter precedes peak activity in the working-side superficial masseter. The dashed vertical lines indicate peak activity in the reference muscle, the working-side superficial masseter (from Hylander *et al.*, 2000).

Table 11.1. *Summary data of mean W/B ratios*

	Superficial masseter				Deep masseter			
Subject	*n*	Mean	SD	Range	*n*	Mean	SD	Range
Baboon	15	1.9	0.40	1.3 to 2.7	12	1.0	0.22	0.7 to 1.4
Macaque	8	1.4	0.25	1.2 to 1.8	7	1.0	0.20	0.7 to 1.3
Owl monkey	12	1.4	0.22	1.1 to 1.8	4	1.4	0.13	1.2 to 1.5
Thick-tailed galago	7	2.2	0.49	1.5 to 2.8	8	4.4	2.39	2.1 to 9.5
Ring-tailed lemur	9	1.7	0.70	1.3 to 3.1	9	2.4	0.99	1.8 to 3.4

Overall mean values are grand means based on the mean of all experiments for each species. The experiment means are based on all power strokes within a chewing sequence. *n* is the number of individual experiments, not the number of power strokes.

Results

Typically, some subjects simply refused to eat certain foods. All, however, chewed at least one of the hard or tough foods. Therefore, EMG data were recorded from all subjects during episodes of relatively forceful and vigorous mastication. We only present data recorded during the chewing of hard or tough foods so as to confine the EMG analysis to the relatively forceful sequences of mastication.

Among galagos the balancing-side EMGs were often very low in magnitude even though their working-side values were consistently large. When this occurs this causes the W/B ratio to become very large, and in theory the W/B EMG ratio can approach infinity. In order to prevent unusually large ratio values from grossly distorting the overall mean W/B ratios, we employed a cutoff value of 10 for each individual ratio. Thus, any W/B ratios larger than 10 were simply assigned a value of 10.0. We were concerned that if this procedure was not employed, we might erroneously conclude that galagos routinely recruit relatively little balancing-side muscle force when in fact their mean W/B ratios were grossly distorted by one or two unusually large individual ratio values.

W/B EMG ratios

Table 11.1 presents the descriptive statistics of the W/B ratios for sequences of chewing hard or tough foods for baboons, macaques, owl monkeys, galagos, and lemurs.

Superficial masseter

For the superficial masseter the anthropoid W/B ratios are smaller than the W/B ratios for galagos, but not so for lemurs. The grand mean W/B ratios of baboons, macaques, owl monkeys, galagos, and lemurs are 1.9, 1.4, 1.4, 2.2, and 1.7, respectively. Moreover, there is a considerable overlap of the experiment mean values between the anthropoid and strepsirhine species. For example, the individual experiment mean values for anthropoids range from 1.1 to 2.7, whereas for the strepsirhines these values range from 1.3 to 3.4.

Deep masseter

For the deep masseter the anthropoid W/B ratios are unlike the W/B ratios for galagos, and lemurs. The W/B deep masseter ratios for baboons, macaques, owl monkeys, galagos, and lemurs are 1.0, 1.0, 1.4, 4.4, and 2.4, respectively. For anthropoids the mean W/B ratios for each experiment range from 0.7 to 1.5, whereas for the strepsirhines these values range from 1.8 to 9.5. Thus, there is no overlap whatsoever in the experiment mean values. Finally, as a cutoff value of 10 was imposed on the W/B ratios, and only galagos had values exceeding 10, we have underestimated the average W/B values for galagos, and therefore the average differences of the W/B ratios between the galagos and anthropoids.

Percentage EMG activity of the masseter

Figure 11.5 shows percent scaled EMG values for the working-side and balancing-side masseters during the chewing of hard or tough foods. In addition to indicating some of what has already been described for the W/B ratios, this figure demonstrates that on average the working-side muscles range from 55% to 75% for both the superficial and deep masseters. In contrast, the balancing-side muscles are more variable. For the balancing-side superficial masseter the values range from 32% to 55%. The largest range in percentage values, however, is associated with the balancing-side deep masseter. Although anthropoids recruit on average from 58% (owl monkeys) to 65% and 66% (baboons and macaques) of peak EMG activity, galagos and lemurs only recruit 25% and 30% of peak activity of the balancing-side deep masseter.

Masseter firing patterns of peak EMG

Table 11.2 presents the descriptive statistics for the firing patterns of the working-side and balancing-side masseters for baboons, macaques, owl monkeys, galagos, and lemurs during the chewing of hard or tough foods. This table

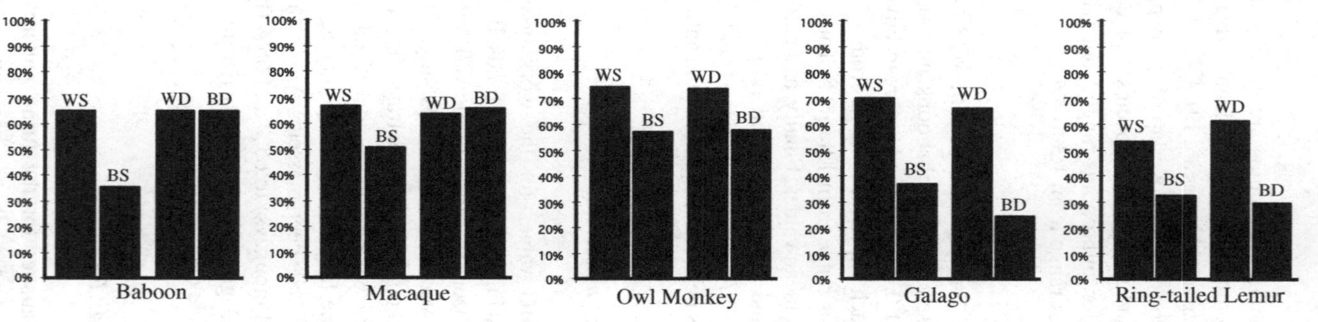

Figure 11.5. Bar graphs of average percent EMG activity for all primates during the chewing of hard and/or tough foods. Abbreviations: WS, working-side superficial; BS, balancing-side superficial; WD, working-side deep; BD, balancing-side deep masseters (from Hylander *et al.*, 2000).

Table 11.2. *Summary data of mean differences in timing (milliseconds) of peak EMG of the working and balancing-side deep and the balancing-side superficial masseters relative to the reference muscle*

	Working deep			Balancing superficial			Balancing deep		
Subject	n	Grand mean	SD	n	Grand mean	SD	n	Grand mean	SD
Baboon	13	47***	16.7	13	17***	14.1	13	−6*	9.1
Macaque	8	65**	17.1	8	17**	8.4	7	−20*	16.9
Owl monkey	4	13*	11.7	12	−1	11.4	4	−11*	9.0
Thick-tailed galago	8	11**	10.4	7	21**	6.3	8	23[a]	11.7
Ring-tailed lemur	9	37**	18.5	9	14**	13.4	9	24[a]	23.0

Overall mean values are grand means based on the mean of all experiments for each species. The experiment means are based on all power strokes within a chewing sequence.

n is the number of individual experiments, not the number of power strokes.

* Significantly greater than zero ($P \leq 0.05$).

** Significantly greater than zero ($P \leq 0.01$).

*** Significantly greater than zero ($P \leq 0.001$).

[a] Working hypothesis is rejected as peak EMG activity of the balancing-side deep masseter precedes peak EMG activity of working-side superficial masseter ($P < 0.01$).

indicates the mean timing differences in milliseconds of peak EMG activity of the working-side deep, the balancing-side superficial, and the balancing-side deep masseters, relative to peak EMG activity of the reference muscle, the working-side superficial masseter.

Positive values indicate that peak EMG of the muscle precedes peak activity of the reference muscle, and negative values indicate the reverse. Table 11.3 consists of the grand means of three additional timing variables for these five groups of primates. These variables are the time intervals between peak EMG of (1) the working-side and balancing-side deep masseters, (2) the working-side deep and the balancing-side superficial masseters, and (3) the balancing-side superficial and deep masseters.

Figures 11.6–11.10 indicate the mean timing differences of the working-side and balancing-side masseter muscles throughout the power stroke for one subject from each of the five groups of primates. In addition to the relative timing of peak EMG values, these figures also indicate the relative timing of 25%, 50%, and 75% of peak EMGs during loading and unloading.

Baboons and macaques

The overall firing patterns for baboons and macaques are remarkably similar, and therefore will be treated together. The data indicate that on average peak EMG of the working-side deep and balancing-side superficial masseters precede

Table 11.3. *Summary data of mean differences in timing (milliseconds) of peak EMG activity of masseter muscles*

	Working deep precedes balancing deep			Working deep precedes balancing superficial			Balancing superficial precedes balancing deep		
Subject	*n*	Grand mean	SD	*n*	Grand mean	SD	*n*	Grand mean	SD
Baboon	13	52***	16.0	12	30***	14.7	12	23***	9.9
Macaque	7	85**	20.8	7	48**	11.5	6	37**	17.6
Owl monkey	4	24*	12.0	4	14†	11.9	4	10*	3.6
Thick-tailed galago	8	−12[a]	14.1	7	−10[b]	10.3	7	−1	11.0
Ring-tailed lemur	9	12**	6.4	9	22**	15.7	9	−9[c]	13.9

Overall mean values are grand means based on the mean of all experiments for each species. The experiment means are based on all power strokes within a chewing sequence.
n is the number of individual experiments, not the number of power strokes.
* Significantly greater than zero ($P \leq 0.05$).
** Significantly greater than zero ($P \leq 0.01$).
*** Significantly greater than zero ($P \leq 0.002$).
† Significantly greater than zero ($P \leq 0.1$).
[a] Working hypothesis is rejected as peak EMG activity of balancing-side deep masseter precedes peak EMG activity of working-side deep masseter ($P \leq 0.05$).
[b] Working hypothesis is rejected as peak EMG activity of balancing-side superficial masseter precedes peak EMG activity of working-side deep masseter ($P \leq 0.05$).
[c] Working hypothesis is rejected as peak EMG activity of balancing-side deep masseter precedes peak EMG activity of balancing-side superficial masseter ($P \leq 0.05$).

the reference muscle by 47 ms and 17 ms for baboons ($P \leq 0.001$), and by 65 ms and 17 ms for macaques ($P \leq 0.01$). Furthermore, peak EMG activity of the working-side deep masseter precedes activity of the balancing-side superficial masseter on average by 30 ms in baboons ($P \leq 0.001$) and 48 ms in macaques ($P \leq 0.01$) (Table 11.3). In contrast, peak EMG of the balancing-side deep masseter occurs after the reference muscle by 6 ms in baboons (mean = −6 ms; $P \leq 0.05$) and 20 ms in macaques (mean = −20 ms; $P \leq 0.05$). Figures 11.6 and 11.7 indicate the timing of the masseters for baboon 5 and macaque 11. Note the considerable similarity between these two figures.

Owl monkeys

Similar to baboons and macaques, the data indicate that on average peak EMG of the working-side deep masseter precedes the reference muscle (mean = 13 ms, $P \leq 0.05$). Unlike macaques and baboons, however, peak EMG of the balancing-side superficial masseter peaks 1 ms after, not well before, the reference muscle. Not surprisingly, this −1 ms mean value is not statistically significant. Similar to baboons and macaques, for owl monkeys peak EMG

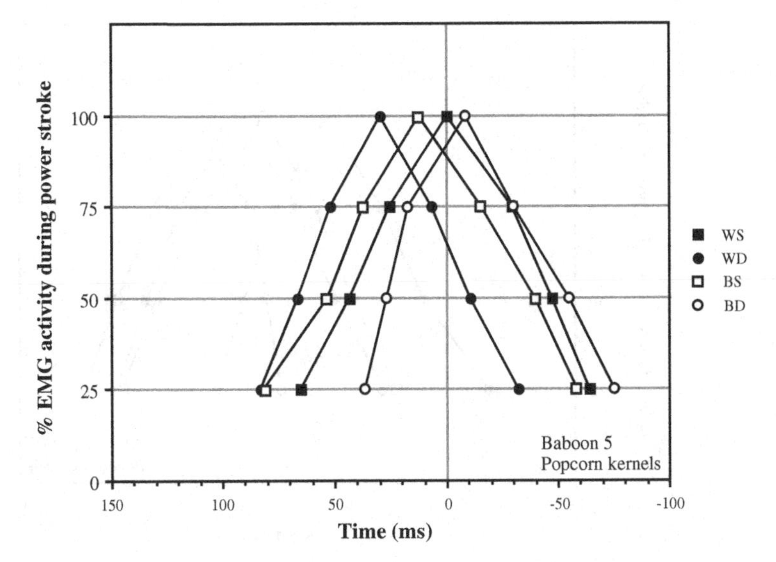

Figure 11.6. Mean values for the timing of average percent EMG activity of the deep and superficial masseters during the chewing of popcorn kernels in baboon 5. The first muscle to reach peak activity is the working-side deep masseter, followed by the balancing-side superficial, working-side superficial, and balancing-side deep masseters. The vertical line indicates peak activity in the reference muscle, the working-side superficial masseter. Positive timing values indicate that a muscle reaches peak activity prior to peak activity of the reference muscle, whereas negative values indicate peak activity after the reference muscle. Symbols and abbreviations as in Figure 11.3 (from Hylander *et al.*, 2000).

activity of the working-side deep masseter precedes activity of the balancing-side superficial masseter (mean = 14 ms; $P \leq 0.05$) (Table 11.3), and peak EMG activity of the balancing-side deep masseter occurs after the reference muscle (mean = -11 ms; $P \leq 0.05$). Finally, Figure 11.8 indicates the relative timing of the various muscles for one of our owl monkeys. This subject exhibits an unusual pattern for the two superficial masseters relative to one another. Note, however, the general similarity in the firing pattern of the deep masseters in all anthropoid subjects (Figs. 11.6, 11.7, and 11.8). The working-side and balancing-side deep masseters are the first and last muscles, respectively, to reach peak EMG activity.

Thick-tailed galagos

The data indicate that on average peak EMG of the working-side deep, balancing-side superficial, and balancing-side deep masseters precede the reference muscle by 11 ms, 21 ms, and 23 ms, respectively. As in the anthropoids, the first two mean values are positive and statistically significant ($P \leq 0.01$), but

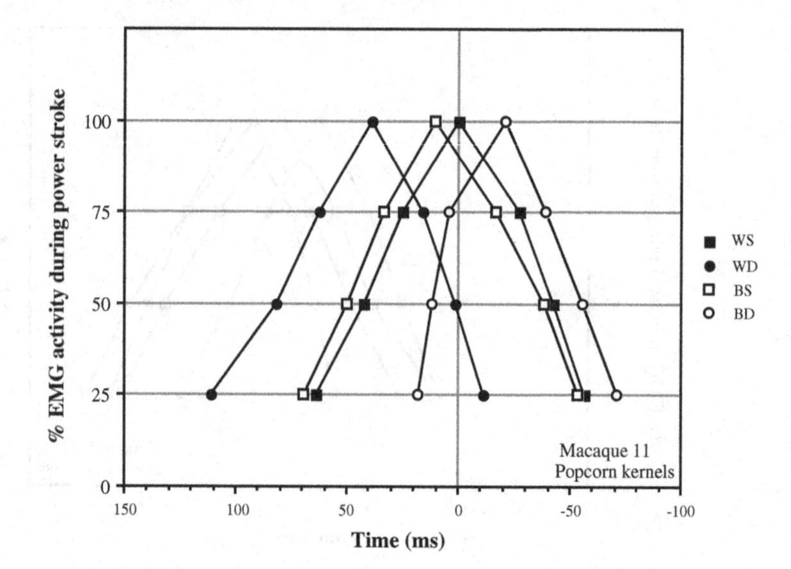

Figure 11.7. Mean values for the timing of average percent EMG activity of the deep and superficial masseters during the chewing of popcorn kernels in macaque 11. The first muscle to reach peak activity is the working-side deep masseter, followed by the balancing-side superficial, working-side superficial, and balancing-side deep masseters. The vertical line indicates peak activity in the reference muscle. Symbols and abbreviations as in Figure 11.3 (from Hylander *et al.*, 2000).

unlike anthropoids, the latter mean value is also positive, demonstrating that in galagos the balancing-side deep masseter is not the last muscle to reach peak activity. Instead, for galagos it is one of the first muscles to reach peak activity. Furthermore, and again unlike the anthropoids, for galagos the balancing-side superficial masseter reaches peak EMG activity prior to the working-side deep masseter (mean $= -10$ ms, $P \leq 0.04$). Thus, in contrast to the firing pattern determined for anthropoids, for galagos the working-side deep masseter is not the first nor is the balancing-side deep masseter the last to reach peak EMG activity. Finally, and once again unlike anthropoids, for galagos the balancing-side deep masseter peaks 1 ms prior to the balancing-side superficial masseter. This mean value is small and not statistically significant (Table 11.3), suggesting that these two muscles reach peak activity more or less at the same time.

Ring-tailed lemurs

The data indicate that on average peak EMG of the working-side deep, balancing-side superficial, and balancing-side deep masseters precede the reference muscle by 37 ms, 14 ms, and 24 ms, respectively. As in anthropoids and galagos the first two mean values are positive and statistically significant

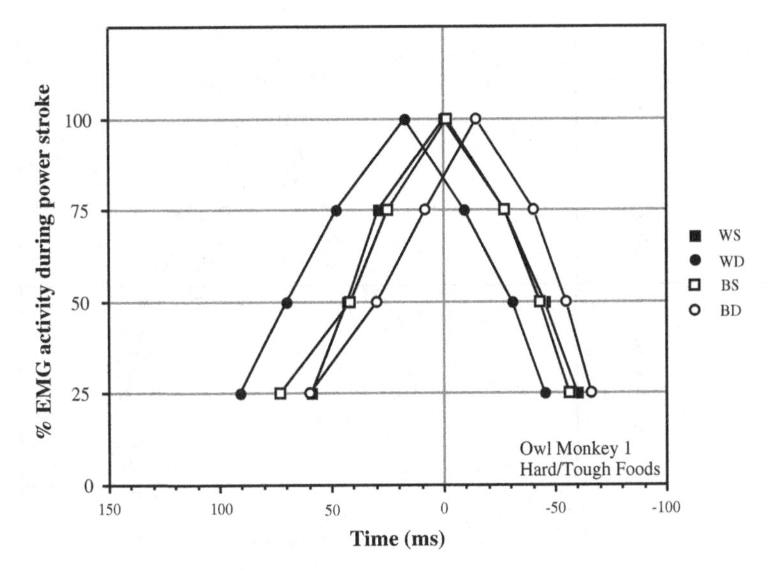

Figure 11.8. Mean values for the timing of average percent EMG activity of the deep and superficial masseters during the chewing of hard/tough foods in owl monkey 1. The first muscle to reach peak activity is the working-side deep masseter, followed by the near simultaneous activity of the working-side and balancing-side superficial masseters, followed by the balancing-side deep masseter. The vertical line indicates peak activity in the reference muscle. Symbols and abbreviations as in Figure 11.3 (from Hylander *et al.*, 2000).

($P \leq 0.01$), but unlike anthropoids, the latter mean value is also positive, demonstrating that in lemurs (and galagos) the balancing-side deep masseter is not the last muscle to reach peak activity. Instead, for lemurs (and galagos) it reaches peak activity early in the power stroke. Furthermore, similar to anthropoids but unlike galagos, for lemurs the balancing-side superficial masseter reaches peak EMG activity after the working-side deep masseter (mean = 22 ms, $P \leq 0.01$). Thus, similar to the firing pattern determined for anthropoids, for lemurs the working-side deep masseter is the first to reach peak EMG activity. Finally and most importantly, and unlike anthropoids, for lemurs the balancing-side deep masseter peaks 9 ms prior to the balancing-side superficial masseter. This mean value is statistically significant (Table 11.3).

Summary of firing patterns

Tables 11.2 and 11.3 indicate quite clearly that for the anthropoids the first muscle to reach peak EMG activity is the working-side deep masseter and

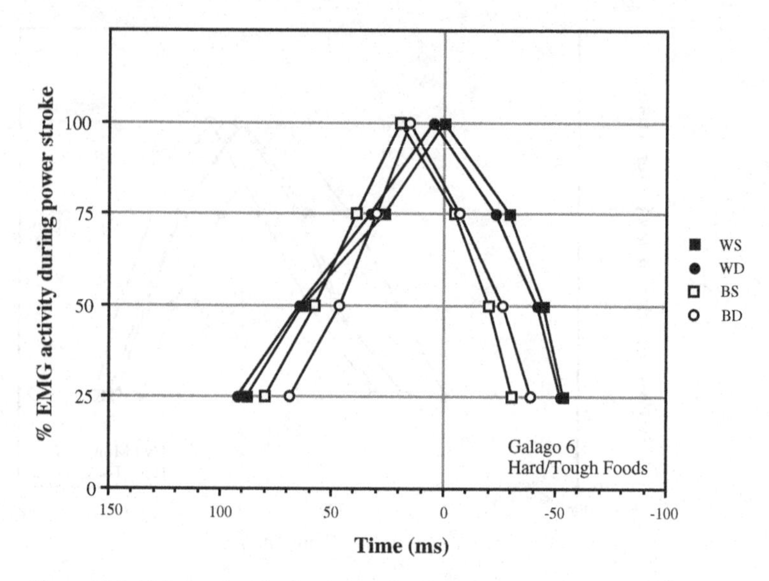

Figure 11.9. Mean values for the timing of average percent EMG activity of the deep and superficial masseters during the chewing of hard/tough foods in galago 6. Unlike baboons, macaques, and owl monkeys, in thick-tailed galagos the balancing-side deep and superficial masseters reach peak activity first. The vertical line indicates peak activity in the reference muscle, which unlike the situation in the above anthropoids, is the last muscle to reach peak activity in thick-tailed galagos. Symbols and abbreviations as in Figure 11.3 (from Hylander *et al.*, 2000).

the last muscle to reach peak activity is the balancing-side deep masseter. In contrast, for galagos, the first muscle to reach peak activity is the balancing-side deep although its timing is not significantly different from that of the balancing-side superficial masseter. For lemurs the first muscle to reach peak activity is the working-side deep masseter. Most importantly, in galagos and lemurs the last muscle to reach peak activity is the reference muscle, the working-side superficial masseter.

Note in Table 11.3 that for the anthropoids the mean values for the three variables listed are all positive and they decrease in value from left to right. Furthermore, with a single exception, the anthropoid values are all statistically significant ($P \leq 0.05$). In contrast, for galagos the mean values for the three variables in Table 11.3 are all negative, and they become less negative, i.e., they increase algebraically, from left to right. Two of these three mean values, however, are not statistically significant. Finally, for lemurs the first two variables are positive as in anthropoids, but the last is negative as in galagos. All three variables are statistically significant.

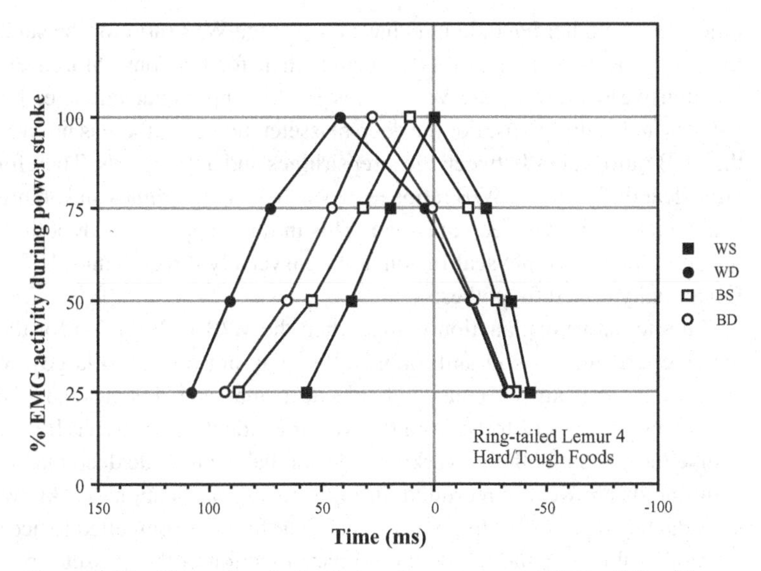

Figure 11.10. Mean values for the timing of average percent EMG activity of the deep and superficial masseters during the chewing of hard/tough foods in lemur 4. Similar to baboons, macaques, and owl monkeys, and unlike thick-tailed galagos, the working-side deep masseter reaches peak activity first. The vertical line indicates peak activity in the reference muscle, which, in contrast to the situation in the above anthropoids, is the last muscle to reach peak activity in ring-tailed lemurs (and thick-tailed galagos). Symbols and abbreviations as in Figure 11.3 (from Hylander *et al.*, 2000).

Discussion

W/B EMG ratios and testing the symphyseal fusion–muscle recruitment hypothesis

The symphyseal fusion–muscle recruitment hypothesis predicts that although anthropoids, galagos, and lemurs recruit about the same percentage of working-side masseter force, and therefore exhibit a similar percentage of working-side masseter EMG activity, anthropoids recruit much more balancing-side muscle force. Therefore, anthropoids should have W/B EMG ratios that are relatively small and similar to one another, whereas the strepsirhines should have W/B ratios that are relatively large. The data in Table 11.1 and Figure 11.5 support certain aspects of this prediction. For example, all species exhibit 55% to 75% of peak average EMG activity of their working-side masseters during forceful mastication. Moreover, galagos have the largest W/B ratios for both the superficial and deep masseters, and all anthropoids have relatively small W/B ratios. On the other hand, lemurs have an average W/B ratio for the superficial masseter that is actually slightly smaller than for baboons. Moreover, there is extensive overlap of the W/B ratios for the superficial masseter between galagos and anthropoids. For the deep masseter, however, there is no overlap of the W/B ratio values between the strepsirhines and anthropoids. Therefore, the only clear difference in W/B ratios between these strepsirhines and anthropoids is associated with the deep masseter. This in turn supports the hypothesis that for anthropoids symphyseal fusion and transversely directed muscle force are functionally linked (hypothesis 2).

An alternative explanation is that when the W/B EMG ratios for the deep masseter are small, as in anthropoids, this may not be due to large contributions of force from the balancing-side deep masseter, but instead is due to small contributions of force from the working-side deep masseter. If so, this of course means that both the working-side and balancing-side deep masseters in anthropoids are weakly recruited during forceful mastication. We know of no such data to support this hypothesis, and therefore are compelled to accept the more plausible hypothesis that the primary function of the masseter muscle is to produce bite force and movement during chewing and biting.

Thus, when primates engage in forceful chewing behaviors, small W/B EMG ratios are best interpreted as indicating relatively large contributions of force from the balancing-side masseter, just as large W/B EMG ratios are best interpreted as indicating relatively small contributions of force from the balancing-side masseter. Furthermore, the results of this EMG analysis are similar to the results of our recent analysis of mandibular corpus bone strains during forceful mastication. This analysis demonstrates that although peak working-side corpus strains are rather similar for macaques, galagos, and owl monkeys, the galagos have much smaller peak balancing-side corpus strains than macaques and owl monkeys. Thus, macaques and owl monkeys have small average W/B mandibular strain ratios whereas galagos have large W/B strain ratios, indicating that compared to galagos, anthropoids recruit more balancing-side muscle force during forceful mastication (Hylander *et al.*, 1998).

Finally, although not directly related to whether the masseter muscles are strongly recruited during forceful mastication, we have established that the magnitude of peak masseter EMG is intensely correlated with peak masseter force during the power stroke of mastication (Hylander and Johnson, 1989, 1993). This finding increases the likelihood that our W/B ratio data do indeed provide a good approximation of the recruitment of relative masseter force during mastication.

In summary, of the three symphyseal fusion hypotheses (hypotheses 1, 2, and 3), the data provide strong support for hypothesis 2, refute hypothesis 3, and seriously weaken hypothesis 1.

Differential firing patterns and the symphyseal fusion–muscle recruitment hypothesis

Previous work on macaques and baboons suggests that they share a common jaw-muscle firing pattern that causes their mandibles to experience wishboning during mastication (Hylander and Johnson, 1994). The working hypothesis predicts that owl monkeys also have this pattern whereas galagos and lemurs do not.

The data in Tables 11.2 and 11.3 indicate quite clearly that the balancing-side deep masseter firing pattern for galagos and lemurs is quite different from that for owl monkeys, macaques, and baboons. Furthermore, the firing pattern for owl monkeys is very similar to the one for baboons and macaques, i.e., in all three anthropoids the last muscle to reach peak activity is the balancing-side deep masseter. Thus, as illustrated in Figures 11.6–11.10, a major difference in firing patterns between the strepsirhines and anthropoids is related to the behavior of the balancing-side deep masseters. Similar to the data on W/B EMG ratios, the firing pattern data support the hypothesis that symphyseal fusion and transversely directed muscle force are functionally linked (hypothesis 2).

Wishboning and the balancing-side deep masseter

In 1994 we discussed the various external forces associated with wishboning of the macaque mandible during mastication (Hylander and Johnson, 1994). These forces (Fig. 11.11) are the laterally directed component to the bite force (F_b) and the opposite laterally directed component to the balancing-side jaw muscle force (F_{mb}). These two force components are probably not equal as residual force associated with the relaxing working-side jaw adductors (F_{mw}) also contributes to wishboning. An important aspect of wishboning is that it occurs late in the power stroke of mastication, and it is due mainly to increased recruitment of the balancing-side deep masseter at a time when force from the balancing-side superficial masseter and medial pterygoids (and also balancing-side lateral pterygoids?) have rapidly declined.[1]

[1] The lateral pterygoid muscles in primates appear to have a biphasic activity during mastication, i.e., they are active during both jaw opening and the power stroke. It has been hypothesized that the medially directed components of force from the balancing-side medial and lateral pterygoids prevents

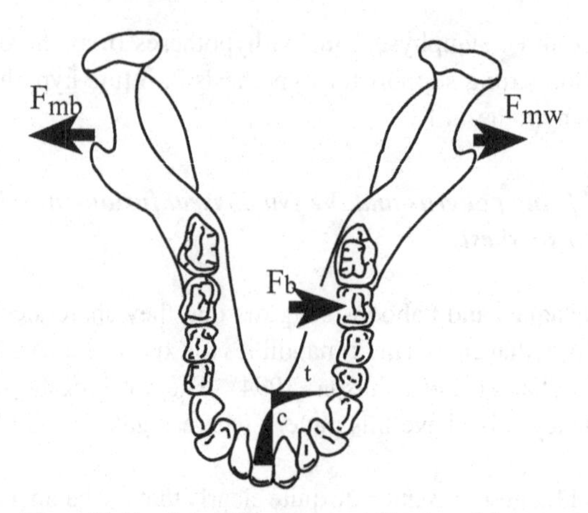

Figure 11.11. Lateral bending or wishboning of the macaque mandibular symphysis. F_b, transverse component to the bite force; F_{mb}, transverse component to the balancing-side jaw-closing muscle force; F_{mw}, transverse component to the working-side jaw-closing muscle force. These three forces wishbone the mandible. The length of the arrows does not indicate the relative magnitude of bite and muscle forces. The darkly shaded areas along the symphysis indicate the distribution of bending stress for a symmetrical curved beam: *t*, tensile bending stress; *c*, compressive bending stress. In contrast to the situation for straight beams, note that the distribution of bending stress is markedly nonlinear. The tensile stress along the surface of the lingual aspect of the symphysis greatly exceeds the compressive stress along the surface of the labial aspect of the symphysis (from Hylander and Johnson, 1994).

The EMG data provide strong evidence for the hypothesis that thick-tailed galagos and ring-tailed lemurs do not significantly wishbone their mandibles during mastication. This is because of two main factors. First, galagos and lemurs recruit relatively low levels of force from their balancing-side deep masseter. Whereas the deep masseter W/B ratio is near unity for anthropoids, it is three to four times larger for lemurs and galagos (2.7 to 4.4., respectively). Second, and perhaps most importantly, unlike anthropoids, galagos and lemurs do not exhibit the late firing of the balancing-side deep masseter. That is, they do not have the wishboning firing pattern. Thus, our EMG data strongly support the hypothesis that symphyseal fusion and transversely directed muscle force in anthropoids are functionally linked. These data do not, however, help us to determine whether symphyseal fusion is an adaptation to strengthen the

the laterally directed component of force of the balancing-side superficial masseter from causing wishboning of the symphysis. This may be why wishboning ordinarily does not occur earlier in the power stroke, a time when all of these muscles are highly active. Instead, wishboning occurs late, at a time when force from the balancing-side medial and lateral pterygoids and superficial masseter are rapidly decreasing and the balancing-side deep masseter force is rapidly increasing or has peaked (Hylander and Johnson, 1994).

symphysis, or to stiffen the symphysis (Hylander *et al.*, 2000; Lieberman and Crompton, 2000). Recently, however, Hogue and Ravosa (2001) and Vinyard *et al.* (in press b) provide a compelling argument that supports the strengthening hypothesis.

Muscle recruitment and firing of the masseter in strepsirhines

Compared to anthropoids, the W/B ratios for lemurs and galagos demonstrate that these strepsirhines recruit much less relative force from their balancing-side deep masseter. Furthermore, these data also suggest that lemurs recruit relatively more balancing-side deep masseter force than do galagos (deep masseter W/B ratio = 4.4 and 2.4 for galagos and lemurs, respectively). This finding is interesting simply because Beecher's data indicate that although the symphysis of *Lemur catta* is unfused, it is much stiffer and stronger than is the unfused symphysis of *Otolemur crassicaudatus* (Beecher, 1977). This leads us to speculate that the differences in mechanical characteristics of the symphysis of ring-tailed lemurs and thick-tailed galagos are functionally linked to differences in muscle force recruitment patterns. Thus, compared to thick-tailed galagos, perhaps the stiffer and stronger symphysis of ring-tailed lemurs is better able to counter increased balancing-side muscle force during mastication.

There are also some interesting differences between galagos and lemurs in the firing patterns of their deep masseters. Our data indicate that unlike nonhuman anthropoids, galagos and lemurs do not have the late firing pattern of the balancing-side deep masseter (Hylander *et al.*, 2000, 2002). Furthermore, galagos do not have the early firing pattern of the working-side deep masseter (Figs. 11.3 and 11.9) (Hylander *et al.*, 2000). These data initially led Ravosa *et al.* (2000) to speculate that the deep and superficial masseters of strepsirhines are tightly coupled in their firing patterns.

Our data on ring-tailed lemurs, however, demonstrate conclusively that the firing patterns of superficial and deep masseters in strepsirhines are uncoupled. In ring-tailed lemurs the working-side deep and superficial masseters do not fire simultaneously as the working-side deep masseter reaches peak activity very early during jaw closing, and the working-side superficial masseter reaches peak activity late in the power stroke (Hylander *et al.*, 2002). Moreover, in lemurs the balancing-side deep and superficial masseters also do not fire simultaneously as the balancing-side deep masseter reaches peak activity early during jaw closing, well before the balancing-side superficial. Finally, recent work on sifakas demonstrates that the superficial and deep masseters also fire asynchronously on both the working and balancing sides during mastication (Hylander *et al.*, 2003).

Conclusions

The functional morphology of the craniofacial region of primates continues to attract a significant amount of research interest. One problem that has received a considerable amount of attention is related to the adaptive significance of symphyseal fusion in mammals. This is a particularly interesting problem as symphyseal fusion has evolved independently in many different mammalian lineages. In spite of many morphological and comparative studies, there is little agreement as to the ultimate reasons for symphyseal fusion. Therefore, we have focused our research on the more immediate or proximate factors that can be plausibly linked to the evolution of symphyseal fusion, and thus to determine if certain types of biomechanical situations or loading patterns are associated with symphyseal fusion.

In this study we tested several hypotheses about how balancing-side jaw-adductor muscles are recruited in primates, and in turn are possibly linked to symphyseal fusion. These hypotheses predict that anthropoids have W/B EMG ratios that are relatively small and similar to one another, and strepsirhines have W/B ratios that are relatively large.

Our data indicate that galagos have the largest W/B ratios for both the superficial and deep masseters, and anthropoids have relatively small W/B ratios. On the other hand, lemurs have W/B ratios for the superficial masseter that are actually smaller than those for baboons, although their W/B ratios for the deep masseter are relatively large. Thus, the only sharp distinction in W/B EMG ratios between thick-tailed galagos and ring-tailed lemurs as compared to anthropoids is associated with the deep masseter. That is, during mastication the balancing-side deep masseter of anthropoids is much more active than is the balancing-side deep masseter of thick-tailed galagos and ring-tailed lemurs. Furthermore, the analysis of masseter firing patterns indicates that whereas baboons, macaques, and owl monkeys possess the deep masseter wishboning firing pattern, galagos and lemurs do not. Thus, the W/B ratios and firing patterns of the deep masseter support the hypothesis that symphyseal fusion and transversely directed muscle force are functionally linked in anthropoids.

Acknowledgments

This study was supported by the Department of Biological Anthropology and Anatomy, Duke University. It was also supported by research grants from NIH (MERIT Award DE04531 to WLH; DE05595 to MJR and WLH; DE05663 to CEW and WLH) and NSF (SBR-9420764 and BCS-0138565 to WLH; BNS-91-00523 to CFR and WLH). Thanks to Dr. Don Schmechel for providing us access to the galagos in his care, and to the veterinary staff at the Duke University Primate Center. This is Duke University Primate Center publication #762.

References

Anapol, F. and Herring, S. W. (2000). Ontogeny of histochemical fiber types and muscle function in the masseter muscle of miniature swine. *Amer. J. Phys. Anthropol.*, **112**, 595–613.

Beecher, R. M. (1977). Function and fusion at the mandibular symphysis. *Amer. J. Phys. Anthropol.*, **47**, 325–336.

(1979). Functional significance of the mandibular symphysis. *J. Morphol.*, **159**, 117–130.

Daegling, D. J. and Hylander, W. L. (2000). Experimental observations, theoretical models, and biomechanical inference in the study of mandibular form. *Amer. J. Phys. Anthropol.*, **112**, 541–551.

Dechow, P. C. and Hylander, W. L. (2000). Elastic properties and masticatory bone stress in the macaque mandible. *Amer. J. Phys. Anthropol.*, **112**, 553–574.

Herring, S. W. (1993). Epigenetic and functional influences on skull growth. In: *The Skull. Vol. 1: Development*, ed. J. Hanken and B. K. Hall. Chicago, IL: University of Chicago Press. pp. 153–206.

Herring, S. W. and Teng, S. (2000). Strain in the braincase and its sutures during function. *Amer. J. Phys. Anthropol.*, **112**, 575–593.

Hogue, A. S. and Ravosa, M. J. (2001). Transverse masticatory movements, occlusal orientation, and symphyseal fusion in selenodont artiodactyls. *J. Morphol.*, **249**, 221–241.

Hylander, W. L. (1977). In vivo bone strain in the mandible of *Galago crassicaudatus*. *Amer. J. Phys. Anthropol.*, **46**, 309–326.

(1984). Stress and strain in the mandibular symphysis of primates: a test of competing hypotheses. *Amer. J. Phys. Anthropol.*, **64**, 1–46.

Hylander, W. L. and Johnson, K. R. (1989). The relationship between masseter force and masseter electromyogram during mastication in the monkey *Macaca fascicularis*. *Arch. Oral Biol.*, **34**, 713–722.

(1993). Modelling relative masseter force from surface electromyograms during mastication in non-human primates. *Arch. Oral Biol.*, **38**, 233–240.

(1994). Jaw muscle function and wishboning of the mandible during mastication in macaques and baboons. *Amer. J. Phys. Anthropol.*, **94**, 523–547.

(1997). In vivo bone strain patterns in the zygomatic arch of macaques and the significance of these patterns for functional interpretations of craniofacial form. *Amer. J. Phys. Anthropol.*, **102**, 203–232.

Hylander, W. L., Johnson, K. R., and Crompton, A. W. (1987). Loading patterns and jaw movements during mastication in *Macaca fascicularis*: a bone-strain, electromyographic, and cineradiographic analysis. *Amer. J. Phys. Anthropol.*, **72**, 287–314.

(1992). Muscle force recruitment and biomechanical modeling: an analysis of masseter muscle function in *Macaca fascicularis*. *Amer. J. Phys. Anthropol.*, **88**, 365–387.

Hylander, W. L., Ravosa, M. J., Ross, C. F., and Johnson, K. R. (1998). Mandibular corpus strain in primates: further evidence for a functional link between symphyseal fusion and jaw-adductor muscle force. *Amer. J. Phys. Anthropol.*, **107**, 257–271.

Hylander, W. L., Ravosa, M. J., Ross, C. F., Wall, C. E., and Johnson, K. R. (2000). Symphyseal fusion and jaw-adductor muscle force: an EMG study. *Amer. J. Phys. Anthropol.*, **112**, 469–492.

Hylander, W. L., Vinyard, C. J., Wall, C. E., Williams, S. H., and Johnson, K. R. (2002). Recruitment and firing patterns of jaw muscles during mastication in ring-tailed lemurs. *Amer. J. Phys. Anthropol.*, Suppl. **34**, 88.

(2003). Convergence of the wishboning jaw-muscle activity pattern in anthropoids and strepsirhines: the recruitment and firing patterns of jaw muscles in *Propithecus verreauxi*. *Amer. J. Phys. Anthropol.*, Suppl. **35**, 120.

Lieberman, D. E. and Crompton, A. W. (2000). Why fuse the mandibular symphysis? A comparative analysis. *Amer. J. Phys. Anthropol.*, **112**, 517–540.

Ravosa, M. J. (1996). Mandibular form and function in North American and European Adapidae and Omomyidae. *J. Morphol.*, **229**, 171–190.

Ravosa, M. J. and Hylander, W. L. (1994). Function and fusion of the mandibular symphysis in primates: stiffness or strength? In: *Anthropoid Origins*, ed. J. G. Fleagle and R. F. Kay. New York, NY: Plenum Press. pp. 447–468.

Ravosa, M. J. and Simons, E. L. (1994). Mandibular growth and function in *Archaeolemur*. *Amer. J. Phys. Anthropol.*, **95**, 63–76.

Ravosa, M. J., Noble, V. E., Johnson, K. R., Kowalski, E. M., and Hylander, W. L. (2000). Masticatory stress, orbital orientation, and the evolution of the primate postorbital bar. *J. Hum. Evol.*, **38**, 667–693.

Ross, C. F. and Hylander, W. L. (1996). In vivo and in vitro bone strain in the owl monkey circumorbital region and the function of the postorbital septum. *Amer. J. Phys. Anthropol.*, **101**, 183–215.

(2000). Electromyography of the anterior temporalis and masseter muscles of owl monkeys (*Aotus trivirgatus*) and the function of the postorbital septum. *Amer. J. Phys. Anthropol.*, **112**, 455–468.

Scapino, R. P. (1981). Morphological investigations into functions of the jaw symphysis in carnivorans. *J. Morphol.*, **167**, 339–375.

Spencer, M. (1998). Force production in the primate masticatory system: electromyographic tests of biomechanical hypotheses. *J. Hum. Evol.*, **34**, 25–54.

Türker, K. S. (2002). Reflex control of human jaw muscles. *Crit. Rev. Oral Biol. Med.*, **13**, 85–104.

van Eijden, T. M. G. J. and Turkawski, S. J. J. (2001). Morphology and physiology of masticatory muscle motor units. *Crit. Rev. Oral Biol. Med.*, **12**, 76–91.

Vinyard, C. J., Williams, S. H., Wall, C. E., Johnson, K. R., and Hylander, W. L. (2001). Deep masseter recruitment patterns during chewing in callitrichids. *Amer. J. Phys. Anthropol.*, Suppl. **32**, 156.

Vinyard, C. J., Ravosa, M. J., Wall, C. E., Williams, S. H., Johnson, K. R., and Hylander, W. L. (in press a). Jaw-muscle function and the origins of primates. In: *Primate Origins and Adaptations: a Mulitdisciplinary Perspective*, ed. M. J. Ravosa and M. Dagosto. New York, NY: Kluwer.

Vinyard, C. J., Williams, S. H., Wall, C. E., Johnson, K. R., and Hylander, W. L. (in press b). Jaw-muscle electromyography during chewing in Belanger's treeshrews (*Tupaia belangeri*). *Amer. J. Phys. Anthropol.*

Wall, C. E., Vinyard, C. J., Johnson, K. R., Williams, S. H., and Hylander, W. L. (2002). A preliminary study of phase II occlusal movements during chewing in *Papio*. *Amer. J. Phys. Anthropol.*, Suppl. **34**, 161.

Weijs, W. A. (1994). Evolutionary approach of masticatory motor patterns in mammals. In: *Biomechanics of Feeding in Vertebrates*, ed. V. L. Bels, M. Chardon, and P. Vandewalle. Berlin: Springer-Verlag. pp. 282–320.

Williams, S. H., Vinyard, C. J., Wall, C. E., and Hylander, W. L. (2003). Symphyseal fusion in anthropoids and ungulates: a case of functional convergence? *Amer. J. Phys. Anthropol.*, Suppl. **35**, 226.

12 *Hind limb drive, hind limb steering? Functional differences between fore and hind limbs in chimpanzee quadrupedalism*

YU LI
University of Bristol

ROBIN HUW CROMPTON, WEIJIE WANG, RUSSELL SAVAGE, MICHAEL M. GÜNTHER
University of Liverpool

Introduction

There are relatively few biomechanical studies of the quadrupedal walking of primates, especially that of the apes. This is almost certainly because of the extreme difficulty of obtaining records of voluntary (and hence naturalistic) behavior, of even the most common elements of a species' locomotor repertoire, since only that which happens to occur on the measuring equipment used to gather force data (and/or within the field of view of the equipment used to gather motion data) can be analyzed. For this reason, and because of the additional danger of the subjects (or the weather!) damaging equipment, the literature primarily contains studies of trained or guided animals (most often subadult), which are usually based on small sample sizes.

The primates as a whole are generally agreed to be hind-limb dominated (e.g., Martin, 1990), and Rollinson and Martin (1981) used this phenomenon to explain the diagonal gaits which are another remarkable characteristic of primate quadrupedalism. However, their study did not consider the force exerted during walking. One of the first comprehensive studies to do so, by Kimura *et al.* (1979), compared the chimpanzee and orangutan to three nonhominoid species. Their findings led them to propose that while both primates and nonprimates primarily use the forelimbs to control the direction of movement (i.e., they are both "forelimb steered"), primates bear a greater proportion of vertical

Shaping Primate Evolution, ed. F. Anapol, R. Z. German, and N. G. Jablonski. Published by Cambridge University Press.

forces on the hind limb and are "hind limb driven." In contrast, bearing more force on the forelimbs, other mammals are "forelimb driven." This argument was further elaborated by Kimura (1992) with special reference to high-speed locomotion.

Reynolds (1985a) addressed himself to explaining the phenomenon of hind limb domination with respect to vertical force, but suggested that the distinction in force-bearing pattern is not the result of a change in the position of the center of gravity, as might follow from Rollinson and Martin (1981). Reynolds (1985b) further distinguished force characteristics of the "inside" and "outside" limbs during chimpanzee quadrupedalism.[1] Demes *et al.* (1994) followed this distinction, but disputed the "hind limb drive" proposition of Kimura and colleagues, on the grounds that their own study, based on a larger and more diverse study sample, appeared to show that primates and nonprimates share a similar pattern of sagittal (fore–aft) forces: forelimbs account for a large proportion of braking forces but hind limbs a large proportion of accelerative forces. Less dispute exists with respect to Kimura and colleagues' (1979) other concept that primates are "forelimb-steered," although some authors regard it as "right for the wrong reasons." For example, Schmitt (1999) conjectures that a critically important role for the forelimb in directional changes (steering) may be inferred from the reduction of forelimb supporting forces in primates.

This chapter reports a study of common chimpanzee (*Pan troglodytes*) quadrupedal gait. As a nonprimate comparison, equivalent data were also collected from a police dog under the verbal control of its trainer, who ensured that the dog only walked. The focus is on the functional difference between the fore and hind limbs in chimpanzees, in comparison to the dog.

Materials and methods

The coordinate system used in this study is based on that of the Kistler 9281B force plate, used for all kinetic recordings. The vertical is defined as axis Z, pointing downwards, and the direction of locomotion of the subject is defined as Y. The remaining axis, perpendicular to both Z and Y, and satisfying a right-hand system, is then X. In the later discussion, vertical force and Z force, and sagittal direction or Y direction, therefore have the same meaning.

[1] Some explanation of these terms is necessary. In essence, great apes do not tend to make co-linear footfalls, so that if the left hind limb, for example, swings to cross in front of the left forelimb from its lateral side, it will be called the "outside" hind limb, and the left forelimb will be known as the "inside" forelimb. The right hind limb will then tend to cross the right forelimb from the medial side, making it the "inside" hind limb.

Kinetic data for chimpanzees were obtained from subjects on a moated enclosure at Chester Zoo (the North of England Zoological Society). The Kistler 9281B force plate was mounted flush with the ground surface, on a subframe set into in a one-tonne concrete block, which intersected the path used by chimpanzees to enter the enclosure from their inside accommodation. Data were fed directly to hard disk via an AD converter card running under DIA/DAGO (Aachen: GfS), and data were sampled under software control at 250 Hz. Two genlocked Panasonic F10 VHS video cameras were set to record kinematic data as described in Li *et al.* (1996), giving a sample rate of 50 fields per second. Observers made no attempt to affect the timing, direction, or mode of the subjects' locomotion: therefore all records are casual, random, and voluntary. The majority of the forces recorded are from locomotion along the long axis of the force plate, so that sagittal force is represented by coordinate Y.

The number of potential subjects is at least 20; but of the total population a couple of individuals in the group are arthritic or otherwise do not move in a normal manner. No visually abnormal records were retained. Although these subjects are kept in captivity, they are free to range on a large moated island enclosure, and all the records reported here resulted from voluntary locomotion. No contact with the animals occurs. It was not possible to obtain body weights from these animals, or to identify individuals from the video records, so we have employed only dimensionless parameters.

The police dog on which our nonprimate comparison is based (a second dog provided only limited data as it was unwilling to walk over the runway or force plate) was of the Alsatian (German shepherd) breed and had a body mass of 18 kg. The data were collected at the Merseyside Police animal training grounds, and the dogs were highly trained to respond to verbal commands. In this case, the force plate was set flush with a wooden pathway, and two synchronized Kodak Ektapro Motion Analyzer cameras were set orthogonally in front and to the side of the pathway respectively. The sample frequency of both cameras and force plate was set at 500 Hz. Because we have data for only one dog, our comparisons of chimpanzee data with the dog were restricted as far as possible to cases where differences were qualitative.

All force records were scrutinized for any tendency for net acceleration or deceleration over the gait cycle. Fully two-thirds of the recorded data had to be excluded from analysis for this reason. Kinetic data for chimpanzee were smoothed with a digital filter, with a setting of $f_s/f_c = 7$, so that the cut-off frequency (f_c) is 250/7, or about 36, where 250 is the sample rate (f_s) (see Winter, 1979). A Fourier analysis method (Alexander and Jayes, 1980) was then applied to the force-plate records so that the curves could be expressed in the form of numerical parameters. The Fourier coefficients of fore and hind limbs were then compared statistically. The direction cosine, the traces of the center of pressure, and the free moments (couples) about the vertical axis were also calculated and compared between the fore and hind limbs.

As indicated above, due to the lack of exact data for body masses of the chimpanzees, in the data analysis only dimension-free parameters such as the ratio of Fourier coefficients and direction cosines, or variables unaffected by weight such as the trace of the center of pressure, are reported. For example, in the Fourier analysis, all the coefficients are expressed as a ratio in which Za_1 (coefficient a_1 of the vertical force) is used as the dominator. For the free moment about the vertical axis, the peak vertical force is used to correct the body-weight factor.

Given the relatively large area of the force plate (40 × 60 cm), recording of both fore and hind limbs in a single record was a reasonably likely event, especially since hind-limb contact with the substrate was usually made not far away from that of the ipsilateral forelimb. Since hind-limb contact occurs when the ipsilateral forelimb is still in its stance phase, overlap will occur. If more than two limbs touched the force plate at the same time, the sequence was discarded. However, although double contact records have their disadvantages, they provide unique information concerning forelimb–hind-limb interactions during the gait cycle, and are particularly important in parameters such as direction cosines and free moments.

It was generally observed that at the stage when the hind limb had made contact with the force plate and its vertical force had begun to rise, that of the forelimb was already declining sharply towards zero. Thus it was possible to separate the two partly overlaid curves of double contact records in a manner similar to the procedure of Demes *et al.* (1994). However, the same is not the case for sagittal forces, since during the period of overlap, they operate in the opposite direction for fore and hind limbs. Sagittal forces from double contacts were therefore excluded from analysis.

While a detailed treatment of Alexander and Jayes' (1980) technique for generating Fourier coefficients from force curves should be sought in their publication, a very brief description may be appropriate here. Figure 12.1 shows a typical sagittal force curve. The value of the independent variable in the analysis, time, is defined as a unit, which represents the whole period of stance. This time unit is divided into six equal intervals by points 0, 1/6, 2/6, 3/6, 4/6, 5/6 and 1 (i.e., 100%) along the curve. As indicated in Figure 12.1, each of these data points has its dependent value, $f(t)$. The coordinates of these points are therefore $\{0, f(0)\}$; $\{1/6, f(1/6)\} \ldots \{1, f(1)\}$. For any force or force-related parameters in locomotion, at the start and end of stance, we have $f(0) = f(1) = 0$. The limited numbers of points used here represents the curve under study,

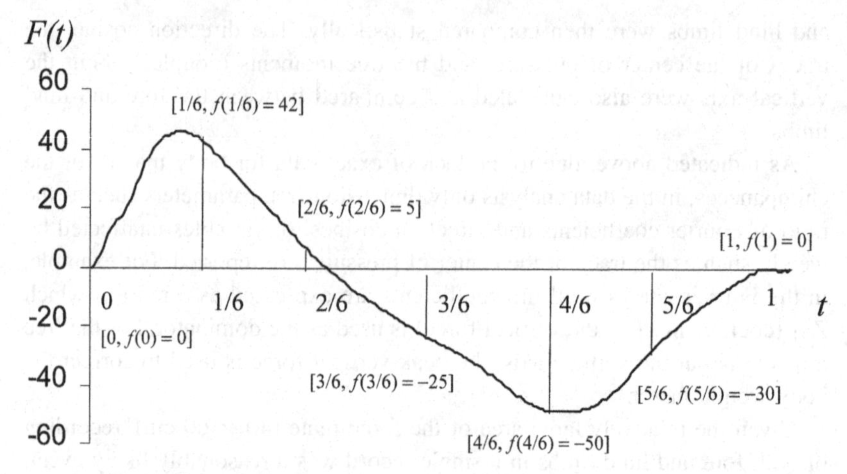

Figure 12.1. Construction of the Fourier coefficients. Horizontal axis represents time, the vertical any kinetic variables (in this case, force).

and the Fourier coefficients of the curve are:

$$a_1 = [f(1/6) + \sqrt{3}f(2/6) + 2f(3/6) + \sqrt{3}f(4/6) + f(5/6)]/6 \quad (12.1)$$

$$a_3 = [-f(1/6) + f(3/6) - f(5/6)]/3 \quad (12.2)$$

$$a_5 = [f(1/6) - \sqrt{3}f(2/6) + 2f(3/6) - \sqrt{3}f(4/6) + f(5/6)]/6 \quad (12.3)$$

$$b_2 = \sqrt{3}[-f(1/6) - f(2/6) + f(4/6) + f(5/6)]/6 \quad (12.4)$$

$$b_4 = \sqrt{3}[f(1/6) - f(2/6) + f(4/6) - f(5/6)]/6 \quad (12.5)$$

All five coefficients are a linear combination of the function values, $f(t)$ of the five points. Value a_1 represents the weighted mean of the data points, in which the middle point (3/6) has the heaviest weight (1/3), points at 1/6 and 5/6 have a weight of 1/6, and so on. If the curve changes its sign between [0,1] however, a_1 measures curve asymmetry about the point (0.5, 0). If the curve changes sign around $t = 0.5$, as indeed the sagittal force often does, coefficient b_2 is proportional to the simple mean of the absolute value of the curve. If a function has the same sign over its domain [0,1], b_2 represents asymmetry about the vertical line $t = 0.5$. Coefficient a_3 only involves three points, dividing the curve into three parts, and contrasts the values in the middle part with those in the other two. Parameters b_4 and a_5 express high-frequency characteristics of the curve, which are less important in this particular study and are therefore not discussed here: they are included for completeness.

The fact that a force or force-like curve (i.e., a curve defined in [0,1], $f(0) = f(1) = 0$, with a low frequency) can be represented by a Fourier function provides us a simple and practical tool for analysis. Compared with the general form of the Fourier equation, $f(t)$ has only odd coefficients for sine terms and even coefficients for cosine terms. The other coefficients (including constant term a_0) are all zero because of the properties of the force and the way in which this Fourier analysis is arranged. This fact greatly simplifies the curve and provides an objective, numerical means of analyzing forces and moments.

In order to give a quantitative expression of the direction along which a force is exerted by a limb against the ground, regardless of its magnitude, we have calculated direction cosines. The direction cosine (*DC*) is defined as a ratio of the force along a certain axis over the force vector. For example, *DCx*, the direction cosine of the X force is:

$$DCx = \frac{Fx}{\sqrt{Fx^2 + Fy^2 + Fz^2}} \tag{12.6}$$

DCy and *DCz* can be defined in a similar way. By definition, $DCx^2 + DCy^2 + DCz^2 = 1$.

Results

Gait pattern

The dog studied in our experiments was controlled verbally by its handler to ensure that it did not run. Its gait was a diagonal-sequence diagonal-couplets walk, but we do include one or two records of what amounted to a walking trot, *sensu* Hildebrand (1966). The walking gaits observed for chimpanzees were primarily diagonal-sequence, diagonal-couplets (*sensu* Hildebrand, 1966; Rollinson and Martin, 1981). For both chimpanzees and the dog, kinetic data (see the force/direction cosine data below) show that the hind limb landed before the same-side forelimb lifted. Kinematic data also show that the duty factors of the hind limbs were larger than 0.5. However, particularly at slow speeds, forelimb and opposite hind limb strikes were sometimes close enough in sequence as to be almost indistinguishable.

The kinetic records obtained in this study for chimpanzees fall into three categories. They include 23 single forelimb contacts, 22 single hind-limb contacts, and 31 fore- and hind-limb double contacts. The records are mostly from a continued "forward cross type" gait (Kimura *et al.*, 1979), also referred to as a diagonal gait (Rollinson and Martin, 1981), except for a small number which result from slow running, and in which they galloped. There are 8 running (galloping) records for forelimbs and 11 for hind limbs. Chimpanzees occasionally engage in bipedal or tripedal walking. In the latter, both hind limbs and the "outside" (Reynolds, 1985b) forelimb were nearly always used to support

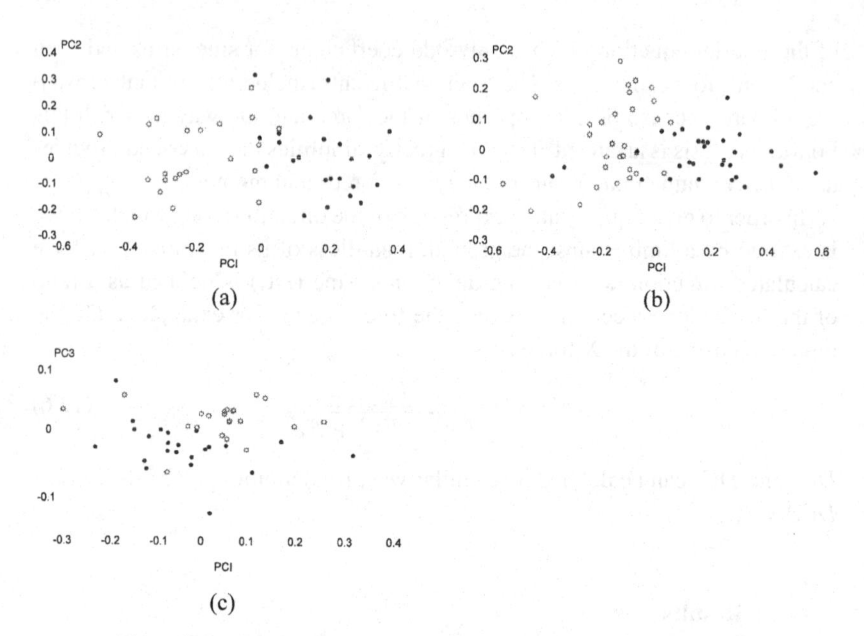

Figure 12.2. Principal components of fore- and hind-limb force in chimpanzees. Dark circles represent forelimb records; stars, hind-limb records. (a) Fourier coefficients of 23 fore- and 22 hind-limb vertical and sagittal force records; (b) Fourier coefficients of 31 fore- and 31 hind-limb vertical force records; (c) Fourier coefficients of 23 fore- and 22 hind-limb sagittal force records.

the body, which leant towards the side of the "inside" hind limb, leaving the "inside" forelimb free to carry objects. Only on one occasion (out of eight) did a chimpanzee carry an object with its "outside" forelimb: this led to a running rather than walking gait.

Force

Figure 12.2 shows a plot of the principal components of the Fourier coefficients of the force curves (in each case, the first three PCs count for over 85% of all variation). Figures 12.2a and 12.2c use separated force records (23 forelimb and 22 hind-limb records). While Figure 12.2a includes nine coefficients (Ya_1, Ya_3, Ya_5, Yb_2, Yb_4, Za_3, Za_5, Zb_2, Zb_4), Figure 12.2c only includes the five Y coefficients. As mentioned above, Za_1 is used with other coefficients to construct ratios. For example, when discussing Za_3, we are actually referring to the ratio Za_3/Za_1. Figure 12.2b has 62 records from the 31 double contacts, but only four Z force coefficients are included. In all three plots, the

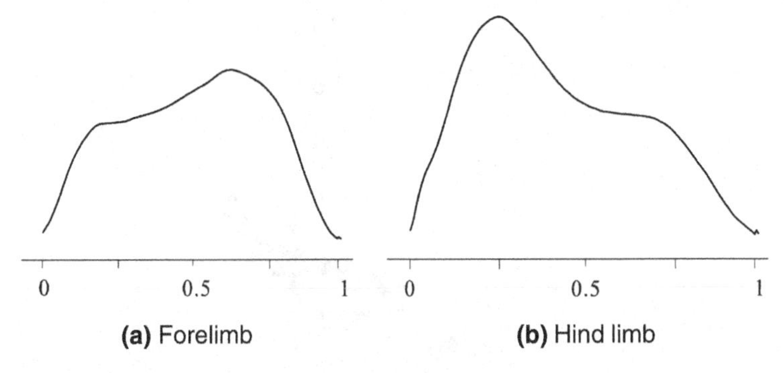

Figure 12.3. Typical vertical forces in chimpanzee quadrupedal walking. Vertical axis, force in N; horizontal axis, time as a proportion of the gait cycle.

fore- and hind-limb forces are well separated, showing that forelimb forces differ from those of hind limbs in both vertical and sagittal components. In all three cases, the variable with largest loading, on the axis that separates fore and hind limbs, is either Zb_2, or Yb_2 in cases where Z coefficients are not included. Following Alexander and Jayes (1980), a ratio constructed using a_1 and b_2 will indicate the degree of symmetry of a curve. For vertical forces, where coefficient a_1 (Za_1) is by definition always positive, a positive b_2 indicates that the curve peaks to the right, and vice-versa, zero values indicating curve symmetry.

A pair of typical force curves for the fore and hind limbs is shown in Figure 12.3. This figure indicates that the forelimb has the characteristic of $b_2 > 0$, while it can be seen from Figure 12.2 that for the hind limb $b_2 < 0$. These results suggest that in chimpanzee quadrupedal walking, vertical force is applied and withdrawn in a different manner by the fore and hind limbs. Forelimbs apply force gradually to the support, and force reaches a peak only after mid-stance, or even shortly before contact is lost. In contrast, hind limbs rapidly increase vertical force, so that it peaks shortly after touchdown; the force then declines gradually as the subject progresses forwards.

Figure 12.2c shows the plot of the first and third principal components (PC1 and PC3) of the Fourier coefficients of the Y force. The best separation occurs across a parallel with PC3 = −PC1. Coefficient a_1 loads heaviest on PC1 but a_3 and b_2 as well as a_1 load heavily on PC3, complicating interpretation. In order to better demonstrate the characteristics of the sagittal force, two new variables were created, which are plotted against each other in Figure 12.4. The first is the ratio of two force integrals on time, the deceleration section of the curve (D) over the acceleration section (A), known as the impulse. The second is the

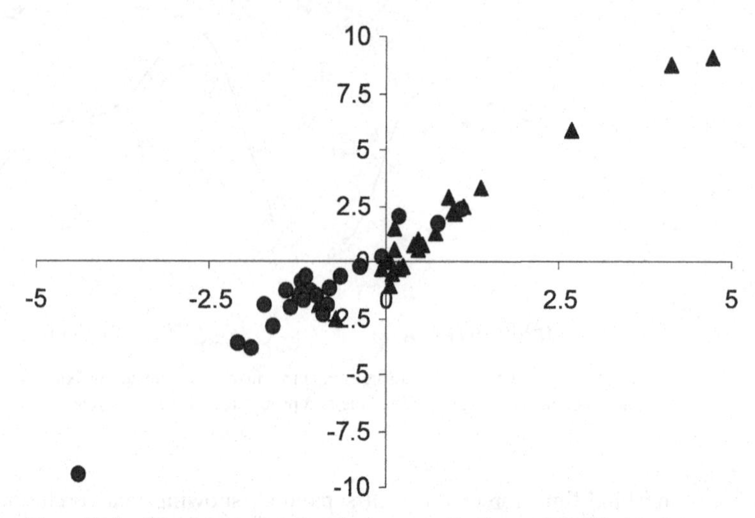

Figure 12.4. Distinctions in sagittal forces exerted by fore and hind limbs of chimpanzees. The horizontal axis represents the ratio of time of deceleration over time of acceleration; vertical axis, the ratio of impulse of the force (force integrals on time), the deceleration component (D) over the acceleration component (A). Dark circles show hind-limb records; triangles, forelimb records. It is apparent from the spread of the symbols against the vertical axis that the chimpanzees were neither decelerating nor accelerating as they crossed the forceplate. Both time ratio and force impulse ratio have been logged.

ratio of deceleration time over acceleration time. This figure shows clearly that the forelimb tends to have a relatively large deceleration component, and small acceleration component, with respect both to duration and force-time integral, and vice-versa for the hind limb.

Effects of speed on the force curve

Table 12.1 is a list of the correlation coefficients between the ratio of Fourier coefficients and duration of stance, and this duration of stance is inversely related to speed of locomotion. Among the most statistically significant values in Table 12.1 are those for the correlation of speed with coefficient Za_3, both fore and hind limb giving significance levels of $P < 0.01$. Alexander and Jayes (1980) explained that a negative value for Za_3 indicates a depression in the vertical force curve at mid-stance, while a positive value indicates a peak. In Figure 12.5a, for the forelimb Za_3 is negatively correlated with time as a proportion of gait cycle (period in stance), and all the data points form a unified entity. In contrast, Figure 12.5b, for the hind limb indicates no single simple

Table 12.1. *Correlation coefficients between Fourier coefficients and support duration*

	Y		Z	
	Forelimb ($n = 23$)	Hind limb ($n = 22$)	Forelimb ($n = 54$)	Hind limb ($n = 53$)
a_1	0.2309	−0.3755	–	–
a_3	0.2342	−0.3306	−0.4906**	−0.5132**
a_5	−0.1237	0.1577	−0.0779	−0.0074
b_2	−0.1900	−0.1459	−0.0534	0.1371
b_4	−0.4742*	0.0186	0.3293	0.1825

*$P < 0.05$
**$P < 0.01$

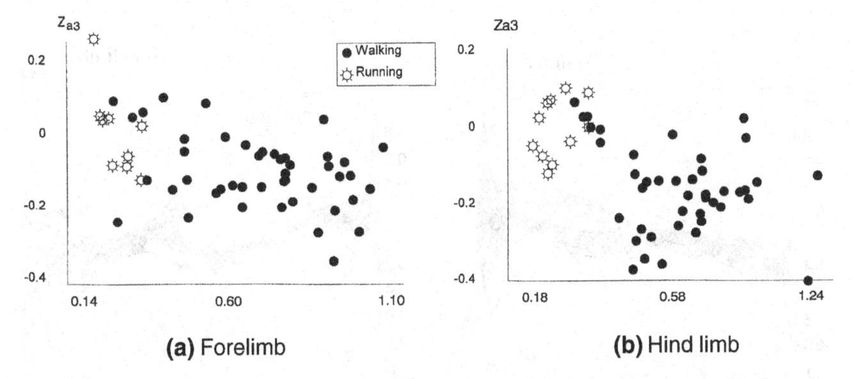

Figure 12.5. Effects of speed on shape of vertical force in chimpanzees. Horizontal axis, time as a proportion of the gait cycle; vertical axis, value of coefficient Za_3. Dark circles represent walking, stars running.

relationship of Za_3 and stance period. Rather, to the left we observe a high-speed band where data points, mainly from running, show little indication of correlation between the two axes. However, further to the right, middle- and low-speed data blocks do show positive correlation ($r = 0.398$, $P < 0.02$), if a probable outlier at the bottom-right of the graph is excluded.

Direction cosine of the ground reaction forces

Figure 12.6 shows the sagittal direction cosines (*DCy*) of the forelimb and hind limb. It can be seen that the direction cosines of the forelimb fall within a narrower band of variation. In the case of the forelimb, records show a steady

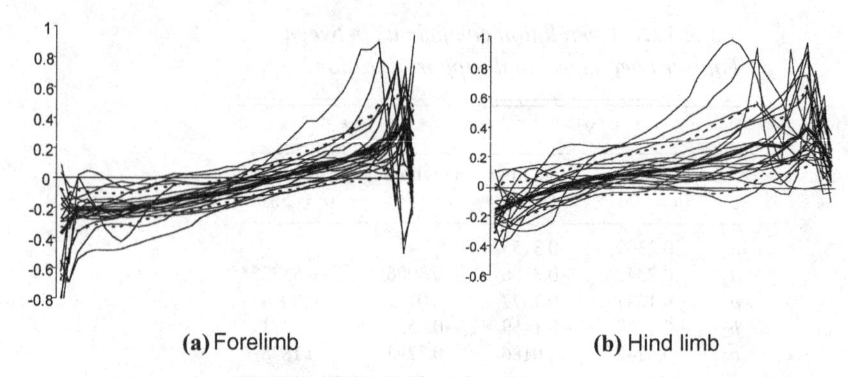

Figure 12.6. Direction cosine of sagittal force in chimpanzees. Thin curves show individual records. The thick dark line shows the mean and the dotted lines ± one standard deviation. Negative numbers are decelerative and positive are accelerative.

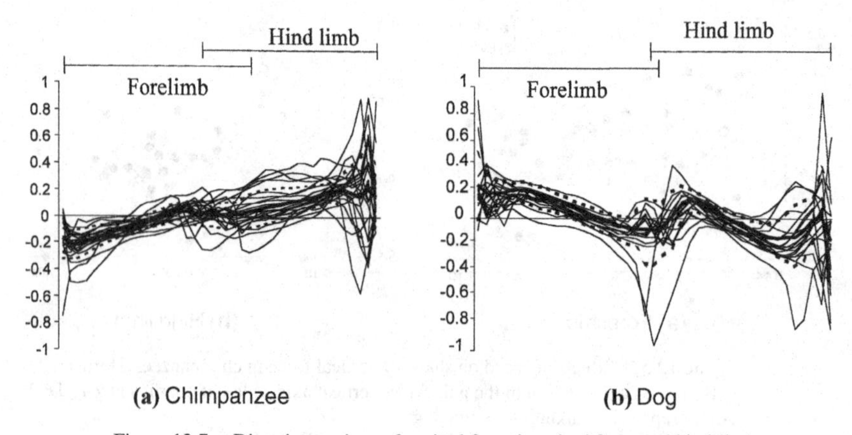

Figure 12.7. Direction cosines of sagittal force in paired fore- and hind-limb contacts for chimpanzee (a) and dog (b). Thin curves show individual records. The thick dark line shows the mean and the dotted lines ± one standard deviation.

change from negative (deceleration) to positive (acceleration), most changing sign half way through stance. In contrast, the hind-limb curves show much more variation, particularly in late stance.

Figure 12.7a (for chimpanzees) shows the same parameter but includes double contacts of a forelimb with the ipsilateral hind limb. Again, it shows that the forelimb curves have less variability. (Since traces in Figure 12.7a represent contiguous fore- and hind-limb records, any random effect would be expected to affect both limbs equally.)

Figure 12.7b shows that overall the dog has less variability in the sagittal force direction cosine. However, since Figure 12.7b reports a single subject, inter-individual variation is absent. Together, Figures 12.7a and 12.7b suggest that the dog has equal variability in fore- and hind-limb direction cosines, but chimpanzees greater variability in the hind limb.

Trace of the center of pressure under the fore and hind limbs

The paths of the center of pressure for chimpanzee fore (Figure 12.8a) and hind limbs (Figure 12.8b) differ also, but in a manner opposite to the case for direction cosines. Figure 12.8c shows a trace for a double record, with both limbs together (H and F), and includes the period of their double support (D). For the hind limb, the path of the center of pressure forms a slightly curved trace, starting at the heel and gradually moving forwards. The high density of time interval markers (separated by 1/250 second for the chimpanzee) shows that velocity is initially low. Motion of the center of pressure accelerates towards the midpoint of hind-limb contact (indicated by an increase in the distance between markers) and through to the end of hind-limb stance. This pattern of forwards movement somewhat resembles the path of the center of pressure in bipedal walking of human children (Li *et al.*, 1996).

From Figure 12.8a, and 12.8c region F, it is apparent that in chimpanzees the range of displacement of the forelimb is small compared to that of the hind limb (30–50 mm compared to 150–200 mm). Further, as Figure 12.8a indicates, the forelimb trace does not proceed consistently forwards. After establishment of a firm ground contact, the center of pressure moves anteriorly for a short period – about 10% of forelimb support time for both Figures 12.8a and 12.8c, with displacements of 10 mm (Fig. 12.8a) to 20 mm (Fig. 12.8c) – before returning to the starting position in the sagittal direction (over a further 10% of contact). The center of pressure then moves forwards again. In Figure 12.8c, the forelimb trace is disturbed by ipsilateral hind-limb contact from about 60% of the whole support period. In Figure 12.8a, where no hind-limb contact was made with the force plate, the forelimb center of pressure can be traced completely, and it is clear that many secondary movements are superimposed on the primary, anterior, displacement.

It cannot be said that the fore–aft center of pressure traces shown in Figures 12.8a and 12.8c are "typical" since a great deal of variation exists between records, and a much larger dataset would thus be required to establish "typical" patterns. Nevertheless, clear, sharp, transverse displacements are common to all the records, whether they be directed medially or laterally and with no

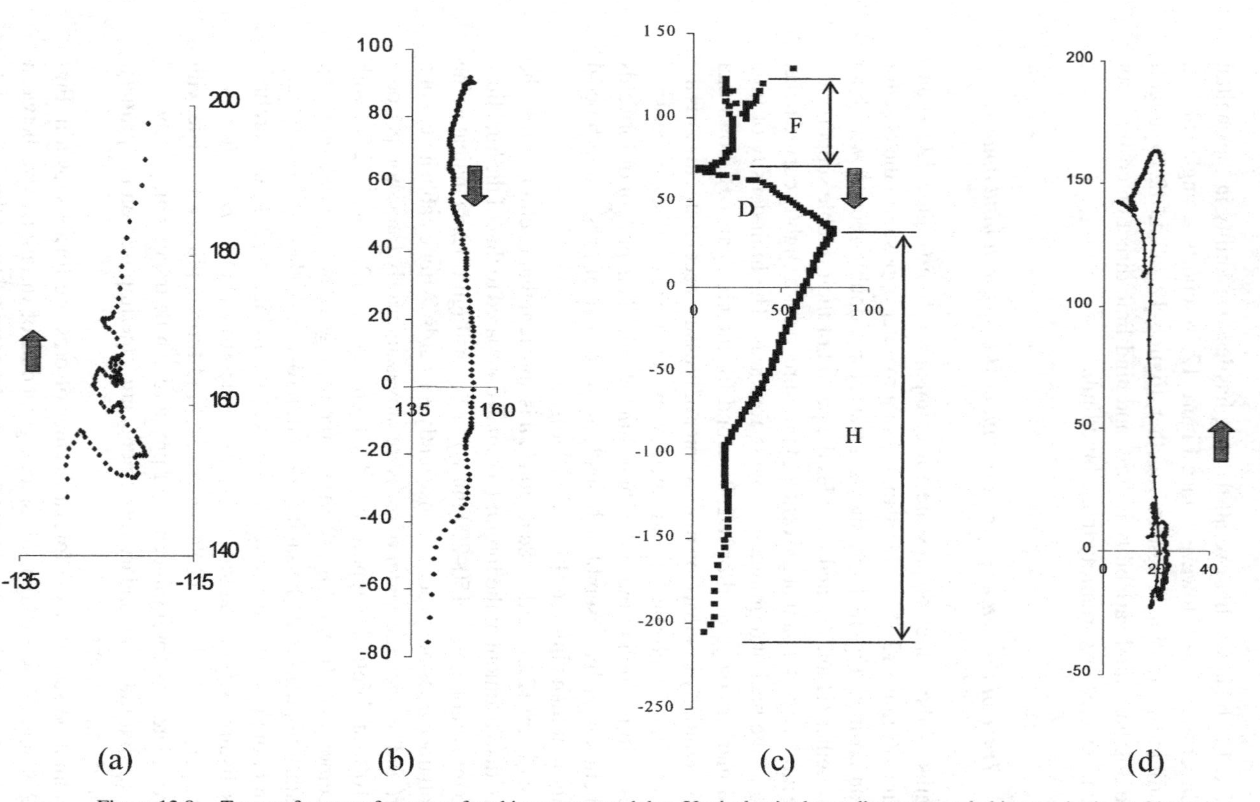

Figure 12.8. Traces of center of pressure for chimpanzees and dog. Vertical axis shows distance traveled in mm, horizontal lateral displacements. The density of markers indicates the speed of displacement of the center of pressure. The arrows in the charts show the direction of movement: (a) Chimpanzee: single forelimb contact; (b) Chimpanzee: single hind-limb contact; (c) Chimpanzee: paired fore- and hind-limb contact. (F: forelimb phase; D: double-contact phase; H: hind-limb phase); (d) Dog, fore and hind limbs. (It was not possible to annotate this chart to distinguish hind limb and forelimb without obscuring too many datapoints.)

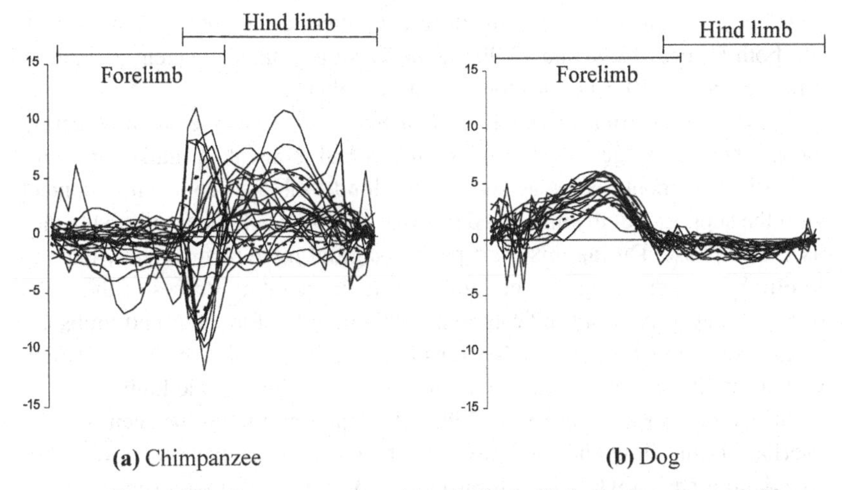

Figure 12.9. Free moment about the vertical axis. Vertical: value of moment (Nm × 10^{-3}) per unit body weight (N); horizontal: time as a proportion of the gait cycle. Directions of moments are affected by both the coordinate system and whether the left or right limb is involved. Therefore, the direction of the moments was assessed with reference to the accompanying video records and cannot be read directly from these plots.

apparent influence of the side (left or right) nor position (inside or outside) of the limb.

In contrast with findings for the chimpanzee, Figure 12.8d indicates that for the dog the traces of center of pressure for the fore and hind limbs are similar. Hind-limb contact is made some 10–13 cm behind the forelimb. Hind-limb traces are short (30–40 mm), and displace in the direction of movement. Some very small transverse motion is apparent in the hind limb.

Free moments about the vertical axis

When an animal's limb touches the ground, apart from vertical, sagittal, and transverse forces, a free moment is also applied to the supporting surface about the vertical axis, M_Z. Figure 12.9 shows its values. In the absence of exact data for chimpanzees, their body mass was estimated as 90% of the largest hind-limb vertical force; measured data were available for the dog. Thus, in Figure 12.9a chimpanzee moments are given as the measured torque value divided by 90% of the peak hind-limb vertical force. Figure 12.9b shows the same parameter for the dog. Here the vertical axis represents measured values divided by the (known) body weight (use of the same estimate as for the chimpanzee would have been inappropriate given the different pattern of bearing vertical forces). For both Figures 12.9a and 12.9b, the dimension of the dependent variable is mm (i.e., Nm × 10^{-3} per newton of body weight).

It can be seen from Figure 12.9a that in chimpanzees the forelimb acting alone tends to produce only a very small vertical moment. A marked increase in vertical moment occurs as soon as the hind limb also comes into contact with the support, but the moment decreases sharply when the forelimb finishes its stance phase. During this short period, sagittal force records show that the forelimb is accelerating but the hind limb is decelerating (traces of the paths of the center of pressure indicate that at this time the fore and hind limbs are separated by about 8 cm) and they thus form a force couple the value of which will normally be greater than the moment produced by a single limb.

In contrast, in the dog (Fig. 12.9b, a walking trot) it can be seen that the overlap of hind-limb and forelimb contacts is very short, and therefore corresponds to a period when forces on both limbs are small; moreover, the hind- and forelimb contacts are close to co-linear with each other and thus also with horizontal force vectors, which are primarily influenced by sagittal forces.

When the sign of the torque in chimpanzee diagonal gaits is analyzed together with the synchronous video records, it is apparent that free moments generated by the hind limbs are mostly laterally directed (17.6% in the medial and 82.4% in the lateral direction). For the chimpanzee forelimb, the direction seems more random (35.3% medial and 64.7% lateral). In the dog, however, the forelimb produces a moment in the medial direction, but the hind limb a smaller one in the lateral direction. Thus, while chimpanzees exert only a small moment with the forelimb compared to the hind limb, the pattern in the dog is exactly the reverse, although the ground contact area offered by the dog's fore paw is not much bigger than that offered by its hind paw.

Discussion

Demes *et al.* (1994) state that there is so much variation in primate gaits that it is impossible to identify a "typical" set of kinematics for any group. However, biomechanical principles suggest that studies of force should as far as possible sample *movement on a level surface* (so that the beginning and end of a gait cycle are not differentiated by a change in potential energy), and *smooth movement*, i.e., movement that does not show an imbalance of acceleration or deceleration after a gait cycle (so that there is no change in kinetic energy). The first condition is possible to approximate in most cases, but our previous experience leads us to expect that the latter condition will be difficult to achieve when ape subjects are under the influence of human experimenters. Thus, firm conclusions need as far as possible to be based on studies of voluntary movement, as were our data on chimpanzees. Therefore it is on these data that we place the greater stress in this discussion, using the dog as a comparator primarily when distinctions from the chimpanzee case are qualitative rather than merely quantitative.

Again, while not disputing that primates bear a greater proportion of vertical forces on the hind limb, Demes *et al.* (1994) argue that nonprimates as well as primates have a larger proportion of accelerative force in the hind limb, and conclude that the concept of hind-limb drive as a primate specialization is "a misconception." There are some difficulties with their argument, since their own data for trotting cats show that the accelerative impulse exerted by the forelimb is much larger than that exerted by the hind limb: 1.410 bw·s (body-weight seconds) versus 1.093 bw·s. Moreover, since the mean braking impulse is 1.530 bw·s but the mean propulsive impulse only 1.026 bw·s, the sequences were predominantly recorded during deceleration, so that the motion was not "smooth."

Let us turn now to our own data. Figures 12.2 and 12.3 confirm the finding of previous studies (Kimura *et al.*, 1979; Reynolds, 1985b; Demes *et al.*, 1994) that the shapes of both vertical and sagittal forces differ between the fore and hind limbs of chimpanzees. A considerably greater proportion of vertical force in chimpanzees is clearly borne by the hind limb, and the manner in which these forces are applied also differs between the fore and hind limbs. But Figure 12.4 further shows that the chimpanzee forelimb has a relatively large deceleration component, and a small acceleration component. The reverse applies to the hind limb, both with respect to duration of acceleration/deceleration in sagittal forces and their force–time integral. Our data show that the forelimb has a smaller accelerative role than the hind limb, in both chimpanzees and dogs, although the distinction is greater in chimpanzees. The forelimb acts mostly in deceleration. Demes *et al.* (1994, in their Figure 12.11) summarize ratios of propulsive over braking impulses in a variety of primates and nonprimates from their own studies and from the literature, and since most primates and nonprimates show net braking in the forelimb, and net propulsive impulses in the forelimb, conclude that it is inappropriate to describe primates as hind-limb driven.

We however suggest that in normal walking both decelerating and accelerating forces are in a sense "driving" locomotion. Decelerative forces in animal locomotion are not an exact parallel to "braking" forces in a vehicle, which release mechanical power as heat. Both in quadrupedal and in bipedal walking, accelerating forces increase the velocity of the center of mass in the direction of locomotion, while decelerating forces have the function of changing the pattern of fluctuation in potential/kinematic energy of the body (internal, external, or both). While only accelerative forces generate power, decelerative forces are also essential to function, in controlling the pattern of power output. But even if we treat only accelerative forces as the "driving" force, Demes and colleagues' (1994) own data offer some support for the concept of nonprimates being relatively "forelimb-driven" (in comparison to primates) in that, as we have noted, their data for cats show that the forelimb has a much larger accelerative impulse than the hind limb (as indicated above: 1.410 bw·s vs. 1.093 bw·s) and their chimpanzee data similarly suggest net acceleration over the gait cycle as the subjects crossed the force platform. We lack the data required to determine the extent to which our results for chimpanzees are generalizable. However, to the extent to which they are generalizable to primates as a whole, since our Figure 12.4 shows that the integral of absolute force over time is high in the chimpanzee hind limb, but low in the forelimb, functional distinctions between the fore and hind limbs of primates do appear to be in an opposite *direction* to those in nonprimates.

Table 12.1 and Figure 12.5 suggest that the forelimb and hind limb of the chimpanzee also respond in different ways to an increase in speed. Referring to the formula used to calculate coefficient a_3, where $a_3 = [-f(1/6) + f(3/6) - f(5/6)]/3$, we see that the middle term has the opposite sign to the first and last non-zero terms, so that a_3 will provide information on the extent to which the mid-portion of the curve differs from the earlier and later portions of the curve.

In this light, Figure 12.5a indicates that high speed slightly increases vertical forces exerted by the forelimb at mid-stance. However, for the hind limb, high values of a_3 (and thus a peak at mid-stance) occur in the high-speed block (to the left), but (to the right) the middle- and slow-speed blocks show a tendency for mid-stance forces to be *reduced* at higher speed.

This complex relationship bears some resemblance to our findings for human bipedal walking and running (Li *et al.*, 1996). Our study found that running gave larger Za_3 values than walking (and thus a peak at mid-stance). In walking, however, Za_3 became lower with a higher speed, producing a deeper valley at mid-stance. Thus, from this perspective, the chimpanzee hind limb, even in quadrupedal locomotion, shares common functional characteristics with human bipedalism. The chimpanzee forelimb however has a much simpler force–speed relationship, suggesting a simpler functionality.

Again, Figures 12.6 and 12.7 show that the function of the chimpanzee forelimb resembles that seen in both the fore and hind limb of the dog, in that the direction cosines for the sagittal force show a narrow band of variation. It is the hind limb of the chimpanzee which is unique, in showing a wide band of variation. (Although outliers certainly contribute to this variation, a core of very variable direction cosines confirms that this is a systematic variation.) Thus, in the chimpanzee, changes in the direction of accelerative and decelerative force (that is, steering) are brought about much more by the hind limb than by the forelimb, while forelimbs and hind limbs play a much more similar role in dogs.

A distinction between hind-limb and forelimb function was also apparent in the traces of the center of pressure (Fig. 12.8), where the hind-limb center of pressure traces a clear, slightly curved line, starting at the heel, and moving slowly forwards with increasing speed. This pattern somewhat resembles the function of the foot in human bipedalism, where Carrier *et al.* (1994) show that as the center of pressure moves anteriorly, the length of the lever formed between the center of pressure and the ankle joint becomes larger. As a result, the functional length of the foot becomes longer, so that it is more effective in accelerating the body, without requiring dramatic increases in the angular velocity of the ankle joints. In contrast, the short, rather randomly directed traces formed by the chimpanzee forelimbs (where, in "knuckle walking," contact is often made along the middle phalanges alone) showed no such function.

Figure 12.9 shows, further, that chimpanzees and dogs apply vertical moments in an opposite manner. Chimpanzees exert only a very small moment via the forelimb, but larger moments via the hind limbs or via the combination of fore and hind limbs, while the dogs exerted vertical moments predominantly via the forelimb.

In humans, the vertical moment acts to reinforce armswing in balancing the trunk (Li *et al.*, 2001), changing from medial to lateral before mid-stance, and the lateral moment peaks at 5–10 Nm $\times$ 10^{-3} per newton of body weight. A direct comparison of these values cannot be made with the present study, because the conditions of the two studies (bipedal in humans and quadrupedal in chimpanzees) are very different. Nevertheless, in chimpanzees, relative values are lower, but more variable. Eighty percent of hind-limb records indicate laterally directed moments. In contrast, the dog's forelimbs exerted moments in the medial direction, but the hind limbs moments in the lateral direction.

The analysis of vertical moments in quadrupedalism is more difficult than for bipedalism. For bipedalism, moments exist either during the stance of a single foot, or during double-support (both feet). In quadrupedalism, however, moments exist during both contralateral and ipsilateral double support. In the double-support period for ipsilateral limbs, moments may be in either direction, but have high absolute values and considerable variation, since a hind limb landing lateral to its same side forelimb (an outside hind limb), will produce an opposite moment to the inside hind limb. However, in the dog studied in detail, which used a diagonal gait, the forelimb–hind-limb double-support period was short, and at this time forces on both limbs were small. Further, hind-limb contacts in the dogs were made co-linear with the forelimb and with the horizontal force vectors, which are dominated by sagittal forces. Thus, the distance between the two forces is small, the forces are small, and the resulting moment is thus also small.

The larger moment exerted by the forelimb of dogs suggests a likely effect on the angular speed of the trunk and head, and hence control of direction. In chimpanzees the greater moments are exerted by the hind limbs, and thus the hind limbs have the effect of balancing or affecting the direction of the body.

Conclusions

Thus, for chimpanzees, while the large, primarily accelerative sagittal forces, as well as the consistent forward progress of the hind-limb center of pressure, are evidence of hind-limb driving, the large vertical moment and complicated direction cosine pattern which characterize the hind limb also appear to be clear indications of hind-limb steering. We do not exclude the forelimb from a role in the steering process, but suggest its role may exist largely during swing, in the selection of touchdown points. Even then, the force or torque required to balance the direction change for the swinging forelimb would be provided primarily by the stance hind limbs.

Acknowledgments

We thank the North of England Zoological Society for permission to study the animals in their care and the Merseyside Police Animal Training Unit for their patient help in studying their dogs. The Biotechnology and Biological Sciences Research Council, the Natural Environment Research Council and the Leverhulme Trust funded this study. The Kodak Motion Analyzer was borrowed from the Joint Research Council's equipment loan pool. Finally, RHC thanks the editors for inviting him to participate in honoring Charles Oxnard's contributions to biological anthropology and, of course, thanks Eleanor and Charles for many years of hospitality, inspiration, and friendship.

References

Alexander, R. McN. and Jayes, A. S. (1980). Fourier analysis of forces exerted in walking and running. *J. Biomech.*, **13**, 383–390.

Carrier, D. R., Heglund, N. C., and Earls, K. D. (1994). Variable gearing during locomotion in the human musculoskeletal system. *Science*, **265**, 651–653.

Demes, B., Larson, S. G., Stern, J. T., Jr., Jungers, W. L., Biknevicius, A. R., and Schmitt, D. (1994). The kinetics of primate quadrupedalism: "hind limb drive" reconsidered. *J. Hum. Evol.*, **26**, 353–374.

Hildebrand, M. (1966). Analysis of symmetrical gaits of tetrapods. *Folia biotheor.*, **6**, 9–22.

Kimura, T. (1992). Hind limb dominance during primate high-speed locomotion. *Primates*, **33**, 465–476.

Kimura, T., Okada, M., and Ishida, H. (1979). Kinesiological characteristics of primate walking: its significance in human walking. In: *Environment, Behavior, and Morphology: Dynamic Interactions in Primates*, ed. M. E. Morbeck, H. Preuschoft, and N. Gomberg. New York, NY: Fischer.

Li, Y., Crompton, R. H., Alexander, R. McN., Günther, M. M., and Wang, W. J. (1996). Characteristics of ground reaction forces in normal and chimpanzee-like bipedal walking by humans. *Folia Primatol.*, **66**, 137–159.

Li, Y., Wang, W. J., Crompton, R. H., and Günther, M. M. (2001). Arm swing, vertical moment, and lateral forces in human normal walking. *J. Exp. Biol.*, **204**, 47–58.

Martin, R. D. (1990). *Primate Origins and Evolution*. London: Chapman and Hall.

Reynolds, T. R. (1985a). Mechanics of increased support of weight by the hind limb in primates. *Amer. J. Phys. Anthropol.*, **67**, 335–349.

(1985b). Stresses on the limbs of quadrupedal primates. *Amer. J. Phys. Anthropol.*, **67**, 351–362.

Rollinson, J. and Martin, R. D. (1981). Comparative aspects of primate locomotion, with special reference to arboreal cercopithecines. *Symp. Zool. Soc. Lond.*, **48**, 377–427.

Schmitt, D. (1999). Compliant walking in primates. *J. Zool. Lond.*, **248**, 149–160.

Winter, D. (1979). *Biomechanics of Human Movement*. New York: Wiley.

Part IV
Theoretical models in evolutionary morphology

Though I had always been interested in the use of mathematical methods of one kind or another, in the earlier days these interests were almost always for data analysis. The models of function with which we worked in the early days were simplistic in the extreme, scarcely deserving of the term "model" – in fact they were little more than functional anatomical inferences. At later stages I used models for assessing mechanical efficiency, but these mainly employed what is now a quite old-fashioned technique – photoelastic analysis – and they were extremely limited – e.g., to two dimensions only, and only to isotropic situations. Only much more recently, in collaboration with others (e.g., O'Higgins), have I come to use better modeling methods (such as finite elements).

However, in a completely surprising way, through a stimulus applied by Sydney Brenner and his invitation to present a model of mtDNA evolution (which I did not do) at a workshop hosted by the International Institute for Advanced Studies in Kyoto, I have come to be involved in true evolutionary modeling (mimicking species evolution and individual lineages) with Dr. Ken Wessen. Though not included as a section in this book, Dr. Wessen's work (in which I have been pleased to share) is reported in my own final chapter.

At this point, however, it is an especial pleasure to recognize, through the following sections, kinds of modeling that I scarcely envisaged in those earlier days. Thus the following chapters on modeling the origins and mechanics of bipedalism, and the mechanics of mastication, are extremely relevant.

Charles Oxnard

13 *Becoming bipedal: how do theories of bipedalization stand up to anatomical scrutiny?*

NINA G. JABLONSKI, GEORGE CHAPLIN
California Academy of Sciences

Introduction

In a volume dedicated to Charles Oxnard and his many contributions to anthropology and zoology, it is only fitting that some papers address the topic of primate, including human, locomotion. A quick survey of Oxnard's long list of publications reveals his abiding interest in the topic – the vast majority of these (and that's a lot!) relate to the evolution of locomotor complexes in primates. Within this area, one of the topics that has captured his attention consistently over the years has been the evolution of human bipedalism and its anatomical correlates. Although Oxnard has never published his own theory on the evolution of human bipedalism, he was one of the first to determine that the australopithecine pelvis was anatomically distinct from that of apes and humans, and that its anatomical uniqueness implied a unique mode of locomotion in the group that had previously not been appreciated (Oxnard, 1973, 1975). To his lasting credit, he never stooped to the level of his detractors on this subject nor gloated when he was shown to be correct by numerous independent sources years later (e.g., Susman *et al.*, 1984).

Theories of bipedalization

The selective factors that operated to promote habitual bipedalism in the ancestors of the human lineage are still hotly debated. As McHenry pointed out years

Shaping Primate Evolution, ed. F. Anapol, R. Z. German, and N. G. Jablonski. Published by Cambridge University Press.

ago, this area of research continues to simultaneously fascinate and frustrate scholars because of its signal importance to human evolution and because we shall never *see* what really happened (McHenry, 1982). Lewin (1999, p. 98) recently remarked that "the plethora of hypotheses offered to explain the evolution of bipedalism reflects both the fertility of ideas among anthropologists and the difficulty of using available evidence to discriminate between them." The end result of bipedalization is clear from paleontological and comparative anatomical evidence; the process is not.

Many different selective agents have been singled out over the years for their role in inaugurating and developing bipedal posture and locomotion in hominins. Rose (1991) summarized these appropriately into four categories. Here we update his list and expand the number of categories to five. The first of these, forelimb pre-emption, comprised carrying behaviors including the carrying of infants (Etkin, 1954; Iwamoto, 1985), food (Hewes, 1961; Lovejoy, 1981; Gebo, 1996), and tools (Bartholomew and Birdsell, 1953; Washburn, 1967; Marzke, 1986), and the throwing of tools and other missiles (Fifer, 1987). The second category, social behavior, included hypotheses concerning the importance of threat displays (Livingstone, 1962; Wescott, 1967; Jablonski and Chaplin, 1993), aggression (Kortlandt, 1980), evasion (Reynolds, 1931), vigilance and sentinel behaviors (Dart, 1959; Day, 1977; Ravey, 1978), sexual display (Guthrie, 1970; Montgomery, 1988), and nuptial gifts (Parker, 1987). The third (and most popular) category includes hypotheses that emphasize the importance of feeding in different contexts, including arboreal gathering (Hunt, 1991, 1994, 1996), terrestrial gathering (DuBrul, 1962; Jolly, 1970; Wrangham, 1980; Rose, 1984; Wundram, 1986; Sarmiento, 1998), aquatic gathering (Hardy, 1960; Morgan, 1982; Veerhaegen, 1985), terrestrial predation (Geist, 1978; Carrier, 1984; Merker, 1984), and terrestrial scavenging (Szalay, 1975; Shipman, 1986; Sinclair *et al.*, 1986). A fourth category embraces those hypotheses that suggest that the evolution of bipedalism was inevitable because it was more efficient as a mode of locomotion in long-distance foraging (Rodman and McHenry, 1980; Isbell and Young, 1996) or because it was superior to quadrupedalism with respect to thermoregulation, especially during high activity levels in hot environments (Wheeler, 1984, 1991). A final miscellaneous category comprises selection pressures such as walking on snow or ice (Kohler, 1959), iodine deficiency (Marett, 1936), or those theories that posit a combination of selective factors (Napier, 1963; Sigmon, 1971; Rose, 1984; Day, 1986). One recent theory proposes that no selective pressure was necessary in the evolution of human bipedalism, because bipedalism itself constitutes the primitive mode of posture and locomotion, from which all quadrupeds evolved (de Sarre, 1994).

Anatomical predictions of selected hypotheses

In this chapter, we seek to assess the likely validity of some of these proposed selective agents by assessing them with respect to whether the anatomies of early fossil hominins meet the anatomical predictions of their respective hypotheses. The hypotheses that we have chosen to evaluate represent three of the five major sets of proposed selective agents: forelimb pre-emption (especially carrying of food), social behavior (threat displays), and feeding (particularly arboreal gathering). In previous publications, we have considered and rejected bipedalization hypotheses based on biomechanical or thermoregulatory inevitability (Jablonski and Chaplin, 1993; Chaplin *et al.*, 1994). We have chosen to evaluate the hypotheses of those authors who have at least made an attempt to be specific and explicit about the anatomical predictions that would follow from their theories, or who have made an effort to explicitly reconcile the observed anatomy of early bipeds with their hypotheses of bipedalization (e.g. Lovejoy, 1981; Jablonski and Chaplin, 1993; Hunt, 1994, 1996).

In Table 13.1, we summarize these hypotheses and include information about the duration of the proposed selective agent and whether the proposed activity involves the handling of an external load.

In terms of their anatomical consequences, the three major hypotheses under discussion differ primarily in their reconstructions of the duration and nature of bipedal activity, and secondarily in the nature and magnitude of any external load that would be handled during the activity. Both the carrying hypothesis of Lovejoy (1981) and the feeding hypothesis of Hunt (1994, 1996) propose a prolonged duration for the bipedal activity: walking in the case of the former and standing in the case of the latter. The threat-display hypothesis proposes that the bipedal activities that propelled the adoption of habitual bipedalism were of short duration. With respect to the nature of the external loads, the carrying hypothesis involves the portage of "significant amounts" of food (Lovejoy, 1981, p. 345) and the threat-display hypothesis entails the occasional "brandishing [of] a branch or other object" (Jablonski and Chaplin, 1993, p. 270). The arboreal-feeding hypothesis proposes a "feed-as-you-go strategy" (Hunt, 1994, p. 196) that does not entail the handling of any significant external load, but it does imply that one or, occasionally, both forelimbs would be maintained in an elevated position for prolonged periods.

Evaluating the hypotheses

The fossil record is now revealing tantalizing glimpses of what appear to have been the earliest phases of human evolution six to seven million years ago – in

Table 13.1. *Hypothesized selective agents for the origin of human bipedalism and the anatomical predictions for early fossil hominins following from them. For each proposed agent, information is provided pertaining to the duration of the activity, nature of the external load handled during the activity if any, the predicted anatomy for the earliest hominin, and the observed hominin anatomy consistent with the agent. Information from the original citation(s) is given in quotations, with a specific page reference. When this information has not been provided explicitly by authors, we have used other information in the papers to derive these predictions, and have presented these without quotation marks.*

Selective agent	Duration of activity	External load	Predicted anatomy for earliest hominin	Observed early hominin anatomy consistent with theory	Key reference
Food carrying	"extended periods of upright walking" (p. 345)	"significant amounts of food" (p. 345)	Truncal erectness; pelvis and lumbar spine adapted for walking (wide pelvis and lumbar lordosis)	"shift toward greater molar dominance" (p. 342); "reduction and effective loss of canine dimorphism" (p. 346); "other forms of dimorphism were apparently being accentuated" (p. 346)	Lovejoy (1981)
Arboreal feeding	"feeding occupies a greater portion of the time-budget . . . than any other behavior" (p. 184)	no external load (small fruits eaten on the spot)	Truncal erectness, long forelimbs; lumbar lordosis permitting prolonged standing	"arboreal arm-hanging/bipedalism and terrestrial bipedalism during small-fruit collecting" (p. 198) with forelimb and trunk adapted for hanging and pelvis and lower limb adapted for standing	Hunt (1994)
Threat displays	"frequent" (p. 273) but of short duration	Occasional "brandishing a branch or other object" (p. 270)	Stature sufficient to be easily visible when standing or running erect; anatomy consistent with "frequent bipedal standing", followed by "changes in the structure and function of the trunk, pelvic girdle, lumbar spine and gluteal muscles and in the neural mechanisms of postural control" (pp. 263–264)	"lessened sexual dimorphism in canine tooth size but relatively larger dimorphism in stature" (p. 272)	Jablonski and Chaplin (1993)

the form of *Sahelanthropus tschadensis* (Brunet *et al.*, 2002), and over four million years ago in *Ardipithecus ramidus* (White *et al.*, 1994). Unfortunately, the locomotor anatomy of these earliest hominins is not known, so we have as yet no glimpse of what the earliest bipedal skeleton looked like. What we do have, however, are a few well-preserved skeletons of later hominins, the most complete of which are the middle Pliocene AL-288 of *Australopithecus afarensis* ("Lucy") and the early Pleistocene KNM-WT 15000 of *Homo ergaster* (the "Turkana Boy"). The anatomy of these skeletons and some other partial skeletons can be used to test predictions generated by theories concerning the origins of habitual bipedalism.

Food carrying

Hypotheses which envision carrying as the agent of selection which initiated bipedalization entail several anatomical requirements, including truncal erectness, a vertebral column adapted for carrying loads, and an ability for the protohominin to engage in "consistent, extended periods of upright walking" (Lovejoy, 1981, p. 345). Studies of the biomechanics of the modern human skeleton have shown that bipedal walking involves coordinated movements of the spine and pelvis, with the "spinal engine" driving the pelvis and lower limb forward (Gracovetsky, 1985; Gracovetsky and Farfan, 1986). The degree of lordosis in the lumbar spine is the main factor that influences the conversion of the extensor power developed by the intrinsic back muscles to axial torsion necessary to rotate the pelvis in walking (Gracovetsky, 1985). Lumbar lordosis also promotes stability during bipedal stance and locomotion by placing the center of gravity of the body directly over the hip joints (Shapiro, 1993). In humans, the lordosis permits the pelvis to rotate backward opposite to the side of lateral flexion of the spine; thus, alternate rotations of the pelvis are caused by alternating lateral flexion of the spine (Farfan, 1995). Chimpanzees and Old World monkeys occasionally engage in bipedal walking, but the lack of a lordosis makes it difficult for them to rotate their pelves. This contributes to the great lateral pelvic movements observed when these animals undertake facultative bipedal walking (Jenkins, 1972; Gracovetsky, 1985; Hayama *et al.*, 1992; Li *et al.*, 1996). Similar gait patterns are observed in humans with fused lumbar spines, in which axial rotation cannot be developed between individual lumbar segments to drive the pelvis forward in walking. Significantly, in monkeys habituated to bipedalism over the course of years, a lordosis develops to varying degrees, along with other modifications of the lower limb skeleton, and their gait becomes more human-like (Gracovetsky and Farfan, 1986;

Hayama *et al*., 1992; Nakatsukasa *et al*., 1995). Therefore, consistent walking requires development of a lordosis.

When an external load is introduced into consideration as in the carrying hypothesis, the situation becomes more complex because the load must be lifted, brought close to the body, and then transported. The load being carried exerts an additional axial load on the spine, which is then compounded by the problem of moving with the load (Gracovetsky and Farfan, 1986). The additional axial load stiffens the intervertebral joint so that the amount of potential axial rotation developed between the joints is reduced (Gracovetsky and Farfan, 1986). With increased weight, spinal flexion works to reduce the compression of the intervertebral joints, but at the expense of the lordosis, the development of torque in the lumbar spine, and stride length (Farfan, 1978; Gracovetsky, 1985). One can immediately recognize here the short-stepped gait of a person carrying a heavy load. These facts have clear implications for the viability of the carrying hypothesis. Modern humans are endowed with anatomical features such as short forelimbs and a highly flexible spine with permanent lordosis that effectively reduce the moment of the external load (Farfan, 1978). The lifting of heavy objects in humans is made possible by the very large weight-bearing surface of the vertebrae, especially in the lower lumbar region, and the recruitment of a large hip extensor musculature, the glutei (Shapiro, 1993; Farfan, 1995). These anatomical provisions are absent in the African apes, hence their limited abilities to carry external loads without placing a hand on the ground (Farfan, 1995). The anatomical situation in early hominins with respect to the existence and degree of the lumbar lordosis thus becomes critical.

Australopithecines such as *A. afarensis* AL-288 exhibited an erect trunk and were capable of some form of walking (Susman *et al*., 1984; Latimer, 1991). Based on studies of the pelvis of AL-288, Berge (1991, 1994) argued that the bipedalism practiced by *A. afarensis* differed from that of modern humans in exhibiting less stability of the lower limb during walking and, thus, shifting the trunk laterally from one limb to another with large rotatory movements of the trunk and shoulders by way of balancing compensation. When the vertebrae of this specimen and others with relatively complete thoracolumbar vertebral series such as Stw-H8/H41 are examined, however, it appears that the australopithecine vertebral column was adapted to human-like lordosis and stable balance of the trunk over the pelvis (Sanders, 1998). The relatively small size of the lumbar vertebral centra in these specimens, however, indicates that vertical forces through the vertebral column were transmitted differently than in modern humans or that smaller increments of compression passed across their centra (Sanders, 1998). The logical deduction in terms of function is that a vertebral column with this configuration was not adapted for carrying anything but a very light load. A similar conclusion was made from the anatomical observations of the early *Homo* skeleton, KNM-WT 15000 (Latimer and Ward, 1993).

The paleontological evidence thus suggests that australopithecines engaged in bipedal walking, but that they did not carry heavy loads at the same time. Whether they could engage in "consistent, extensive walking" is also somewhat questionable. The relatively short legs of *Australopithecus* did not preclude a modern form of bipedalism (Kramer, 1999), but they did not guarantee it either. Crompton and colleagues (1998) have recently demonstrated that the limb proportions of AL-288 are compatible with both a modern-human-like gait and a compliant, "bent-hip, bent-knee" gait. In the latter case, the mechanical ineffectiveness of the gait would have resulted in an increase in core body heat, and the authors speculate that any selection pressure acting to favor such a gait would have had to be intense to overcome what amounts to a full doubling of energy costs (Crompton *et al.*, 1998). Thus, at the very best, early bipeds were walking far but empty-handed, or at worst weren't walking far at all.

Arboreal feeding

We now turn our attention to feeding hypotheses, and the arboreal-feeding hypothesis in particular, to see if these fare any better. These hypotheses also require truncal erectness, forelimbs adapted to harvesting and manipulation of food items, stability in bipedal standing, and probably a low energy expenditure in feeding if it is to be undertaken for a long time. Hunt (1994, 1996), the most recent proponent of the feeding hypothesis, describes a situation in modern chimpanzees in which harvesting of small food items is undertaken in bipedal postures on the ground and in low trees, and extends this scenario to the putative prehominins. As Hunt (1994, 1996) points out, the wide pelvis of *A. afarensis* is consistent with truncal erectness, as is the anatomy of the forelimbs with an adaptation to the harvesting of small food items.

The real question posed by the feeding hypothesis is whether prolonged standing could be endured by the earliest hominins. One of the important consequences of bipedal feeding postures in which the forelimbs are raised above the level of the shoulder joint is that this activity shifts the body's center of gravity and imposes high physical stresses on the thoracic and lumbar vertebrae. In bipedal feeding postures, animals with relatively long forelimbs such as modern apes and, presumably, the earliest hominins, are carrying the equivalent of a heavy parcel away from their bodies at the level of their shoulders or above, thus greatly increasing the moment of the load. This is still a problem even if only one forelimb is used in food harvesting, while the other supports part of the body weight. Bipedal feeding in an arboreal context thus results in at least two serious consequences: an increase in the compressive loading of the thoracolumbar spine and an increase in energetic costs due to the need to support the weight of the forelimbs during feeding. It also involves an increased energetic cost in relation to the maintenance of bipedalism in an arboreal context while moving around in foraging and standing in feeding. We have already discussed, in the context of the carrying hypothesis, the difficulties with any bipedalization hypothesis that requires increased compressive loading of the vertebral column: there is no evidence from the paleontological record that such loading could be tolerated. So we must now look at the putative energetics of arboreal bipedal feeding.

Clambering bipedally on arboreal substrates (or their equivalents) is energetically very costly (Elton *et al.*, 1998). Simply standing bipedally on a branch is also relatively expensive in terms of energy because such a posture requires adoption of a compliant posture with flexed hip and knee (and possibly foot) joints, with its attendant metabolic costs. A further issue to be considered here is how females with dependent infants would have managed in this context. With their higher basal metabolic costs (if weaning) and at least one forelimb occupied in stabilizing their infant, they would have been hard pressed to harvest sufficient high-quality food on a regular basis to make bipedal arboreal feeding a worthwhile activity.

We can only conclude, therefore, that the energetic penalties of this posture and mode of locomotion in food harvesting argue against feeding as the primary agent of selection in bipedalization. In fact, feeding in low trees or shrubs favors a morphology in which long, powerful forelimbs operate freely from a trunk that is stabilized in sitting or squatting on the ground, such as in giant ground sloths or – much closer to home – living gorillas. In such postures, long forelimbs served by a powerful retractor musculature operate from an erect trunk and are used to pull food-laden branches closer to the animal's center of gravity, where they are consumed (Jablonski and Chaplin, 2000).

Threat displays

Bipedalization originating as a result of the bipedal threat display appeasement complex (Jablonski and Chaplin, 1993) does not have any strict anatomical requirements because it does not propose that prolonged standing, walking, or carrying were required. The single general anatomical requirement would be that the display be visible, and in this connection taller stature would be an advantage. In addition, the hypothesis has a few behavioral requirements. One of these is that the display be recognized as having a specific meaning. In modern African apes, bipedal displays generally consist of short periods of bipedal standing – in a bent-hip, bent-knee position and often with forelimbs raised – followed by a bipedal charge that is in turn followed by contact with another individual or a bout of quadrupedal charging. These displays are of short duration, but have profound impacts on social relationships and access to resources and mates, especially when they are performed between males. Male–male aggression is common among mammals, and anthropoid primates are unusual among mammals in that they exhibit aggressive displays that do not generally result in serious injury or death (Jablonski and Chaplin, 1993; Jablonski *et al.*, 2002). Despite this, the potential for serious injury needs to be real for a male to achieve dominance (Darwin, 1871; Geist, 1971; Andersson, 1994; Carrier *et al.*, 2002). This is the second important behavioral requirement of the hypothesis.

An important point here is that the duration of time spent in any given activity is not directly proportional to its impact on reproductive success. Lovejoy (1981) and Hunt (1994, 1996, 1998), in their respective propounding of the carrying and arboreal-feeding hypotheses make the point that it was the duration of time spent in a particular bipedal activity that mattered in the adoption of habitual bipedalism. In a similar vein, Videan and McGrew (2002) have recently concluded that the carrying and foraging hypotheses were the most tenable because these activities were the most commonly observed of four (also including vigilance and display) in their behavioral tests of hypotheses of bipedalization on populations of captive chimpanzees and bonobos. This is akin to saying that the activity that one is engaged in for the longest amount of time (e.g., sleeping, for many species) is the most important to one's reproductive success. This is clearly a fallacy. The importance of rare behaviors in determining reproductive success has been well illustrated by Stanford's observations of hunting by male chimpanzees (Stanford, 1995, 1998). Hunting and subsequent meat-sharing are occasional activities, not linked with any known nutritional motivation, that have profound social consequences and effects on the reproductive success of the hunters (Stanford, 1995, 1998). Therefore, we suggest that bipedalization was driven by a different, relatively rare behavior – the bipedal threat display and appeasement behavior complex – that enhanced the reproductive success of particular individuals and ultimately had dramatic consequences for species demography.

We have recently modeled prehominin demographic trends using the demographic profile of the present-day Gombe chimpanzee population as the basis for the model (Jablonski *et al.*, 2002). Among the scenarios we developed was the growth of a chimpanzee population over 50 generations and 500 iterations under simulated conditions comparable to those we proposed for protohominins, that is, a situation in which morbidity and mortality have been reduced by 10% due to the introduction of a more efficacious method of settling potentially violent disputes over resources and mates. In this case, a dramatic rise in the intrinsic rate of population increase was observed (Jablonski *et al.*, 2002). It is clear that a level of increased survivorship as a result of reduced mortality would have a dramatic effect on the intrinsic rate of increase of the population of living chimps at Gombe or that of a putative pre- or protohominin population with broadly comparable life-history parameters. What would have been the effects of such a change in prehominins? We reason that such a change would have relieved the pressure on females to reproduce early (Jablonski *et al.*, 2002). By decreasing mortality due to intraspecific killing, the periods during which somatic growth could occur and before which reproduction must commence in females were extended, as suggested by Stearns and Koella (1986) and Charnov (1991; Charnov and Berrigan, 1993). Early phases of development could therefore "afford" to be prolonged without leading to extinction (Jablonski *et al.*, 2002).

As we have pointed out elsewhere (Jablonski and Chaplin, 1993; Jablonski *et al.*, 2002), one of the important ways in which our hypothesis stands up to anatomical scrutiny is in the concordance between the predictions of the hypothesis and observed levels of hominin sexual dimorphism in canine size and body size. The reduced levels of canine dimorphism observed in *A. afarensis* suggest a reduced importance of sharp canine teeth in fighting, especially between males (Kelley, 1995). The high levels of dimorphism in overall body size in *A. afarensis* suggest that tall stature was important to male reproductive success. We have speculated that this latter effect has had long-lasting consequences for human behavior (Jablonski and Chaplin, 1993), with respect to the kinds of postures that denote confidence or success in modern human societies. In this connection it is significant that a recent study has shown on a sample of over 4000 modern humans that tall men have more reproductive success than men of shorter stature (Pawlowski *et al.*, 2000).

Conclusions

The adoption of habitual bipedal posture and locomotion was the key innovation responsible for the origin of the human lineage. The results of this study suggest that two of the most strongly favored hypotheses for the evolution of habitual bipedalism, the food-carrying hypothesis of Lovejoy (1981) and the arboreal-feeding hypothesis of Hunt (1994, 1996) are untenable because they are not compatible with the anatomical evidence of the fossil record and because they propose activities that are extremely costly in terms of energetic expenditure. This leaves the threat-display hypothesis as the only one compatible with the available evidence.

In our research on the evolution of bipedalism, we have argued that hominin bipedalism evolved as an extension of the bipedal threat display and appeasement behavior complex. This is one of the most important morphological–behavioral complexes shared by the African great apes and humans and provides a mechanism for increasing individual reproductive success by reducing morbidity and mortality due to intra- and intergroup aggression. We have advocated this theory because it meets important criteria for any theory concerning the origin of the process of bipedalization (Jablonski and Chaplin, 1993). First, it is consistent with the paleontological, anatomical, physiological, environmental, and behavioral evidence. Second, it describes a behavioral complex that would clearly lead to greater reproductive success on the parts of those individuals engaging in it. And third, the potential demographic consequences of the theory can be addressed and investigated. Finally, it is not an exclusive theory. The bipedal display–appeasement behavior complex was not the only agent that brought about bipedalization, rather it was the key behavior that gave a kick-start to the process because of its strong positive impact on individual reproductive success. We recognize the importance of a multiplicity of factors in the evolution of modern bipedalism, as emphasized by Rose (1984, 1991), but the primacy of one in initiating the process of becoming bipedal.

Behavioral hypotheses for the origin of hominin bipedalism are often eschewed because they are viewed as untestable, unparsimonious, and unscientific (e.g., Lewin, 1999), or because scientists feel uncomfortable in proposing and defending "soft" explanations for critical evolutionary events. This ignores the critically important role of behavioral innovation in primate evolution, especially as a way for creating new ecological opportunities (Lee, 1991). Behavioral hypotheses are more difficult to test, but they can be evaluated with great rigor and their feasibility objectively judged.

Much remains to be done in the study of the evolution of bipedal posture and locomotion in humans, but we feel confident that by putting more and more evidence together in creative ways – in a manner typical of Charles Oxnard – we will get ever closer to understanding what really happened.

Acknowledgments

As we were developing our theory of the evolution of human bipedalism in the early 1990s, Charles Oxnard's questions and insights never failed to challenge and inspire us. We will be forever grateful to him for his intellectual generosity and friendship.

References

Andersson, M. (1994). *Sexual Selection*. Princeton, NJ: Princeton University Press.

Bartholomew, G. A. and Birdsell, J. B. (1953). Ecology and the protohominids. *Amer. Anthropol.*, **55**, 481–498.

Berge, C. (1991). Size- and locomotion-related aspects of hominid and anthropoid pelves: an osteometrical multivariate analysis. *Hum. Evol.*, **6**, 365–376.

(1994). How did the australopithecines walk? A biomechanical study of the hip and thigh of *Australopithecus afarensis*. *J. Hum. Evol.*, **26**, 259–273.

Brunet, M., Guy, F., Pilbeam, D., *et al.* (2002). A new hominid from Upper Miocene of Chad, Central Africa. *Nature*, **418**, 145–151.

Carrier, D. R. (1984). The energetic paradox of human running and hominid evolution. *Curr. Anthropol.*, **25**, 483–495.

Carrier, D. R., Deban, S. M., and Otterstrom, J. (2002). The face that sank the Essex: potential function of the spermaceti organ in aggression. *J. Exp. Biol.*, **205**, 1755–1763.

Chaplin, G., Jablonski, N. G., and Cable, N. T. (1994). Physiology, thermoregulation and bipedalism. *J. Hum. Evol.*, **27**, 497–510.

Charnov, E. L. (1991). Evolution of life history variation among female mammals. *Proc. Natl. Acad. Sci. U. S. A.*, **88**, 1134–1137.

Charnov, E. L. and Berrigan, D. (1993). Why do female primates have such long lifespans and so few babies? or Life in the slow lane. *Evol. Anthropol.*, **1**, 191–194.

Crompton, R. H., Yu, L., Wang, W.-J., Günther, M., and Savage, R. (1998). The mechanical effectiveness of erect and "bent-hip, bent-knee" bipedal walking in *Australopithecus afarensis*. *J. Hum. Evol.*, **35**, 55–74.

Dart, R. A. (1959). *Adventures with the Missing Link*. London: Harper.

Darwin, C. (1871). *The Descent of Man, and Selection in Relation to Sex*. London: John Murray.

Day, M. H. (1977). Locomotor adaptation in man. *Biol. Hum. Affairs*, **42**, 149–151.

(1986). Bipedalism: pressures, origins and modes. In: *Major Topics in Primate and Human Evolution*, ed. B. Wood, L. Martin, and P. Andrews. Cambridge: Cambridge University Press. pp. 188–202.

de Sarre, F. (1994). The theory of initial bipedalism: on the question of human origins. *Riv. Biol/Biol. Forum*, **87**, 237–258.

DuBrul, E. L. (1962). The general phenomenon of bipedalism. *Amer. Zool.*, **2**, 205–208.

Elton, S., Foley, R. A., and Ulijaszek, S. J. (1998). Habitual energy expenditure of human climbing and clambering. *Ann. Hum. Biol.*, **25**, 523–531.

Etkin, W. (1954). Social behavior and the evolution of man's faculties. *Amer. Nat.*, **88**, 129–142.

Farfan, H. F. (1978). The biomechanical advantage of lordosis and hip extension for upright activity: man as compared with other anthropoids. *Spine*, **3**, 336–342.

(1995). Form and function of the musculoskeletal system as revealed by mathematical analysis of the lumbar spine. *Spine*, **20**, 1462–1474.

Fifer, F. C. (1987). The adoption of bipedalism by the hominids: a new hypothesis. *Hum. Evol.*, **2**, 135–147.

Gebo, D. L. (1996). Climbing, brachiation, and terrestrial quadrupedalism: historical precursors of hominid bipedalism. *Amer. J. Phys. Anthropol.*, **101**, 55–92.
Geist, V. (1971). *Mountain Sheep: a Study in Behavior and Evolution*. Chicago, IL: University of Chicago Press.
(1978). *Life Strategies, Human Evolution, Environmental Design*. New York, NY: Springer.
Gracovetsky, S. (1985). An hypothesis for the role of the spine in human locomotion: a challenge to current thinking. *J. Biomed. Eng.*, **7**, 205–216.
Gracovetsky. S. and Farfan, H. (1986). The optimum spine. *Spine*, **11**, 543–573.
Guthrie, R. D. (1970). Evolution of human threat display organs. *Evol. Biol.*, **4**, 257–302.
Hardy, A. (1960). Was man more aquatic in the past? *New Scientist*, **7**, 642–645.
Hayama, S., Nakatsukasa, M., and Kunimatsu, Y. (1992). Monkey performance: the development of bipedalism in trained Japanese monkeys. *Acta Anat. Nippon*, **67**, 169–185.
Hewes, G. (1961). Food transport and the origin of hominid bipedalism. *Amer. Anthropol.*, **63**, 687–710.
Hunt, K. D. (1991). Mechanical implications of chimpanzee positional behavior. *Amer. J. Phys. Anthropol.*, **86**, 521–536.
(1994). The evolution of human bipedality: ecology and functional morphology. *J. Hum. Evol.*, **26**, 183–202.
(1996). The postural feeding hypothesis: an ecological model for the evolution of bipedalism. *S. Afr. J. Sci.*, **92**, 77–90.
(1998). Ecological morphology of *Australopithecus afarensis*: traveling terrestrially, eating arboreally. In: *Primate Locomotion: Recent Advances*, ed. E Strasser. New York, NY: Plenum Press. pp. 397–418.
Isbell, L. A. and Young, T. P. (1996). The evolution of bipedalism in hominids and reduced group size in chimpanzees: alternative responses to decreasing resource availability. *J. Hum. Evol.*, **30**, 389–397.
Iwamoto, M. (1985). Bipedalism of Japanese monkeys and carrying models of hominization. In: *Primate Morphophysiology, Locomotor Analyses, and Human Bipedalism*, ed. S Kondo. Tokyo: University of Tokyo Press. pp. 251–260.
Jablonski, N. G. and Chaplin, G. (1993). Origin of habitual terrestrial bipedalism in the ancestor of the Hominidae. *J. Hum. Evol.*, **24**, 259–280.
(2000). Do theories of bipedalization stand up to anatomical scrutiny? *Amer. J. Phys. Anthropol.*, **30** (Suppl.), 187.
Jablonski, N. G., Chaplin, G., and McNamara, K. J. (2002). Natural selection and the evolution of hominid patterns of growth and development. In: *Human Evolution through Developmental Change*, ed. N. Minugh-Purvis and K. J. McNamara. Baltimore, MD: Johns Hopkins University Press. pp. 189–296.
Jenkins, F. A. (1972). Chimpanzee bipedalism: cineradiographic analysis and implications for the evolution of gait. *Science*, **178**, 877–879.
Jolly, C. J. (1970). The seed-eaters: a new model of hominid differentiation based on a baboon analogy. *Man*, **5**, 1–26.
Kelley, J. (1995). Sexual dimorphism in canine shape among extant great apes. *Amer. J. Phys. Anthropol.*, **96**, 365–389.
Kohler, W. (1959). *The Mentality of Apes*. New York, NY: Vintage Books.

Kortlandt, A. (1980). How might early hominids have defended themselves against large predators and food competitors? *J. Hum. Evol.*, **9**, 79–112.

Kramer, P. A. (1999). Modeling the locomotor energetics of extinct hominids. *J. Exp. Biol.*, **202**, 2807–2818.

Latimer, B. (1991). Locomotor adaptations in *Australopithecus afarensis*: the issue of arboreality. In: *Origine(s) de la Bipédie Chez les Hominidés*, ed. Y. Coppens and G. Senut. Paris: CNRS. pp. 169–176.

Latimer, B. and Ward, C. V. (1993). The thoracic and lumbar vertebrae. In: *The Nariokotome* Homo erectus *Skeleton*, ed. A. Walker and R. Leakey. Cambridge, MA: Harvard University Press. pp. 266–293.

Lee, P. C. (1991). Adaptations to environmental change: an evolutionary perspective. In: *Primate Responses to Environmental Change*, ed. H. O. Box. New York, NY: Chapman and Hall. pp. 39–56.

Lewin, R. (1999). *Human Evolution: an Illustrated Introduction*. Malden, MA: Blackwell.

Li, Y., Crompton, R. H., Alexander, R. M., Günther, M. M., and Wang, W. J. (1996). Characteristics of ground reaction forces in normal and chimpanzee-like bipedal walking by humans. *Folia Primatol.*, **66**, 137–159.

Livingstone, F. B. (1962). Reconstructing man's Pliocene pongid ancestor. *Amer. Anthropol.* **64**, 301–305.

Lovejoy, C. O. (1981). The origin of man. *Science*, **211**, 341–350.

Marett, J. R. de la H. (1936). *Race, Sex, and Environment*. London: Bale.

Marzke, M. W. (1986). Tool use and the evolution of hominid hands and bipedality. In: *Primate Evolution*, ed. J. G. Else and P. C. Lee. Cambridge: Cambridge University Press. pp. 203–209.

McHenry, H. M. (1982). The pattern of human evolution: studies on bipedalism, mastication, and encephalization. *Annu. Rev. Anthropol.*, **11**, 151–173.

Merker, B. (1984). A note on hunting and hominid origins. *Amer. Anthropol.*, **86**, 112–114.

Montgomery, G. D. (1988). Rhythmical man: an alternative hypothesis to ape-human evolution. *Spec. Sci. Tech.*, **11**, 153–159.

Morgan, E. (1982). *The Aquatic Ape*. London: Souvenir.

Nakatsukasa, M., Hayama, S., and Preuschoft, H. (1995). Postcranial skeleton of a macaque trained for bipedal standing and walking and implications for functional adaptation. *Folia Primatol.*, **64**, 1–29.

Napier, J. R. (1963). The locomotor functions of hominids. In: *Classification and Human Evolution*, ed. S. L. Washburn. New York, NY: Viking. pp. 178–189.

Oxnard, C. E. (1973). *Form and Pattern in Human Evolution*. Chicago, IL: University of Chicago Press.

(1975). *Uniqueness and Diversity in Human Evolution*. Chicago, IL: University of Chicago Press.

Parker, S. T. (1987). A sexual selection model for hominid evolution. *Hum. Evol.*, **2**, 235–253.

Pawlowski, B., Dunbar, R. I. M., and Lipowicz, A. (2000). Tall men have more reproductive success. *Nature*, **403**, 156.

Ravey, M. (1978). Bipedalism: an early warning system for Miocene hominoids. *Science* **199**, 372.

Reynolds, E. (1931). The evolution of the human pelvis in relation to the mechanics of the erect posture. *Papers Peabody Mus. Amer. Archaeol. Ethnol.*, **11**, 255–334.

Rodman, P. S. and McHenry, H. M. (1980). Bioenergetics and the origin of hominid bipedalism. *Amer. J. Phys. Anthropol.* **52**, 103–106.

Rose, M. D. (1984). Food acquisition and the evolution of positional behaviour: the case of bipedalism. In: *Food Acquisition and Processing in Primates*, ed. D. J. Chivers, B. A. Wood, and A. Bilsborough. New York, NY: Plenum Press. pp. 509–524.

(1991). The process of bipedalization in hominids. In *Origine(s) de la Bipédie Chez les Hominidés*, ed. Y. Coppens and G. Senut. Paris: CNRS. pp. 37–48.

Sanders, W. J. (1998). Comparative morphometric study of the australopithecine vertebral series Stw-H8/H41. *J. Hum. Evol.*, **34**, 249–302.

Sarmiento, E. E. (1998). Generalized quadrupeds, committed bipeds, and the shift to open habitats: an evolutionary model of hominid divergence. *Amer. Mus. Novit.*, **3250**, 1–28.

Shapiro, L. (1993). Functional morphology of the vertebral column in primates. In: *Postcranial Adaptation in Nonhuman Primates*, ed. D. L. Gebo. DeKalb, IL: Northern Illinois University Press. pp. 121–149.

Shipman, P. (1986). Scavenging or hunting in early hominids: theoretical framework and tests. *Amer. Anthropol.*, **88**, 27–43.

Sigmon, B. A. (1971). Bipedal behavior and the emergence of erect posture in man. *Amer. J. Phys. Anthropol.*, **34**, 55–60.

Sinclair, A. R. E., Leakey, M. D., and Norton-Griffiths, M. (1986). Migration and hominid bipedalism. *Nature*, **324**, 307–308.

Stanford, C. B. (1995). Chimpanzee hunting behavior and human evolution. *Amer. Sci.*, **83**, 256–261.

(1998). *Chimpanzee and Red Colobus: the Ecology of Predator and Prey*. Cambridge, MA: Harvard University Press.

Stearns, S. C. and Koella, J. C. (1986). The evolution of phenotypic plasticity in life-history traits: predictions of reaction norms for age and size at maturity. *Evolution*, **40**, 893–913.

Susman, R. L., Stern J. T., and Jungers, W. L. (1984). Arboreality and bipedality in the Hadar hominids. *Folia Primatol.*, **43**, 113–156.

Szalay, F. (1975). Hunting–scavenging protohominids: a model for hominid origins. *Man*, **10**, 420–429.

Veerhaegen, M. (1985). The aquatic ape theory: evidence and a possible scenario. *Med. Hypotheses*, **16**, 17–21.

Videan, E. N. and McGrew, W. C. (2002). Bipedality in chimpanzee (*Pan troglodytes*) and bonobo (*Pan paniscus*): testing hypotheses on the evolution of bipedalism. *Amer. J. Phys. Anthropol.*, **118**, 184–190.

Washburn, S. L. (1967). Behaviour and the origin of man. *Proc. Roy. Anthropol. Inst.*, **3**, 21–27.

Wescott, R. W. (1967). The exhibitionistic origin of human bipedalism. *Man*, **2**, 630.

Wheeler, P. E. (1984). The evolution of bipedality and loss of functional body hair in hominids. *J. Hum. Evol.*, **13**, 91–98.

(1991). The thermoregulatory advantages of hominid bipedalism in open equatorial environments: the contribution of increased convective heat loss and cutaneous evaporative cooling. *J. Hum. Evol.*, **21**, 107–115.

White, T. D., Suwa, G., and Asfaw, B. (1994). *Australopithecus ramidus*, a new species of early hominid from Aramis, Ethiopia. *Nature*, **371**, 306–312.
Wrangham, R. W. (1980). Bipedal locomotion as a feeding adaptation in gelada baboons, and its implications for hominid evolution. *J. Hum. Evol.*, **9**, 329–331.
Wundram, I. J. (1986). Cortical motor asymmetry and hominid feeding strategies. *Hum. Evol.*, **1**, 183–188.

14 *Modeling human walking as an inverted pendulum of varying length*

JACK T. STERN, JR., BRIGITTE DEMES
Stony Brook University

D. CASEY KERRIGAN
University of Virginia

Symbols and abbreviations

A_c	axial (centripetal or centrifugal) acceleration of the mass
ΔC	forward translation of the substrate contact point
COM	center of mass
D	instantaneous horizontal distance traveled by the mass since $t = 0$
D_h	the total horizontal distance traveled by the mass during one swing of the inverted pendulum, i.e., the step length
f	stride frequency
F_c	centripetal force required to keep the mass on a circular path
GRF_v	vertical component of the ground reaction force
H	instantaneous value of the height of the point mass (or COM)
H_{MDS1}	height of the point mass (or COM) at the middle of the first double-support phase
H_{MDS2}	height of the point mass (or COM) at the middle of the second double-support phase
H_{MSS}	height of the point mass (or COM) at the middle of the single-support phase
ΔH	maximum vertical excursion of the point mass (or COM) from MDS to MSS
I	moment of inertia of the mass about the pivot point of the inverted pendulum
L_c	the axially directed force exerted on the virtual stance limb (VSL) by the mass
$-L_c$	the axially directed force exerted on the mass by the virtual stance limb (VSL)
M	mass
MDS	middle of double-support phase

Shaping Primate Evolution, ed. F. Anapol, R. Z. German, and N. G. Jablonski. Published by Cambridge University Press.

MSS middle of single-support phase
R the length of the virtual stance limb (from substrate contact-point to the point mass or its surrogate) at any moment in time
R_{MDS} length of the virtual stance limb at MDS
R_{MSS} length of the virtual stance limb at MSS
S stature
t time
t_{MSS} half the duration of the swing of the inverted pendulum
V velocity of the mass
V_c axial (centripetal or centrifugal) velocity of the mass
V_o orbital (tangential) velocity of the mass
$\overline{V_h}$ desired average horizontal velocity of the mass
V' walking speed/stature
VSL the virtual stance limb (i.e, the massless rod of the inverted pendulum)
W gravity vector acting on the mass
W_c centripetal component of the gravity vector
W_o tangential component of the gravity vector
Z_2 the minimum value of GRF_v near MSS
ω angular velocity of the virtual stance limb
θ the angle between the virtual stance limb and vertical
θ_0 initial value of θ at the beginning of the virtual stance limb's swing (MDS1), thus half the total angular excursion of the VSL

Introduction

During walking, the potential energy of the human body is greatest at the middle of single support (MSS), whereas the kinetic energy is greatest near the middle of double support (MDS) (Cavagna *et al.*, 1976). For this reason, various authors have seen fit to analogize the stance phase of human bipedalism to an inverted pendulum. In this case, the walking human is represented by a point mass affixed to the upper end of a massless rod of length approximating that of the extended lower limb (Fig. 14.1). This rod has been called the "virtual stance limb" (VSL) by Lee & Farley (1998), and we shall adopt this terminology. Only two forces act on the mass: (1) gravity (W), which can be resolved into a tangential (W_o) and an axial (W_c) component, and (2) the axial force ($-L_c$) exerted on the mass by the VSL. W_c will be centripetally directed (i.e., toward the axis of rotation, which is the point of contact of the VSL with the ground), and will exceed $-L_c$ by an amount sufficient to supply the centripetal force required to keep the mass on a circular path. The higher the angular velocity (ω) of the mass, the greater is the requirement for such centripetal force ($M\omega^2R$) and, consequently, the smaller is the magnitude of $-L_c$. The magnitude of $-L_c$ will drop to 0 if the mass is moving so quickly that all of W_c must be used to supply centripetal force (i.e., if W_c=$M\omega^2R$).

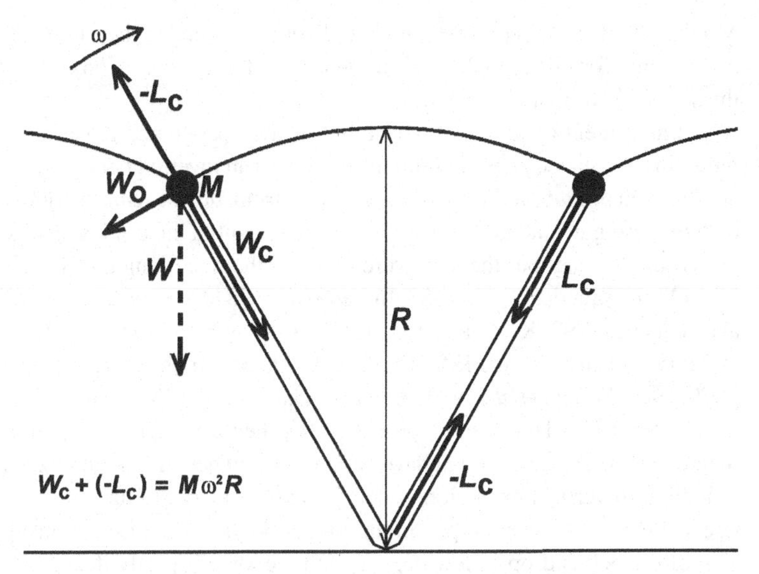

Figure 14.1. Forces acting on the mass (left) and virtual stance limb (or massless rod) of an inverted pendulum. See list of symbols and abbreviations.

Because the VSL is massless, it can be treated as being in equilibrium under the force applied to it by the mass (L_c) and an equal but opposite force applied by the ground ($-L_c$). It is not possible for the mass to have a higher angular velocity than satisfies the equation $W_c = M\omega^2 R$ because this would require that the VSL pull the mass centripetally. Such a pull could only be exerted if the VSL were affixed to the ground, a condition not met by the lower limb of a walking human. With this in mind, Alexander (1976, 1984, 1992a), Helene (1984), and Kram *et al.* (1997) used an inverted pendulum model to estimate a maximum possible speed of human walking. Alexander (1991) also used such a model for calculating the work needed to redirect the center of mass from its downward path at the end of one step to its upward path at the beginning of the next.

If an inverted pendulum is given an initial tangential velocity, its upswing will be slowed by W_o, and its downswing will be accelerated by this same force. The angular velocity will be greatest at the onset and end of the swing; it will be minimum in the middle. The required centripetal force ($M\omega^2 R$) must follow the same pattern. Since $-L_c$ is inversely related to the need for centripetal force, $-L_c$ will be smallest at the onset and end of the swing, and largest in the middle. Thus, so will the ground reaction force. This result, so very different from the actual situation for a bipedal human, in which the vertical component of the ground reaction force attains a local *minimum* at MSS, was obtained by

Mochon & McMahon (1980a) and could not be avoided in their model (1980b) by allowing for rise onto the ball of the foot followed by flexion of the knee in the second half of stance phase.

Authors mentioned above, and others, have recognized the limitations of modeling the stance phase of human walking as an inverted pendulum. Primary among such limitations is that markers on the trunk do not actually follow the set of "compass-gait" intersecting arcs that would result if the trunk's distance from the ground throughout the step were equal to the fixed length of the extended lower limb (Saunders *et al.*, 1953; Inman *et al.*, 1981; Alexander, 1992a; Minetti and Saibene, 1992; Kerrigan *et al.*, 1995). Although the trunk is indeed lowest at MDS and highest at MSS (Fischer, 1899; Saunders *et al.*, 1953; Murray *et al.*, 1966; Waters *et al.*, 1973; Alexander and Jayes, 1978; Inman *et al.*, 1981; Alexander, 1984; Thorstensson *et al.*, 1984), the total vertical excursion is less than would be predicted if it behaved as an inverted pendulum with fixed R. This reduction in vertical excursion is due in small part to a failure of the trunk to rise as high as the inverted pendulum model predicts because the unsupported side of the pelvis drops a few degrees and the knee is partly flexed (Gard and Childress, 1997, 1999). Of greater significance is the failure of the trunk to fall as far as the model predicts. This is caused almost entirely by plantar flexion of the trailing ankle near the end of stance phase (Kerrigan *et al.*, 2000), with some minor contribution made by pelvic rotation about the vertical axis (Kerrigan *et al.*, 2001).

Not only is the vertical excursion of the trunk during human walking less than predicted by the inverted-pendulum model, the points of intersection of the hypothetical arcs are rounded, so that the vertical motion of the trunk follows a path that is described or pictured as sinusoidal (Saunders *et al.*, 1953; Murray *et al.*, 1966; Lamoreux, 1971; Waters *et al.*, 1973; Inman *et al.*, 1981; Thorstensson *et al.*, 1984; Whittle, 1997). More importantly, the same is true of the calculated location of the center of mass (COM) (Fischer, 1899; Cavagna *et al.*, 1976; Simon *et al.*, 1977; Shimba, 1984; Iida and Yamamuro, 1987). Most authors refer to this sinusoidal oscillation as a function of time, and both the path of a trunk marker (Cappozzo, 1981; Whittle, 1997) and that of the calculated location of the COM (Crowe *et al.*, 1995; Whittle, 1997) have been shown to closely follow curves that can be described as sine functions of time. However, authors who have used interrupted-light photography (Saunders *et al.*, 1953; Murray *et al.*, 1966; Inman *et al.*, 1981) show figures illustrating sinusoidal vertical oscillation of the trunk with horizontal distance as the abscissa. In fact, were vertical oscillation of the trunk a sine function of horizontal distance, its path would be so close to that dictated by a sine function of time that it is highly unlikely experimental methods could distinguish the two possibilities, or either from the possibility that the vertical motion is a sine function of angular excursion of the lower limb. Regardless, in order for any such a path to be followed, the distance between the COM and the point of foot contact with the ground (hereafter referred to as virtual stance limb length) must be longer at both the beginning and end of the step than at mid-stance. In other words, the VSL should be both collapsible and extensible (Siegler *et al.*, 1982; Pandy and Berme, 1988a, 1988b; Alexander, 1992b; Lee and Farley, 1998). Various authors (Saunders *et al.*, 1953; Cavagna *et al.*, 1976; Alexander, 1992a; Minetti and Saibene, 1992; Lee and Farley, 1998) offer explanations for how this might occur.

If one were to accept that the vertical path of the COM is a sine function of time, it would enable ready prediction of how the vertical component of the ground reaction force changes with time. At any moment the vertical acceleration of the COM is the second derivative of its height (H) vs. time (t) curve, which itself has the shape of a sine curve from $-\pi/2$ to $3\pi/2$ over the interval from MDS1 to MDS2. In terms of the nomenclature used in Figure 14.2, at any moment (t) during this interval

$$H = H_{\mathrm{MDS}} + \Delta H \cdot [\sin(\pi t/t_{\mathrm{MSS}} - \pi/2) + 1]/2 \tag{14.1}$$

and

$$d^2H/dt^2 = (\Delta H/2)(\pi/t_{\mathrm{MSS}})^2 \cos(\pi t/t_{\mathrm{MSS}}) \tag{14.2}$$

In other words, a plot of vertical acceleration, and consequently vertical force, against time will have the shape of a cosine function of t, being maximum at MDS (when $t = t_{\mathrm{MSS}}$ or $t = 2\,t_{\mathrm{MSS}}$) and minimum at MSS (when $t = 0$). This is indeed the general shape of the human vertical ground reaction force trace. As it turns out, the best evidence that during walking the height of the COM is a sine function of time is derived by doubly integrating the record of vertical force against time for subjects walking at their preferred speeds, then subjecting the results to Fourier analyses (Crowe *et al.*, 1995). If we accept that $H \propto \sin t$ for all normal walking speeds (as Cappozzo's 1981 study of the motion of a trunk marker suggests), we would require only empirical data relating ΔH to speed of walking in order to reproduce the vertical component of the ground reaction force for any speed.

The usefulness of accepting that $H \propto \sin t$ is limited to calculating vertical velocities and vertical accelerations of the COM. It cannot provide us with a complete picture of how the collapsible/extensible inverted pendulum model moves because it provides no way to calculate length changes in the VSL against time, nor does it enable evaluation of the motion of an inverted pendulum whose mass follows a vertical path that is a sine function of the horizontal distance traveled by the mass or of the angular excursion of the VSL. It is the goal of this chapter to present a method that allows comparison of these alternatives

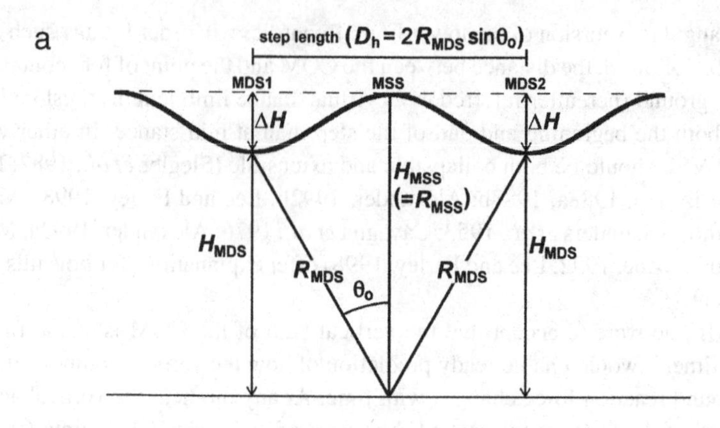

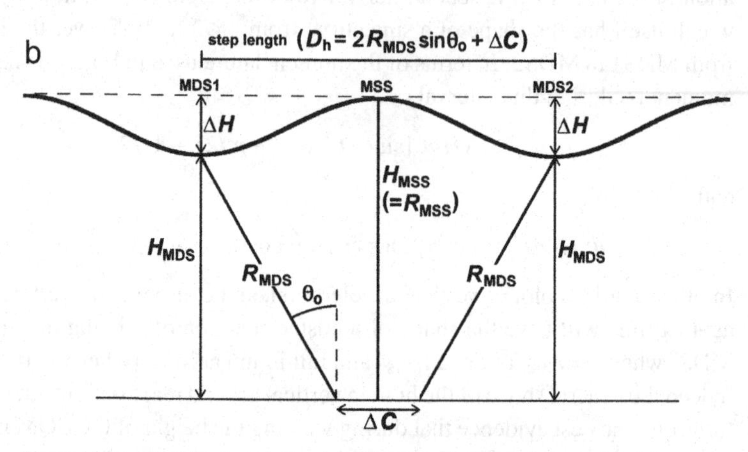

Figure 14.2. (a) A fixed-contact-point (FCP) inverted-pendulum model of the stance limb adjusted to allow for the sinusoidal path of the point mass during walking. The height (*H*) of the mass above the substrate is minimum at the middle of the double-support phases (MDS1 and MDS2). It is maximum in the middle of the single-support phase (MSS). At any location along the path, the length of the line (*R*) connecting the mass to the substrate contact point is a function of θ and *H* corresponding to that location. (b), Adjustment of the inverted-pendulum model to allow for a moving contact point (MCP). For the same angular excursion as illustrated in panel (a), the step length of the MCP model becomes the sum of the contact point's forward translation (ΔC) and that produced by the swing of the virtual stance limb (massless rod).

and also reveals how the length of the VSL, angular velocity of the lower limb, and virtually any other kinematic or kinetic parameter must change with time. A model that reliably predicts dynamic parameters of human walking will allow an evaluation of the energetics of bipedalism as well as size- and speed-related aspects of this gait.

Methods

Virtual stance limb length, R

In a walking human, assuming a fixed contact point (FCP) with the ground, several factors determine the instantaneous value of R: (1) the length of the support-phase lower limb, (2) the motion of the swing-phase lower limb, (3) the motion of the upper limbs, (4) the degree of pelvic tilt in the coronal plane, and (5) the displacement of internal organs (although this is likely to be far less significant in walking than in running). If one models walking as a collapsible/extensible inverted pendulum with the height of the point mass constrained to follow a path that is a sine function of time, the relationship between VSL length (R) and t can be calculated with complete accuracy if one knows the following facts (Fig. 14.2a):

(1) the height of the mass at the middle of the first double-support phase (H_{MDS1})
(2) the height of the mass at the middle of the single-support phase (H_{MSS})
(3) the height of the mass at the middle of the second double-support phase (H_{MDS2}), which can be presumed to be equal to H_{MDS1}
(4) the initial angle (θ_0) that the VSL makes with the vertical
(5) the time course of change in θ from MDS1 to MDS2, which can be presumed to be divided into two symmetric halves, MDS1 to MSS and MSS to MDS2

The greatest challenge to this reconstruction is posed by determining the time course of change in θ. One can easily determine the time from MDS1 to MSS (and MSS to MDS2) given the average horizontal velocity ($\overline{V}_{\mathrm{h}}$) from MDS1 to MDS2 and by assuming that the changes in velocity are symmetric on either side of MSS. The difficulty arises in determining instantaneous values of θ because its value at any moment is dependent on the preceding history of ω. This problem is soluble by computerized numerical integration under a set of initial conditions specifying H_{MSS}, vertical excursion (ΔH) of the mass from MDS to MSS, the initial value of θ, and $\overline{V}_{\mathrm{h}}$.

Lee and Farley (1998) emphasize that previous attempts to model human gait as an inverted pendulum have neglected the very important fact that the point of contact (identified by them as the point of force application) is not stationary but travels from the heel to the toe during a single step. There are two ways of viewing the significance of a moving contact point (MCP). Lee and Farley see it as allowing a reduced angular excursion for any given step length. The reduced angular excursion would result in the COM following a flatter arc of motion and, consequently, having a reduced vertical excursion. For reasons that will become apparent later, we choose to view any forward movement of the contact point as allowing a greater step length for any given angular excursion (Fig. 14.2b). Regardless, allowing for an MCP adds yet another factor that must be considered when attempting to reconstruct the relationship between the length *R* and *t*.

Our method of numerical integration, and our determination of values to be used as initial conditions, are now described.

Initial conditions of numerical integration

Vertical excursion of the center of mass during walking

Although we are interested in vertical excursion of the COM as a function of walking speed, far more data exist on vertical excursion for a trunk marker. A few studies have compared these two parameters. Whittle (1997), using double integration of the vertical ground reaction force to determine vertical excursion of the COM, found it to be 15% smaller than that of the pelvis, although phasic relationships were exactly the same. Employing similar methodology, Saini *et al.* (1998) reported an excursion of the COM 11% greater than that of either a sacral marker or the calculated center of the pelvis. These same authors also used segmental analysis to calculate motion of the COM, and with this technique found vertical excursion of the COM to be 7% smaller than that of a sacral marker, although the difference did not reach statistical significance. Their comparison of the different ways to measure vertical excursion of the COM led them to regard the force-plate method as least reliable, and this was also found to be the case by Thirunarayan *et al.* (1996) in their study of patients with gait disability. Our initial assumption will be that information on vertical excursion of trunk markers can substitute for that of the COM, but we will investigate the possibility that this is incorrect.

Employing the technique described by Kerrigan *et al.* (1995, 2000), we have acquired data on the vertical excursion of a marker attached to skin over the sacrum in 38 female and 22 male subjects walking barefoot across a 10 m walkway. Proper informed consent of all subjects was obtained. Each subject was instructed to walk at four speeds: very slow, slow, comfortable, and fast. The actual walking speed of each trial was determined by determining the average horizontal displacement over time of a body marker over six strides. Our results are presented in Figure 14.3a.

We have been able to generate comparative data by analyzing previous publications on the vertical excursion of a trunk marker (Murray *et al.*, 1966; Waters *et al.*, 1973; Inman *et al.*, 1981, using data from Lamoreux, 1971;

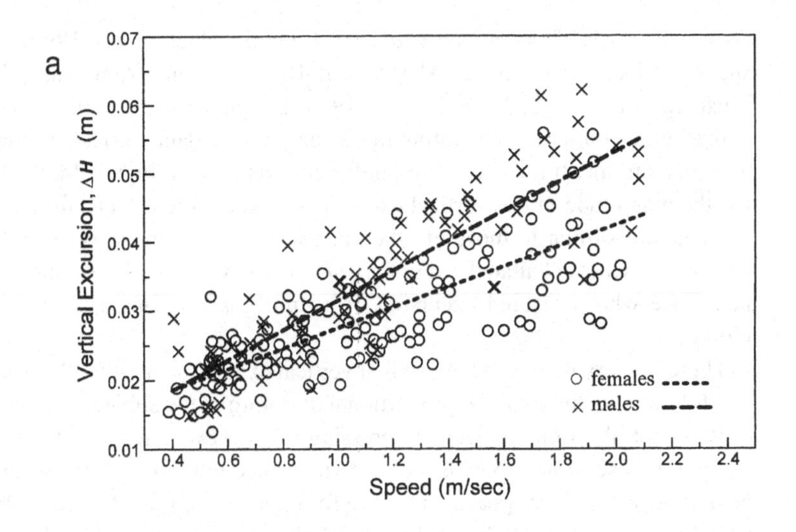

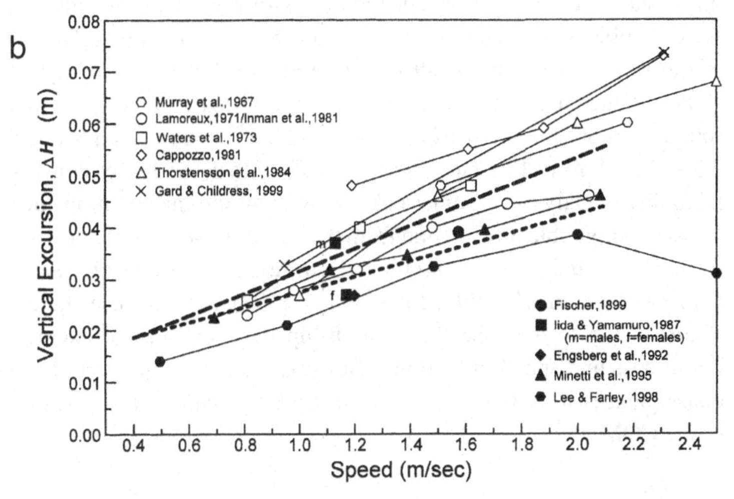

Figure 14.3. (a) Vertical excursion of a marker placed on the skin over the sacrum plotted as a function of speed in 38 female and 22 male subjects, each asked to walk at four different speeds. Of the possible 240 data points, 10 have been excluded as clear outliers. Superimposed onto the plotted data points are best-fit linear regression lines. The formula for females is $\Delta H = 0.01262 + 0.01494 \cdot \textit{Speed}$; for males $\Delta H = 0.00985 + 0.02173 \cdot \textit{Speed}$. The slopes of these regressions are significantly different ($P < 0.001$). (b) Data available from the literature on vertical excursion of trunk markers (open symbols and X) and the calculated position of the COM (closed symbols) plotted as a function of walking speed. The data point from Engsberg *et al.* (1992) represents children. Superimposed onto the literature data are the regression lines for our subjects (female, short-dashed line; male, long-dashed line). There seems to be a tendency for authors who calculated excursion of the COM to find a smaller value for ΔH than those who measured excursion of a trunk marker. Our data fall in the middle.

Cappozzo, 1981; Thorstensson *et al.*, 1984; Gard and Childress, 1999) or the calculated location of the COM (Fischer, 1899; Iida and Yamamuro, 1987; Engsberg *et al.*, 1992; Minetti *et al.*, 1995; Lee and Farley, 1998). There is an inaccuracy built into generating these comparative data because it often required taking measurements from published illustrations. Figure 14.3b shows that the magnitude of our vertical excursions, and the regressions fitting them to speed, are similar to those of other authors. There seems to be a tendency for authors who calculated excursion of the COM to find a smaller value than those who measured excursion of a trunk marker. Our data fall in the middle.

There is every reason to believe that vertical excursion of either the trunk or COM during walking will be proportional to stature of the subject. Stature was available for all of our subjects. Expressing vertical excursion as a fraction of stature for these subjects yields the graph presented in Figure 14.4a. Several of the authors cited above provided stature for their subjects, or data from which mean stature could be derived. Figure 14.4b presents the combined-sex line for our subjects superimposed onto data from the literature. The magnitudes of our relative excursions fall in the middle of those reported by others. Our slope relating relative vertical excursion to speed is very close to what was found by more than half of the other authors (if one excludes the outlier for 2.5 m sec^{-1} from Lee and Farley, 1998). In that our regression was derived from direct analysis of numerical data without the machinations necessary to extract comparable information from the literature, and since vertical excursion of the COM may not be significantly different from that of a sacral marker (Thirunarayan *et al.*, 1996; Saini *et al.*, 1998) we shall use the equation of Figure 14.4b to describe the relationship between vertical excursion of the point mass and speed of walking. However, at one point we will examine the consequences if vertical movement of a sacral marker has overestimated that of the COM.

Height of the center of mass during standing and walking

The most extensive literature review and original study of the location of the COM in the anatomic position is provided by Croskey *et al.* (1922). These authors report that nineteenth-century research (only on male subjects) provided the following estimates of COM location as a percentage of stature: 54.8 (Braune and Fischer), 56.7 (Weber and Weber), 58.0 (Harless), and 59.5 (von Meyer). Croskey *et al.* themselves found values of 56.5% for one group of 20 females, 56.2% for a second group of 50 females, and 55.8% for a group of 50 males. There was no correlation between relative location of the COM and either

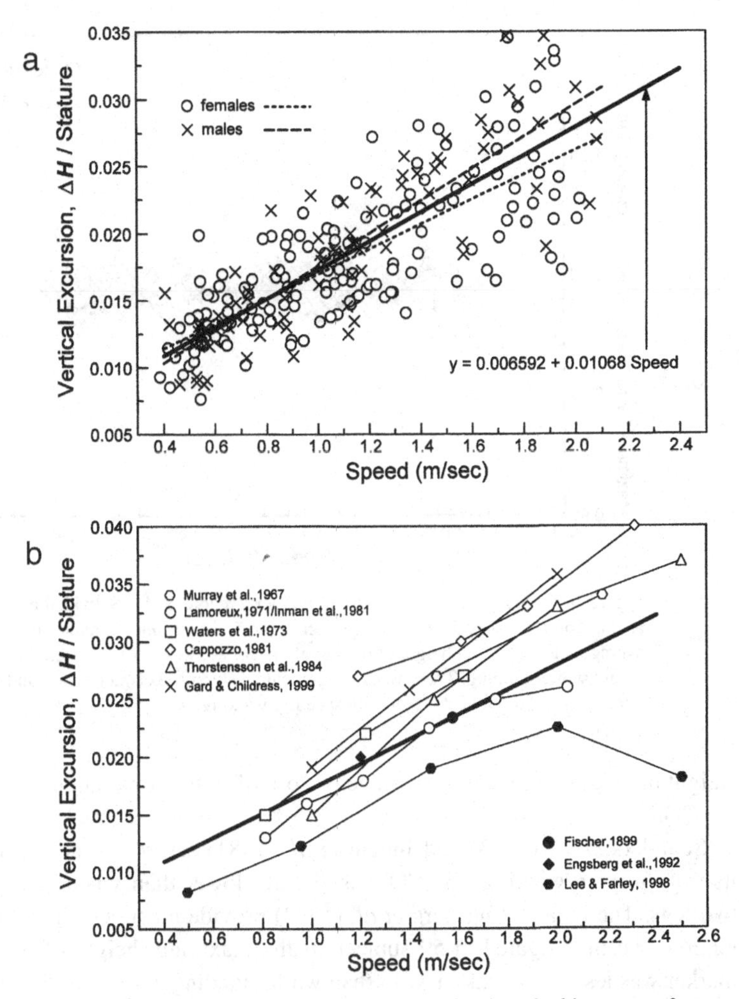

Figure 14.4. (a) Vertical excursion expressed as a fraction of subject stature for a marker placed on the skin over the sacrum plotted as a function of speed in our 38 female and 22 male subjects, each asked to walk at four different speeds. Of the possible 240 data points, 10 have been excluded as clear outliers. Superimposed onto the plotted data points are best-fit linear regression lines. The formula for females is $\Delta H/S = 0.00769 + 0.009276 \cdot Speed$; for males $\Delta H/S = 0.00549 + 0.012091 \cdot Speed$. The slopes of these are significantly different ($P < 0.01$), but we decided to use a combined-sex regression (solid line) that is halfway between those for the two sexes. (b) Data available from the literature on $\Delta H/S$ for trunk markers (open symbols and X) and the calculated location of the COM (closed symbols) plotted against walking speed. Superimposed onto data from the literature is the combined-sex line (thick) for our data. The magnitudes of our relative excursions fall in the middle of those reported by others. Our slope relating relative vertical excursion to speed is very close to more than half the other authors (if one excludes the outlier for 2.5 m sec^{-1} from Lee and Farley, 1998).

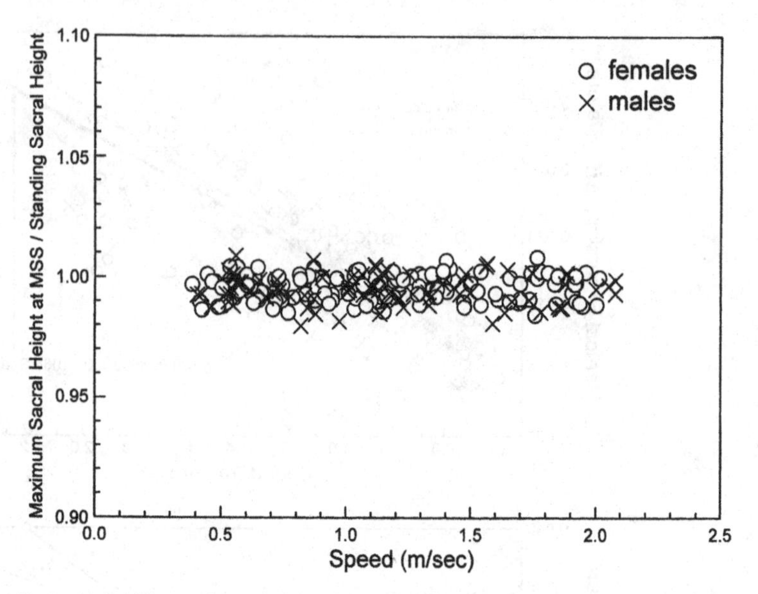

Figure 14.5. The maximum height of the sacral marker at MSS divided by its height at standing plotted against speed of walking for 37 female and 22 male subjects. On average, the maximum height of the sacral marker at MSS is 99.5% (s.e., 0.04%) of its height while standing. There was no significant difference between males and females, nor any correlation whatsoever with speed of walking.

weight or height. We shall use a value of 56% of stature for standing height of the COM.[1]

Saunders *et al.* (1953) and Inman *et al.* (1981) assert that at the peak of its trajectory in walking, the COM is slightly lower than when standing on two legs. The data of Engsberg *et al.* (1992) provide no support for this contention. We investigated in 59 subjects if the maximum height of the sacral marker was less at its peak in MSS than while standing. Figure 14.5 illustrates that the maximum height of the sacral marker at MSS is on average 99.5%

[1] The sacral marker placed by Kerrigan and coworkers is located in the median sagittal plane below the posterior superior iliac spines, at the apex of an equilateral triangle whose base connects these spines. Such a position overlies the fourth or fifth sacral vertebra. For 47 females (including those serving in this study), the average position of this marker was 56.3% (s.e., 0.22) of the distance from the soles to the top of the head. For 32 males (including those serving in this study), the average position of this marker was also 56.3% (s.e., 0.32) of the distance from the soles to the top of the head. These positions are virtually identical to the location of the COM reported in the literature. A conclusion that the COM lies at the level of the lower end of the sacrum is at odds with the findings of Braune and Fischer (1889), who placed the average location in three frozen cadavers at about the second sacral vertebra. However, the authors note that the abdomens of their specimens were "sunken." Possibly this resulted in an artificially high position of the COM.

(s.e., 0.04%) of its height while standing. There was no significant difference between males and females, nor any correlation whatsoever with speed of walking. In our analyses it was assumed that H_{MSS} was 99.5% of the standing height of the COM.

Initial angle of VSL to vertical

It is known that during normal walking, step length, and consequently angular excursion of the limb, increases with speed (Grieve and Gear, 1966; Murray *et al.*, 1966; Andriacchi *et al.*, 1977; Cappozzo, 1981; Inman *et al.*, 1981; Reynolds, 1987). We can evaluate this using measurements of H_{MSS}, ΔH, and step length for our 60 subjects walking at speeds ranging from very slow to fast. From such measurements we can calculate the angular excursion of a line that connects the sacral marker to the ground (see Fig. 14.2a):

$$\theta_0 = \arctan \frac{\textit{step length}/2}{H_{\text{MSS}} - \Delta H} \tag{14.3}$$

These results are presented in Figure 14.6a.

Comparative data are available from three sources:

(1) Fischer (1903) measured lower limb length and step length in 111 males walking overground at freely chosen speeds ranging from 1.5 to 2.1 m sec^{-1}. From these data one can calculate half angular excursion of the lower limb as

$$\theta_0 = \arctan \frac{\textit{step length}/2}{\textit{lower limb length}} \tag{14.4}$$

(2) Dr. Daniel Schmitt (Department of Biological Anthropology and Anatomy, Duke University) has kindly provided us with his direct measurements of the angular excursion of the hip–ankle line in five male and three female subjects walking overground at a range of speeds from 1.1 to 2.6 m sec^{-1}.

(3) Lee and Farley (1998, Fig. 5) provide data gathered from five subjects on the angular excursion of the line connecting the calculated position of the COM and the moving contact point.

Figure 14.6b presents a comparison of our estimates of VSL angular excursion to those of the authors cited above. The magnitudes of our angular excursions are comparable to those found by Schmitt, but are smaller than those we calculated from Fischer's data or, more importantly, those reported by Lee and Farley. Because our sample size is the most extensive, we shall use the relationship suggested by it.

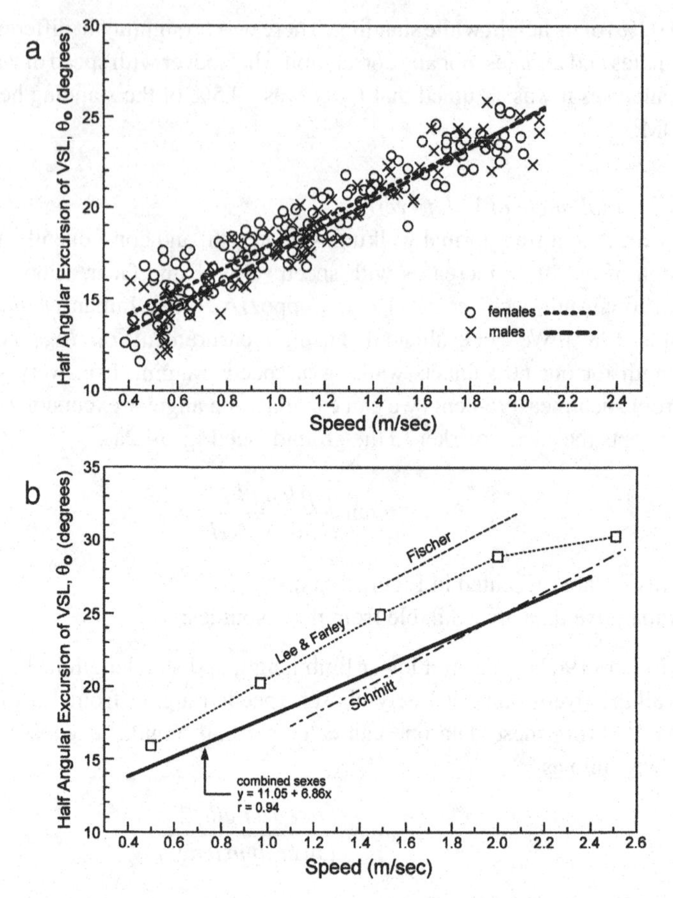

Figure 14.6. (a) Calculated values for half the total angular excursion of the VSL (in this case using a sacral marker as a substitute for the COM) plotted against walking speed for 38 female and 22 male subjects, each walking at four different speeds. Of the possible 240 data points, 10 have been excluded as clear outliers. Superimposed onto the plotted data points are best-fit linear regression lines. The formula for females is $\theta_0 = 11.48 + 6.64 \cdot \mathit{Speed}$; for males $\theta_0 = 10.24 + 7.29 \cdot \mathit{Speed}$. The slopes of these two regressions are not significantly different. (b) A comparison of our data for θ_0 as a function of speed compared to data available in the literature. The thick line represents the best-fit linear regression for all our subjects. The other lines are: (1) the best-fit linear regression line for angular excursion of the lower limb in eight subjects studied by Dr. Daniel Schmitt (Duke University), (2) the best-fit linear regression line calculated from data in Fischer (1903), and (3) data on five subjects taken from Figure 5 in Lee and Farley (1998). Each of the lines from the literature represents a different way of measuring angular excursion. The magnitude of our angular excursions is comparable to that found by Schmitt, who used a line between hip and ankle as the limb axis and measured the angle of this line to vertical. The magnitude of our values is smaller than those reported by Lee and Farley, who calculated the angle of the VSL from analysis of force-plate data.

Earlier we stated that there are two ways of viewing the significance of a moving contact point. Lee and Farley saw it as allowing a reduced angular excursion for any given step length, but it could just as well be viewed as allowing a greater step length for any given angular excursion (see Fig. 14.6b). It is precisely because the angular excursions reported by Lee and Farley using an MCP model are actually greater, not less, than those we derived for our subjects assuming an FCP that we have chosen to regard an MCP as allowing a greater step length for any given angular excursion.

Motion of the inverted pendulum with a fixed contact point (FCP)

The motion of an FCP inverted pendulum that varies in length during its "swing" was calculated by numerical integration on a desktop computer. Input into the program consists only of

(1) stature (*S*), from which H_{MSS} ($= R_{\mathrm{MSS}}$) is calculated as $0.56S \times 0.995$, and
(2) desired average horizontal velocity ($\overline{V}_{\mathrm{h}}$).

Using these values:

(3) the initial angle (θ_0) between the VSL and vertical is calculated as

$$\theta_0 = 11.05 + 6.86\overline{V}_{\mathrm{h}} \qquad \text{[see Fig. 14.6b]} \qquad (14.5)$$

hereafter referred to as the standard condition for the angular excursion-speed relationship

(4) the vertical excursion of the mass from MDS to MSS is calculated as

$$\Delta H = (0.006592 + 0.01068\overline{V}_{\mathrm{h}})S \qquad \text{[see Fig. 14.4a]} \qquad (14.6)$$

hereafter referred to as the standard condition for the vertical excursion–speed relationship

(5) the initial height of the mass at MDS is calculated as

$$H_{\mathrm{MDS}} = H_{\mathrm{MSS}} - \Delta H \qquad (14.7)$$

(6) the initial length of the pendulum (i.e., *VSL*) at MDS as

$$R_{\mathrm{MDS}} = H_{\mathrm{MDS}}/\cos\theta_0 \qquad (14.8)$$

(7) the horizontal distance covered by the "swing" (i.e., the step length) as

$$D_{\mathrm{h}} = 2R_{\mathrm{MDS}}\sin\theta_0 \qquad (14.9)$$

(8) the time it takes for the pendulum to reach MSS as

$$t_{MSS} = 0.5D_h/\overline{V}_h \tag{14.10}$$

The simulation has the requirement that the mass follows a curve of H vs. t that has minima equal to H_{MDS} at $t = 0$ and $t = 2t_{MSS}$ (i.e., at MDS1 and MDS2), a maximum equal to H_{MSS} at t_{MSS}, and has the shape of a sine curve from $-\pi/2$ to $3\pi/2$ over the entire time interval. This is accomplished throughout the simulation by making the instantaneous value of H equal to $H_{MDS} + \Delta H \cdot [\sin(\pi t/t_{MSS} - \pi/2) + 1]/2$.

The calculations performed by the computer program are described in the Appendix. Briefly, knowing the initial angle, angular velocity, and components of the weight vector, the motion of the mass is determined by numerical integration performed at Δt intervals of 0.1 msec.[2] At the end of every interval, the program calculates (1) the new angle of the VSL based on the angular velocity at the beginning of the interval, (2) new angular and tangential velocities resulting from the action of W_o on the mass during the interval, (3) the new components of the weight vector, (4) a change in VSL length caused by the constraint that the height of the mass follows a sinusoidal path, (5) a further change in angular and tangential velocities caused by the requirement that angular momentum remains the same after the change in VSL length, (6) the centripetal velocity and centripetal acceleration of the mass during the interval, (7) the new centripetal force that would be required to maintain the mass on a circular path with its new VSL length and angular velocity, (8) the new outward force that the VSL would exert on the mass if it were following such a circular path, (9) an adjustment to the force that the VSL exerts on the mass to account for the fact that mass has either more or less axial acceleration than required to keep it on a circular path, and finally (10) the force exerted on the VSL by the substrate. The value of the mass is irrelevant since all calculations of force are expressed relative to body weight.

Each simulation begins with the operator specifying the desired average horizontal velocity ($\overline{V}_h$) during the simulated step. During the first iteration of the simulation the mass is assigned an initial angular velocity (at $t = 0$, MDS1) that knowingly will produce a calculated average horizontal velocity greater than desired. At the end of the first iteration, the average horizontal velocity during this invalid swing is used to calculate a new initial angular velocity for a second iteration that will result in an average horizontal velocity closer to $\overline{V}_h$. Further iterations are performed with progressively smaller initial angular

[2] Test simulations carried out with $\Delta t = 0.01$ msec gave results that differed by less than 1% for all numerical output.

velocities until one is identified that produces the desired average horizontal velocity, which is also the one that leads to an *H–t* curve meeting the requirements of shape and location of maxima and minima. For this final simulation, the program outputs values of all relevant parameters at intervals of 0.1 msec.

It is worth noting that this simulation allows determination of the force exerted on the substrate in relation to any other relevant parameter. Thus, it allows determination of the vertical component of the ground reaction as a function of time. The accuracy of the computer program was verified by comparing its moment-by-moment calculation of vertical ground reaction force to that derived by knowing that the vertical acceleration of the mass must be the second derivative of its *H–t* curve (see above). The two methods of calculating vertical ground reaction force were identical to four significant digits. An additional test of the progam consisted of demonstrating that the average of all the individual values for horizontal velocity during each 0.1 msec interval was equal to the input desired value of $\overline{V}_{\text{h}}$.

Two additional variants of the FCP inverted-pendulum model were investigated. In one, *H* was constrained to follow a sine function of the instantaneous horizontal distance (*D*) traveled since the onset of the motion. The results of these simulations will be compared to those in which *H* was assumed to be a sine function of time. In the second variant, *H* was constrained to follow a sine function of the instantaneous angle (θ) of the VSL. The results of simulations in which $H \propto \sin \theta$ were intermediate between the other two, but so close to the variant with $H \propto \sin D$ that they will not be discussed separately.

Motion of the inverted pendulum with a moving contact point (MCP)

As stated above, for any given total angular excursion of the VSL, a moving contact point produces a longer step length (Fig. 14.2b). For any given walking speed, an increase in step length implies an increase in step duration. Since we have empirically measured the vertical excursion of a COM-substitute, an increase in step duration simply stretches out the sine wave in the same way as would an increase in angular excursion. Thus, at any given speed of progression, the vertical ground reaction force, being proportional to the double derivative of the vertical excursion, would be changed in precisely the same way if step duration were increased by a moving contact point or by a greater angular excursion. Other kinematic and kinetic parameters will differ, particularly the curve of *R* vs. *t*. An MCP model was created to allow consideration of its effect on parameters other than vertical force, as well as its interaction with differing assumptions on the sinusoidal nature of *H*.

For the five subjects studied by Lee and Farley (1998), the point of contact traversed an average distance of 0.20 m. For these subjects, the average length of the line from the ground to the greater trochanter was 0.88 m. Martin and Saller (1959) report that such a line is on average 51.5% of a person's stature. This predicts an average height of 1.71 m for Lee and Farley's subjects, and thus a motion of the contact point amounting to 11.7% of stature. We shall assume a value of 0.117 S for ΔC, partitioned so that half this movement occurs prior to MSS and half after MSS. In our MCP model, the contact point moves from its initial position to its final position at a rate that is linearly proportional to time.

Results

Stride frequency

In our FCP model (see Fig. 14.2a), stride length $= 2D_h$ (i.e., $4\,R_{MDS} \sin \theta_0$), and stride duration $= 2D_h/\overline{V}_h$. Stride frequency is the reciprocal of stride duration, and can be calculated as $\overline{V}_h/4R_{MDS} \sin \theta_0$. In turn, $R_{MDS} = (R_{MSS} - \Delta H)/\cos \theta_0$, and R_{MSS} is a fixed percentage of stature. Thus, our model predicts that for any given stature, stride frequency is solely dependent on the preset $\overline{V}_h$ and our choice of formulae relating angular and vertical excursions to speed. If these formulae are appropriate, each simulation should yield a stride frequency close to that found in experimental data for persons walking at the associated $\overline{V}_h$. Grieve and Gear (1966) determined in 20 subjects (mostly adult, ranging in stature from 1.42 m to 1.81 m) that stride frequency (f) is best fit by an equation $f = 64.8V'^{\,0.57}$, where V' is walking speed divided by stature. The average of the coefficients found by Cappozzo (1981) for five male subjects ranging in height from 1.72 m to 1.81 m leads to the equation $f = 63.0V'^{\,0.53}$.

Figure 14.7a shows the scatter of stride-frequency data points generated by our FCP model at 1 cm intervals of stature from 1.42 m to 1.81 m and at 0.2 m sec^{-1} intervals of speed from 0.4 to 2.4 m sec^{-1}. An effect of stature on the relationship between f and V', not mentioned by other authors, is indicated. The best-fit power function for all our generated data is $f = 67.9V'^{0.57}$.

Figure 14.7b presents the best-fit power function curve for our generated data alongside curves representing the power formulae of Grieve and Gear (1966) and Cappozzo (1981), as well as four data points derived from averages for six subjects studied by Minetti *et al.* (1995). Also plotted in this figure is the best-fit power curve of our model altered to allow for an MCP by adding $2\Delta C$ to the stride length, where $\Delta C = 0.117S$. The stride frequencies predicted by our FCP model are as close to experimentally observed values as are the results of one experiment to another. The stride frequencies predicted by the MCP

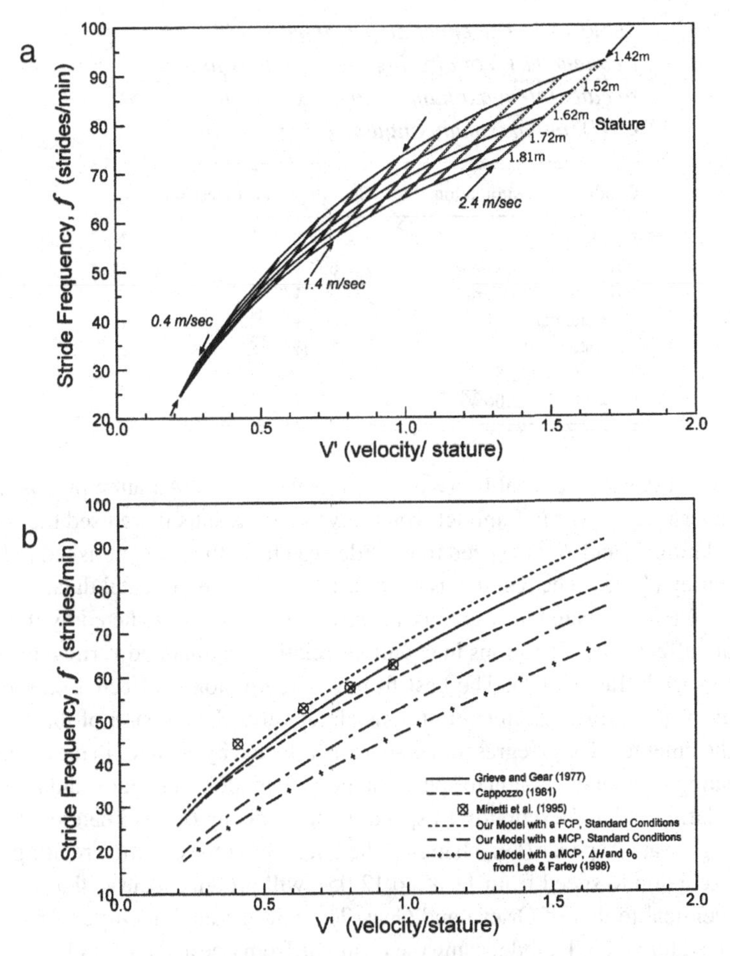

Figure 14.7. (a) Our model's prediction of the relationship between stride frequency and relative speed of walking at statures spanning the range characterizing subjects in the study of Grieve and Gear (1966). Each data point (+) represents the calculated stride frequency generated by our model at 1cm intervals of stature (1.42–1.81 m) and at 0.2 m sec^{-1} intervals of speed (0.4–2.4 m sec^{-1}). An effect of stature on the relationship between f and V', not mentioned by Grieve and Gear, is indicated. (b) A comparison of f–V' curves generated by (1) the power formulae of Grieve and Gear (1966) and Cappozzo (1981), (2) four data points derived from averages for six subjects studied by Minetti *et al.* (1995), (3) the best-fit power regression to values (shown in panel a) generated by our model assuming a fixed pont of contact (FCP), and what we have accepted as the standard relationships between ΔH, θ_0, and speed, from our 60 subjects, (4) the best-fit power regression to values generated by our model assuming a moving point of contact (MCP) and what we have accepted as the standard relationships between ΔH, θ_0, and speed, and (5) the best-fit power regression to values generated by our MCP model but using the relationships between ΔH, θ_0, and speed determined from data published by Lee and Farley (1998). Our FCP model predicts stride frequencies that match closely those observed in actual subjects.

Table 14.1. *The effect on predicted stride frequencies (f) of altering the equations used to calculate maximum vertical excursions (ΔH) and maximum angular excursions (θ_0)*

Conditions of simulation	Best-fit power regression
Standard	$f = 67.9V'^{\,0.57}$
$\Delta H = 140\%$ Standard	$f = 69.2V'^{\,0.58}$
$\Delta H = 60\%$ Standard	$f = 66.7V'^{\,0.56}$
$\theta_0 = \text{standard} - 4°$	$f = 84.5V'^{\,0.48}$
$\theta_0 = \text{standard} + 4°$	$f = 56.3V'^{\,0.62}$
$\theta_0 = 11.05 + 5.86\,\overline{V}_h$	$f = 73.9V'^{\,0.61}$
$\theta_0 = 11.05 + 7.86\,\overline{V}_h$	$f = 62.7V'^{\,0.54}$

model deviate noticeably from experimental values. An attempt was made to determine if the MCP model would give better results if we used the formulae relating θ_0 and ΔH to speed that we derived from the data published in Lee and Farley (1998). The result was even further from experimental data.

We investigated how greatly the stride frequencies predicted by the models are affected by alterations in equations relating angular and vertical excursions to speed (Table 14.1). The best-fit power regression is affected minimally by even major changes in assumptions about vertical excursion of the COM. On the other hand, this regression is strongly affected by changes in the equation relating angular excursion to speed. In the FCP model using our standard formula relating vertical excursion to speed, simply raising our estimate of θ_0 by one degree at all speeds (i.e., changing the intercept of the equation relating angular excursion to speed from 11.05 to 12.05), will produce a curve that is virtually identical to that of Grieve and Gear. The power function curve of f vs. V' derived for an MCP model using the formulae from Lee and Farley (1998) can also be made virtually identical to that from Grieve and Gear, but in this case it requires lowering Lee and Farley's estimate of θ_0 by 7.5 degrees at all speeds.

Horizontal and angular displacement

Having doubly integrated the horizontal ground reaction force to obtain horizontal displacement of the COM, Lee and Farley (1998) found an approximately linear relationship of horizontal displacement with time. This led them to assume that the angle of the VSL also changed linearly with time. In both the FCP and MCP versions of our inverted-pendulum model, the horizontal distance and angle are indeed only "approximately" linear with time (Fig. 14.8). We do not know if this deviation from linearity would have significantly affected the calculations of Lee and Farley.

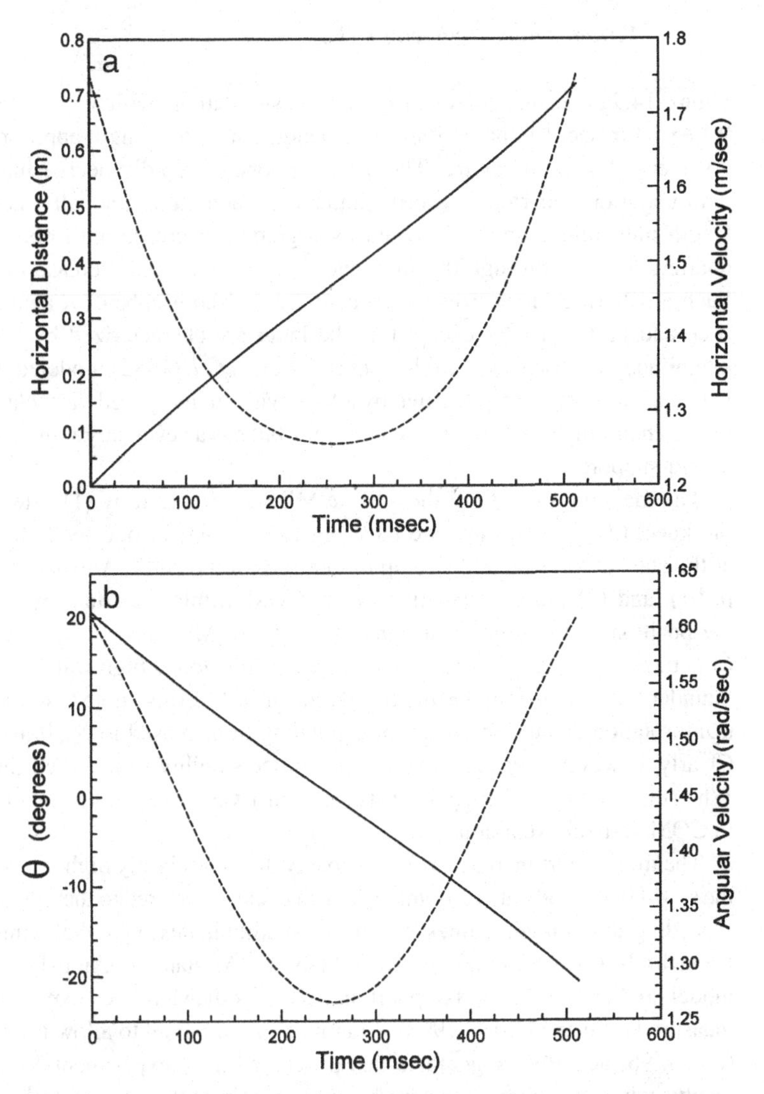

Figure 14.8. Our model's predictions for the time course of horizontal displacement of the point mass (a, solid line), horizontal velocity of the point mass (a, dashed line), angular displacement of the VSL (b, solid line), and angular velocity of the VSL (b, dashed line) in a virtual person 1.78 m tall (that of the senior author prior to aging) walking at 1.4 m sec^{-1} with a fixed point of contact. While changes in horizontal displacement of the mass and angular displacement of the VSL are approximately linear with time, they are not precisely so. This is equally true in simulations with a moving contact point, and may alter the calculations of any models that assume strict linearity.

Virtual stance limb length (R)

Figure 14.9a presents curves of R vs. t for simulations with $\overline{V}_h$ ranging from 0.4 to 2.4 m sec^{-1} in an FCP inverted pendulum with a length appropriate for a subject of 1.78 m stature. The pattern is one of rapidly decreasing length until about one-quarter way through the step, then more gradually increasing length until mid-support, followed by a gradual decrease until about three-quarters the way through the step, and ending by a rapid increase in length. Such a pattern characterizes experimentally derived graphs of R against time (Lee and Farley, 1998), except that the latter are characterized by a smaller magnitude increase in R at mid-support. Siegler *et al.* (1982) modeled walking with each lower limb represented by a VSL with an interposed viscoelastic element. Their graph of R vs. t resembles ours, but has an even larger increase in R at mid-support.

The maximum length of the VSL at MDS is produced by (1) extension of the knee, (2) the fact that "we have big feet... and set our heels down first at the start of the step and rise up on our toes at the end" (Alexander, 1992a, p. 61), and (3) the high position of the COM within the trunk by virtue of the position of the limbs (Cavagna *et al.*, 1976; Minetti and Saibene, 1992; Minetti *et al.*, 1995). All these factors are accentuated at high walking speeds. Saunders *et al.* (1953) and Minetti and Saibene (1992) also note that horizontal trunk rotation creates an effect corresponding to increased lower limb length. Clearly, however, there must be a limit to one's ability to lengthen the VSL. When this limit is reached, gait must change in a way that avoids further increase in COM vertical excursion.

The midsupport increase in R is probably due to a slightly higher position of the COM in the body at this point in the gait cycle, compared to the immediately preceding and following times, because of maximum flexion of the swing-phase lower limb (Cavagna *et al.*, 1966) and elbows. As stated previously, the FCP model predicts a larger mid-support increase in R than has been experimentally measured (Lee and Farley, 1998). Alteration of our model to allow for an MCP (Fig.14.9b) actually exaggerates this difference from experimentally derived results, whereas alteration that restrains the height of the mass to follow a sine function of horizontal distance traveled (Fig. 14.9c) brings it more into line with experimental results regarding this one point.

Vertical ground reaction force

Figure 14.10a presents curves of the FCP model's prediction of the vertical component of the ground reaction force (GRF_v) plotted against time for $\overline{V}_h$

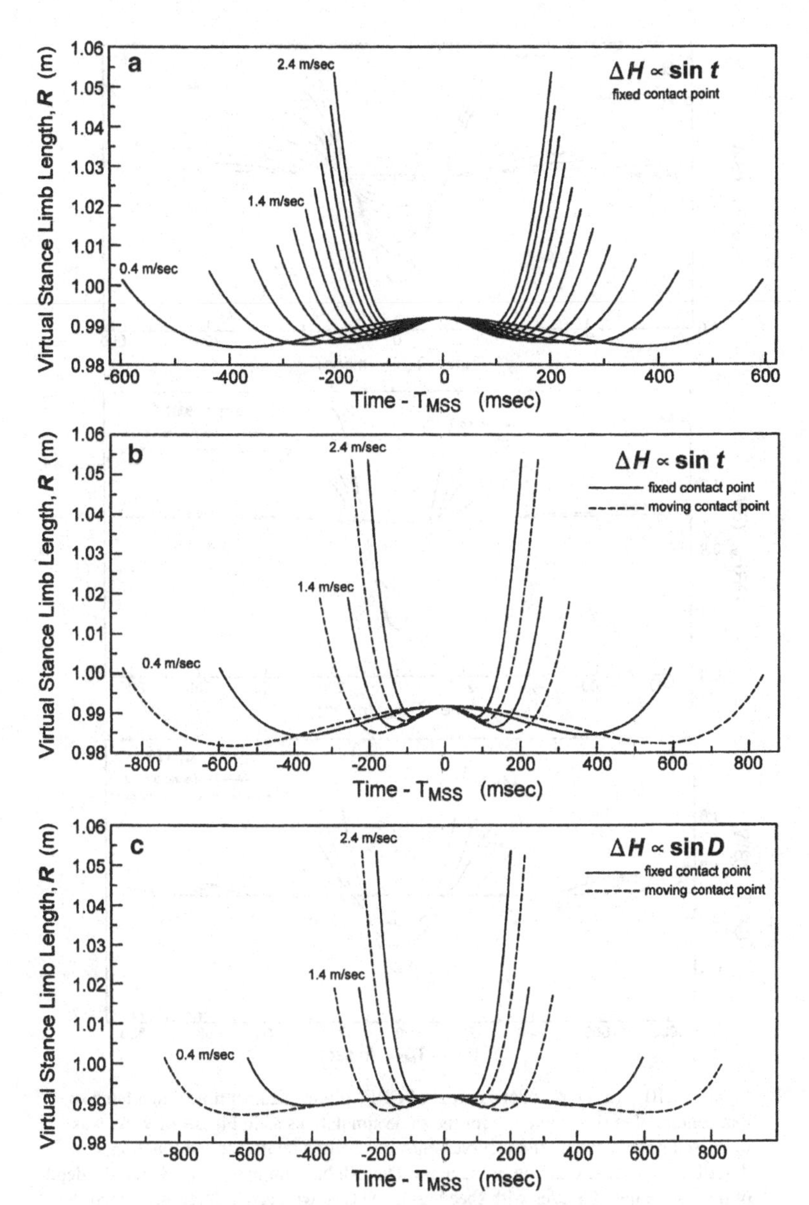

Figure 14.9. (a) Our model's output of virtual stance limb length, R, plotted against time (calculated so that centers of curves coincide) for the simulations of a 1.78 m tall subject walking at values of $\overline{V}_{\mathrm{h}}$ ranging from 0.4 to 2.4 m sec^{-1}. (b) Alteration of the model to allow for a moving contact point. (c) Alteration of the model so as to constrain the height of the mass to follow a sine function of horizontal distance traveled. See text for discussion.

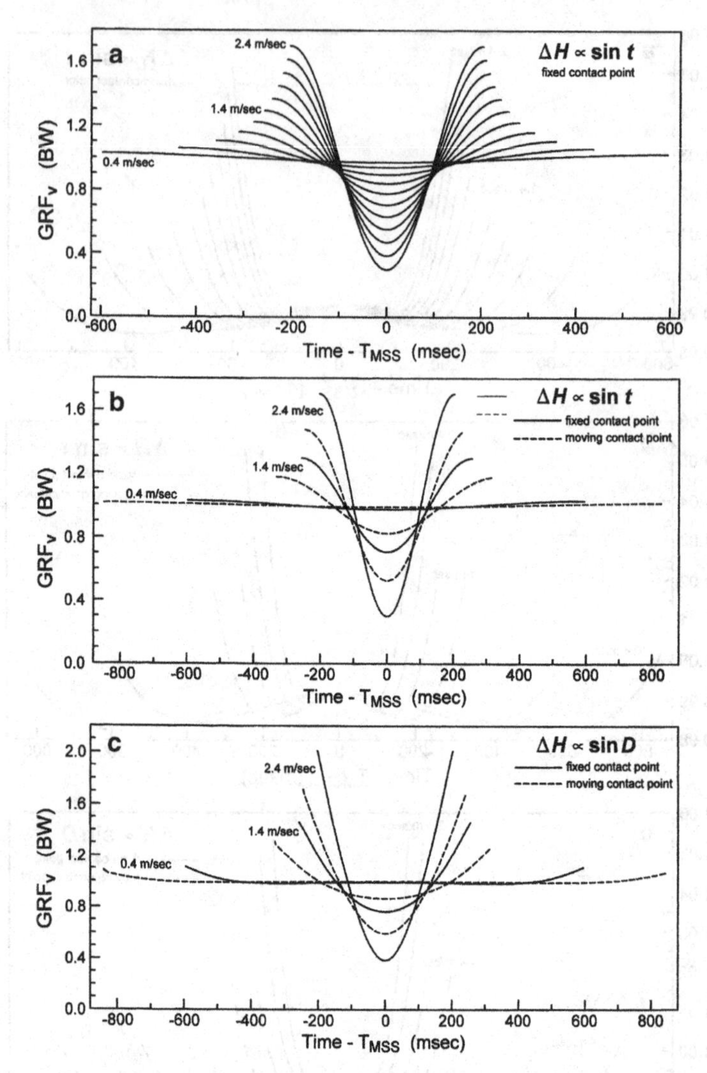

Figure 14.10. (a) Our model's output of GRF_V plotted against time (calculated so that centers of curves coincide) for the same simulations as in Figure 14.9. As was discussed previously, all these curves must be inverted relative to the function describing vertical excursion of the mass. They all have minima at MSS, and the depth of this minimum increases with speed, as is the case with actual force-plate records. (b) Alteration of the model to allow for a moving contact point produces vertical force curves with lower maxima and higher minima. (c) Alteration of the model so that $\Delta H \propto \sin D$ also produces vertical force curves with lower maxima and higher minima, but there is no longer any requirement that the curves be inverted sine waves. In fact, the result is a change in shape that makes them quite unlike force-plate curves recorded experimentally.

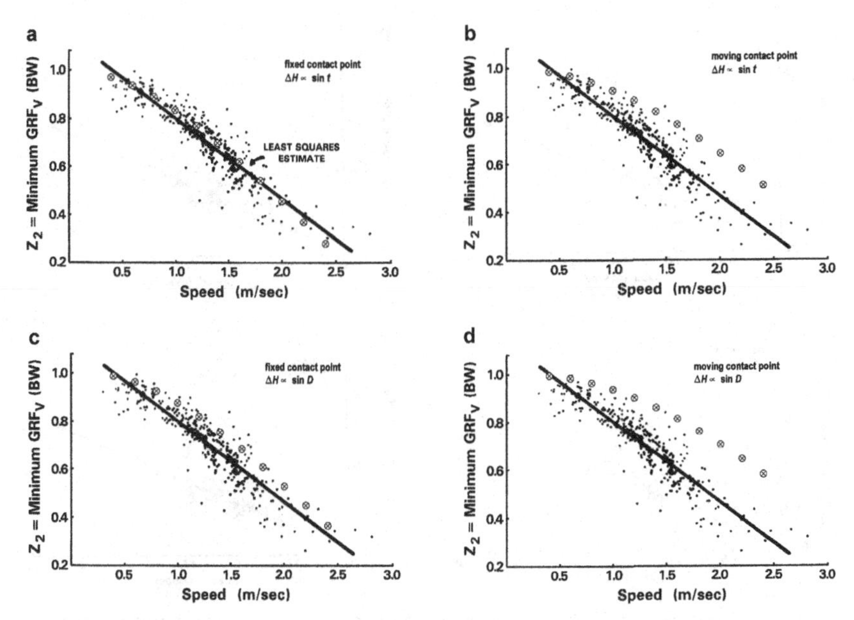

Figure 14.11. Each of the four panels contains the same reproduction of a portion of Figure 5 from Andriacchi *et al.* (1977, with the permission of Elsevier Science) illustrating the relationship between the minimum in GRF_v ($= Z_2$) and walking speed found in 17 subjects with an average stature of 1.73 m. Their data are shown as black dots and their regression line is indicated. In each panel the results of one or another variant of our model are superimposed as ⊗ symbols. (a) values calculated by our FCP model with $\Delta H \propto \sin t$. (b) values calculated by our MCP model with $\Delta H \propto \sin t$. (c) values calculated by our FCP model with $\Delta H \propto \sin D$. (d) values calculated by our MCP model with $\Delta H \propto \sin D$. The results for the FCP model with $\Delta H \propto \sin t$ are very similar to experimental data, and even the curvilinear relationship produced by the model is reflected by a curvilinear trend in the actual data. Our FCP model's predictions assuming that $\Delta H \propto \sin D$ produce results more divergent from actual data, and the results of simulations assuming a moving contact point, fit worst of all.

from 0.4 to 2.4 m sec^{-1} in an inverted pendulum with length appropriate for a subject of 1.78 m stature and $H \propto \sin t$. As was discussed previously, all these curves must be inverted relative to the function describing vertical excursion of the COM. Alteration of the model to allow for an MCP (Fig.14.10b) produces vertical force curves with lower maxima and higher minima. Alteration of the model so that $H \propto \sin D$ (Fig. 14.10c) also produces vertical force curves with lower maxima and higher minima, and alters their shape so as to make them quite unlike experimentally determined force-plate records.

Andriacchi *et al.* (1977) have quantified the relationship between the minimum in GRF_v (which they call Z_2) and speed in 17 subjects with an average stature of 1.73 m. Figure 14.11 presents their results along with the values

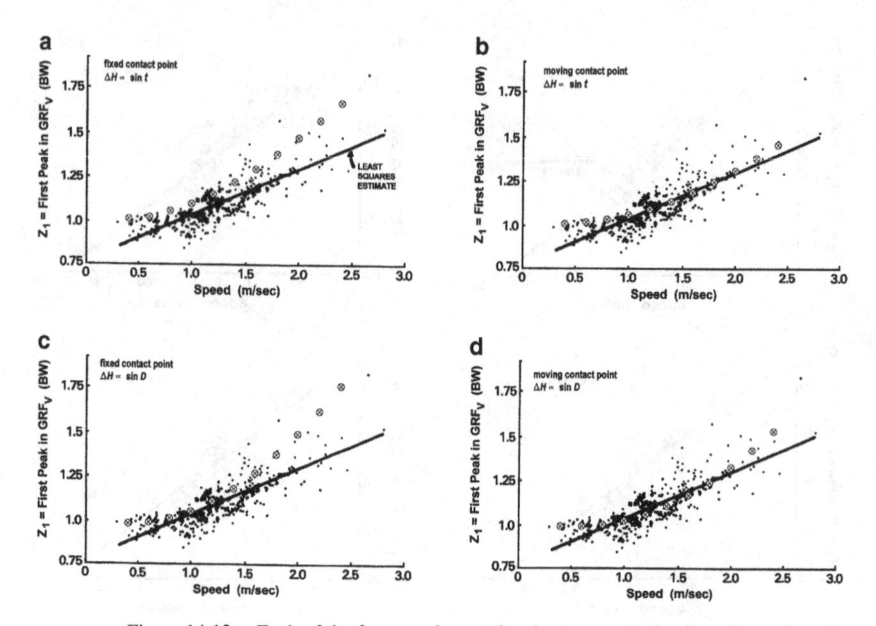

Figure 14.12. Each of the four panels contains the same reproduction of a portion of Figure 5 from Andriacchi *et al.* (1977, with the permission of Elsevier Science) illustrating the relationship between the first maximum in GRF_V (= Z_1) and walking speed found in 17 subjects with an average stature of 1.73 m. Their data are shown as black dots and their regression line is indicated. In each panel the results of one or another variant of our model are superimposed as ⊗ symbols. (a) values calculated by our FCP model with $\Delta H \propto \sin t$. (b) values calculated by our MCP model with $\Delta H \propto \sin t$. (c) values calculated by our FCP model with $\Delta H \propto \sin D$. (d) values calculated by our MCP model with $\Delta H \propto \sin D$. Simulations assuming a moving point of contact produce results that match most closely the experimental ones. Again we note that the curvilinear relationship produced by the model is reflected by a curvilinear trend in the actual data for the first peak in GRF_V. Also, simulations assuming that the vertical excursion is a sine function of time are a bit better than those assuming it follows a sine function of horizontal distance.

generated by our simulations for an inverted pendulum with length appropriate for this stature. The results of our FCP simulation with $H \propto \sin t$ are very similar to the data of Andriacchi *et al.* (1977). We note that the curvilinear relationship produced by the model is reflected by a curvilinear trend in the actual data. Simulations assuming that $H \propto \sin D$ produce results more divergent from actual data, and the results of simulations assuming an MCP fit worst of all.

It is more difficult to determine if the model produces realistic estimates of maximum GRF_V. Our simulation begins and ends in MDS, but attributes all forces to only one limb. We would not expect realistic predicted values for GRF_V outside the period of single support. It is known that during a complete stride,

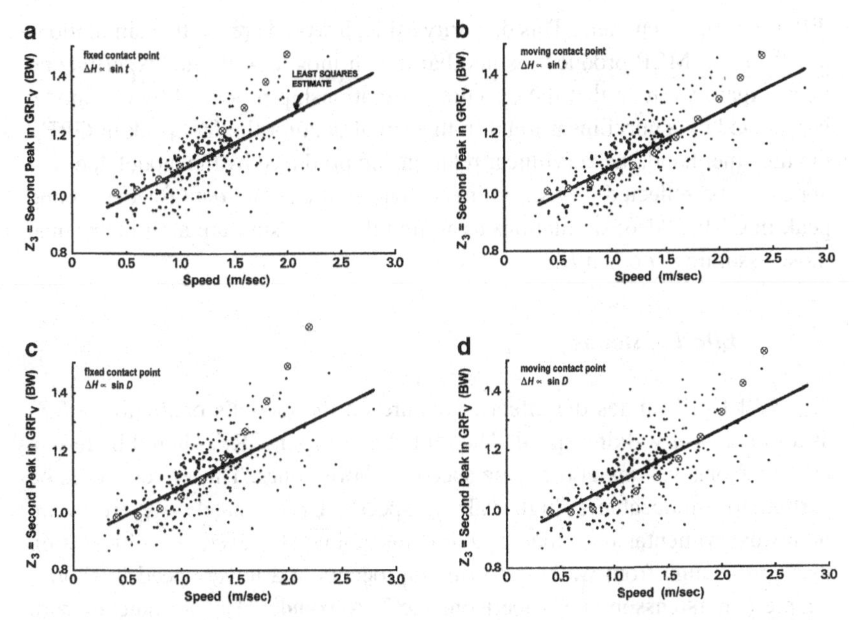

Figure 14.13. Each of the four panels contains the same reproduction of a portion of Figure 5 from Andriacchi *et al.* (1977, with the permission of Elsevier Science) illustrating the relationship between the second maximum in GRF_v ($= Z_3$) and walking speed found in 17 subjects with an average stature of 1.73 m. Their data are shown as black dots and their regression line is indicated. In each panel the results of one or another variant of our model are superimposed as ⊗ symbols. (a) values calculated by our FCP model with $\Delta H \propto \sin t$. (b) values calculated by our MCP model with $\Delta H \propto \sin t$. (c) values calculated by our FCP model with $\Delta H \propto \sin D$. (d) values calculated by our MCP model with $\Delta H \propto \sin D$. Simulations assuming a moving point of contact produce results that match most closely the experimental ones. The curvilinear relationship produced by the model does not appear to be reflected by a curvilinear trend in the actual data for the second peak in GRF_v. Also, simulations assuming that $\Delta H \propto \sin t$ are a bit better than those assuming $\Delta H \propto \sin D$.

the percentage of time spent in double support declines with speed. Combining results summarized in Inman *et al.* (1981) with our own observations, we feel we can roughly approximate the percentage of a stance phase devoted to double support as equal to $(65 - 18.75\overline{V}_h)$. Allotting half of this percentage value to the beginning of the stance phase and half to the end, and knowing that our simulation starts and ends in the middle of double support, we were able to eliminate portions of the GRF_v curves that occurred outside the period of single support. In Figures 14.12 and 14.13 our model's predictions for single-support maximum values of GRF_v are plotted onto the results of Andriacchi *et al.* (1977) for the first and second peaks of GRF_v. One obvious difference is that our model predicts equal first and second peaks, whereas actual data reflect a difference between them. This disparity aside, it would appear that simulations assuming an MCP produce results that match most closely the experimental ones. Again we note that the curvilinear relationship produced by our model is reflected by a curvilinear trend in the actual data for the first peak in GRF_v. On the other hand, the curvilinear relationship produced by the model does not appear to be reflected by a curvilinear trend in the actual data for the second peak in GRF_v. Also, simulations assuming that $H \propto \sin t$ are a bit better than those assuming $H \propto \sin D$.

Effect of stature

Figure14.14 illustrates the effect of stature on the model's predictions of Z_2 as a function of walking speed. The mid-stance dip in GRF_v should be less in taller persons walking at the same speed as shorter ones. This effect should be particularly noticeable at high walking speeds. Unfortunately, we are aware of no experimental data that bear on this point. However, the observation that gait change from walking to running occurs at a lower speed in shorter people (Thorstensson and Robertson, 1987; Alexander, 1992a) matches with this particular result of our model.

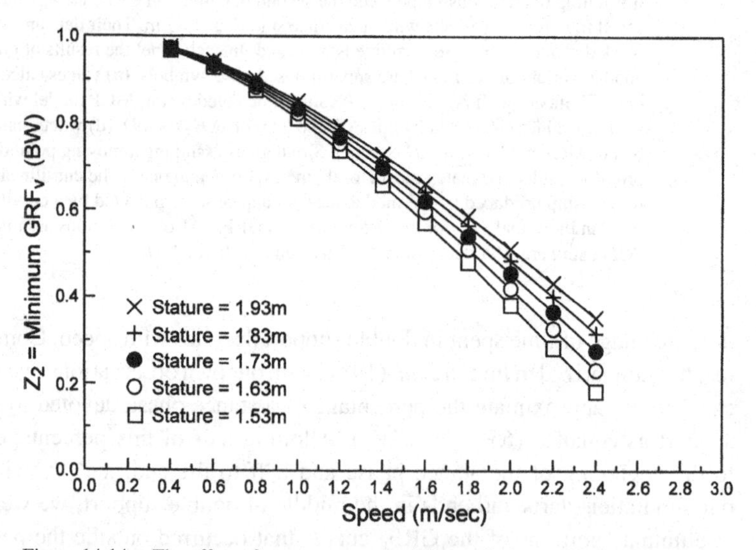

Figure 14.14. The effect of stature on our model's predictions of Z_2 as a function of walking speed. The midstance dip in GRF_v is predicted to be less in taller persons walking at the same speed as shorter ones. The effect should be particularly noticeable only at high walking speeds.

Effect of differing assumptions about vertical excursion of the COM

Whittle (1997), employing double integration of a force-plate record, reported that the vertical excursion of the COM is approximately 85% that of a pelvic marker, and that this reduction occurs both by lowering the MSS peak and by raising the MDS trough. Using the same methodology, Lee and Farley (1998) also reported values for vertical excursion of the COM that are less than other authors have reported for trunk markers (Fig. 14.4b). Saini *et al.* (1998), adopting what they considered to be the more accurate segmental analysis method, found that vertical excursion of the COM was on average only 7% smaller than that of a sacral marker, but the difference was not statistically significant. We investigated the effect of reducing ΔH by 7% and 15%. (It makes little difference if this reduction occurs symmetrically at peaks and troughs, or is applied only at the peak or the trough.) The effect of reducing ΔH on predicted Z_2*–speed* curves (Fig. 14.15) is greatest at high walking speeds. If our standard conditions have overestimated ΔH, a corrected Z_2*–speed* curve would resemble the results of Andriacchi *et al.* (1977) a bit less closely. However, we note that the evidence for any substantial difference between vertical excursion of the COM and that of a sacral marker is based on double integration of force-plate

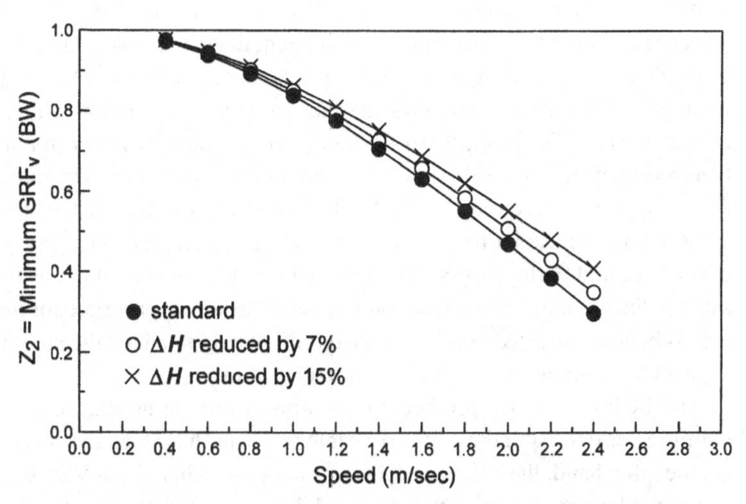

Figure 14.15. The effect on our FCP model's predictions of Z_2 as a function of walking speed if we had assumed the vertical excursion of the mass was either 7% or 15% smaller than indicated by our data on a sacral marker.

records, and this method of assessing COM positional change has been shown to be the least reliable (Thirunarayan *et al.*, 1996; Saini *et al.*, 1998).

Discussion

In this paper we evaluate various ways to model human walking as an inverted pendulum. Experimental data on the relationships between speed, angular excursion of the lower limb, and vertical excursion of a sacral marker are input into models that allow for a fixed point of contact, a moving point of contact, a vertical path of the COM that is a sine function of time, or a vertical path of the COM that is a sine function of distance (or angular excursion of the lower limb). In general, assuming vertical excursion is a sine function of time produces results more in accord with experimental data than does any other assumption about its path. Intuitively, one would expect models using a moving point of contact to be more realistic than those assuming a fixed point of contact. But MCP models did not always yield superior results. The relationship between stride frequency and speed predicted by the FCP model matched quite closely the experimental data of Grieve and Gear (1966). The match would have been nearly perfect if we had increased our estimate of step length by a small amount, as would occur with a total angular excursion 2° greater than our data suggest. In order to achieve a comparable fit with a model that used an MCP, it would be necessary to reduce the estimate of step length by a very substantial amount, as would occur if total angular excursion were decreased by 14–15°. It does not seem possible that our experimental data on angular excursion, which match closely those of D. Schmitt (pers. comm.) and are already lower in magnitude than reported by some other authors, could be so erroneous. The explanation for the apparent inaccuracy of the MCP model is that its associated step length is too great. We used a formula relating ΔC to stature that corresponds to the data in Lee and Farley (1998), but they identified the contact point as being the same as the point of force application. Maybe this is inappropriate for modeling the body as an inverted pendulum. After all, during much of the step, rotation appears to be about the axis of the talocrural joint.

The FCP model also predicts the minimum vertical ground reaction force with greater correspondence to experimental data than does the MCP model. On the other hand, the MCP model seems to predict changes in VSL length that correspond more closely to experimental data, though there are so few of these available that such correspondence may be spurious. Of greater interest is the fact that the MCP model seems to generate predictions for the peaks in vertical ground reaction force that are closer to experimental data.

Our conclusion is that a combination of the two models may turn out to be best of all. The peaks in GRF_V are events occurring early and late in a step. The minimum GRF_V occurs in the middle of a step. It is possible that the early part of a step, when the foot is being brought flat onto the ground, is best modeled by assuming a moving point of contact. The same might be true for the latter part of the step, when the heel is rising. During the middle of the step, when the limb is rotating about the talocrural joint, an FCP model would be best. Such a mechanism may have the reduced step length required to make the MCP model predict more realistic stride frequencies.

The ability to accurately model human walking as an inverted pendulum that varies in length according to the requirement that the height of the mass is a sine function of time implies that the external work of walking is devoted to producing/controlling length changes in the VSL and to redirecting the tangential velocity of the COM from its downward and forward trajectory at the end of one step to an upward and forward trajectory at the beginning of the next step (Cavagna *et al.*, 1976; Minetti and Saibene, 1992; Minetti *et al.*, 1995; Duff-Raffaele *et al.*, 1996). Deviations of the model's predictions from experimental data are largely confined to the second peak at high-speed walking. Cavagna *et al.* (1976) suggested that at high-speed walking the push of the trailing foot interferes with the pendulum mechanism.

Alexander (1976, 1992a) inquired about the maximum possible walking speed if energy considerations were disregarded. Having assigned to an inverted pendulum the dimensions appropriate for an adult human, he calculated the average horizontal speed when the tangential velocity reached the magnitude that would require more centripetal force than could be supplied by gravity, i.e., a tangential velocity that would require a negative substrate-reaction force that pulls on the person. One obvious problem with this approach is that for an inverted pendulum of fixed length, the highest tangential velocities occur at the beginning of its swing, rather than in the middle, when the actual substrate reaction force of walking is minimum. Alexander calculated the maximum horizontal speed of walking to be 3 m sec^{-1}, which is substantially higher than the speed at which humans naturally change from a walk to a run: at ~2 m sec^{-1} (Noble *et al.*, 1973; Thorstensson & Robertson, 1987; Beuter and Lefebvre, 1988; Hreljac, 1993, 1995). This led him to favor an explanation for gait change based on optimization of energetic expenditure.

Our model is associated with a minimum substrate reaction force occurring at MSS. We can carry the simulations up to the speed at which this drops to zero. For a stature of 1.73 m, this occurs at 3.02 m sec^{-1}, still far above the normal walk–run transition. For taller individuals the speed would be even higher (e.g., 3.28 m sec^{-1} for a stature of 1.93 m). Thus, we can confirm Alexander's conclusion that the walk–run transition is not *normally* determined by the maximum possible speed of an inverted pendulum. On the other hand, the ability to avoid negative ground reaction forces may very well limit the speed of "abnormal" walking. Persons can walk at speeds (~4 m sec^{-1}) that are higher than our model predicts as possible under standard conditions. However, as shown above, diminishing the vertical excursion of the COM either by raising the troughs or lowering the peaks of the *H–t* curve causes elevation of the Z_2–*speed* curve (Fig. 14.15), enabling the attainment of higher speeds before Z_2 falls to zero. Race walking corresponds to a lowering of the peaks of the *H–t* curve (Alexander, 1984, 1992a), and we agree it is entirely possible that under conditions of race walking the maximum speed is limited by the ability to avoid negative ground reaction forces.

Several authors (Cavagna *et al.*, 1976; Alexander, 1984, 1992b; Kram *et al.*, 1997) have suggested that the normal walk–run transition may be determined by energy considerations. Strong evidence against this view has been presented by Hreljac (1993, 1995), who proposes that the transition occurs when the stress on muscles that dorsiflex the ankle is perceived as too great. Our model directs attention to one other factor that might play a role. During normal walking it is absolutely essential that VSL length increase with speed (see Fig. 14.9). There must be a limit to which this distance can be increased by normal limb movements. Cappozzo (1981, p. 415) notes that above a speed of 2.4 m sec^{-1} "walking traits undergo considerable modification and assume the characteristic of race walking." It may be that above this speed it is impossible to increase the length of the VSL at MDS to the extent required by extrapolation of the normal relationships between angular excursion, vertical excursion, and speed. If that were so, a person might choose to run. When this choice is made, the sign of ΔH changes in that $H_{MSS} < H_{MDS}$, and H_{MSS} drops well below the standing height of the COM. Altering the sign of ΔH produces GRF_v–*t* curves that are the inverted from those presented in Figure 14.10. In other words, they look like sine curves that are peaked in the middle. Such is the typical pattern of actual substrate reaction forces in running.

While we are gratified that a model of human walking as an inverted pendulum of varying length predicts dynamic parameters of walking very similar to those determined experimentally, we acknowledge inaccuracy of the variables used as input for the model. Accurate data on vertical motion of the actual COM over a wide range of walking and running speeds is badly needed. So are data on the appropriate pivot point to be used at different moments in the step, and on the angular excursion of a line joining this point to the COM as a function of speed. When such data become available, it may be worth extending analysis of our model to predictions of the energetic cost associated purely with altering pendular length, and thereby relating our work to the basic science and clinical problems focusing on this parameter.

Acknowledgments

We thank Dr. Daniel Schmitt (Department of Biological Anthropology and Anatomy, Duke University) for making available to us his data on angular excursion of the lower limb during walking. This research was supported by NSF Research Grant BCS 0109331 (to JTS and BD) and NIH Research Grant HD 01351 (to DCK).

References

Alexander, R. McN. (1976). Mechanics of bipedal locomotion. In: *Perspectives in Experimental Biology, vol. 1*, ed. P. S. Davis. Oxford: Pergamon. pp. 493–504.

(1984). Walking and running. *Amer. Sci.*, **72**, 348–354.

(1991). Energy saving mechanisms in walking and running. *J. Exp. Biol.*, **160**, 55–69.

(1992a). *The Human Machine*. New York, NY: Columbia University Press.

(1992b). A model of bipedal locomotion on compliant legs. *Phil. Trans. Roy. Soc. Lond.*, **338**, 189–198.

Alexander, R. McN. and Jayes, A. S. (1978). Vertical movements in walking and running. *J. Zool. Lond.*, **185**, 27–40.

Andriacchi, T. P., Ogle, J. A., and Galante, J. O. (1977). Walking speed as a basis for normal and abnormal gait measurements. *J. Biomech.*, **10**, 261–268.

Beuter, A. and Lefebvre, R. (1988). Un modèle théorique de transition de phase dans la locomotion humaine. *Can. J. Sport Sci.*, **13**, 247–253.

Braune, W. and Fischer, O. (1889). In: *On the Centre of Gravity of the Human Body*, translated by P. G. J. Maquet and R. Furlong. Berlin: Springer, 1985.

Cappozzo, A. (1981). Analysis of the linear displacement of the head and trunk during walking at different speeds. *J. Biomech.*, **14**, 411–425.

Cavagna, G. A., Thys, H., and Zamboni, A. (1976). The sources of external work in level walking and running. *J. Physiol.*, **262**, 639–657.

Croskey, M. I., Dawson, P. M., Luessen, A. C., Marohn, I. E., and Wright, H. E. (1922). The height of the center of gravity in man. *Amer. J. Physiol.*, **61**, 171–185.

Crowe, A., Schiereck, P., de Boer, R. W., and Keessen, W. (1995). Characterization of human gait by means of body center of mass oscillations derived from ground reaction forces. *IEEE Trans. Biomed. Eng.*, **42**, 293–303.

Duff-Raffaele, M., Kerrigan, D. C., Corcoran, P. J., and Saini, M. (1996). The proportional work of lifting the center of mass during walking. *Amer. J. Phys. Med. Rehabil.*, **75**, 375–379.

Engsberg, J. R., Tedford, K. G., and Harder, J. A. (1992). Center of mass location and segment angular orientation of below-knee amputee and able-bodied children during walking. *Arch. Phys. Med. Rehabil.*, **73**, 1163–1168.

Fischer, O. (1899). Der Gang des Menschen II. Theil: die Bewegung des Gesammtschwerpunktes und die äusseren Kräfte. *Abhandl. K. Sächsischen Gesellsch. Wissensch.*, **43**, 1–130.

(1903). Kinematics of the swing of the leg. In: *The Human Gait,* translated by P. Maquet and R. Furlong. Berlin: Springer, 1987. pp. 315–384.

Gard, S. A. and Childress, D. S. (1997). The effect of pelvic list on the vertical displacement of the trunk during normal walking. *Gait and Posture*, **5**, 233–238.

(1999). The influence of stance-phase knee flexion on the vertical displacement of the trunk during normal walking. *Arch. Phys. Med. Rehabil.*, **80**, 26–32.

Grieve, D. W. and Gear, R. J. (1966). The relationship between length of stride, step frequency, time of swing and speed of walking for children and adults. *Ergonomics*, **5**, 379–399.

Helene, O. (1984). On "waddling" and race walking. *Amer. J. Phys.*, **52**, 656.

Hreljac, A. (1993). Preferred and energetically optimal gait transition speeds in human locomotion. *Med. Sci. Sports and Exerc.*, **25**, 1158–1162.

(1995). Determinants of the gait transition speed during locomotion: kinematic factors. *J. Biomech.*, **28**, 699–677.

Iida, H. and Yamamuro, T. (1987). Kinetic analysis of the center of gravity of the human body in normal and pathological gaits. *J. Biomech.*, **20**, 987–995.

Inman, V. T., Ralston, H. J., and Todd, F. (1981). *Human Walking*. Baltimore, MD: Williams and Wilkins.

Kerrigan, D. C., Viramontes, B. E., Corcoran, P. J., and LaRaia, P. J. (1995). Measured versus predicted vertical displacement of the sacrum during gait as a tool to measure biomechanical gait performance. *Amer. J. Phys. Med. Rehabil.*, **74**, 3–8.

Kerrigan, D. C., Della Croce, U., Marciello, M., and Riley, P. O. (2000). A refined view of the determinants of gait: significance of heel rise. *Arch. Phys. Med. Rehabil.*, **81**, 1077–1080.

Kerrigan, D. C., Riley, P. O., Lelas, J. L., and Della Croce, U. (2001) Quantification of pelvic rotation as a determinant of gait. *Arch. Phys. Med. Rehabil.*, **82**, 217–220.

Kram, R., Domingo, A., and Ferris, D. P. (1997). Effect of reduced gravity on the preferred walk–run transition speed. *J. Exp. Biol.*, **200**, 821–826.

Lamoreux, L. W. (1971). Kinematic measurements in the study of human walking. *Bull. Prosthet. Res.*, **10–15**, 3–84.

Lee, C. R. and Farley, C. T. (1998). Determinants of the center of mass trajectory in human walking and running. *J. Exp. Biol.*, **201**, 2935–2944.

Martin, R. and Saller, K. (1959). *Lehrbuch der Anthropologie*. Band II Stuttgart: Gustav Fischer.

Minetti, A. E., Capelli, C., Zamparo, P., di Prampero, P. E., and Saibene, F. (1995). Effects of stride frequency on mechanical power and energy expenditure of walking. *Med. Sci. Sports and Exerc.*, **27**, 1194–1202.

Minetti, A. E. and Saibene, F. (1992). Mechanical work rate minimization and freely chosen stride frequency of human walking: a mathematical model. *J. Exp. Biol.*, **170**, 19–34.

Mochon, S. and McMahon, T. A. (1980a). Ballistic walking. *J. Biomech.*, **13**, 49–57.

(1980b). Ballistic walking: an improved model. *Math. Biosci.*, **52**, 241–260.

Murray, M. P., Kory, R. C., Clarkson, B. H., and Sepic, S. B. (1966). Comparison of free and fast speed walking patterns of normal men. *Amer. J. Phys. Med.*, **45**, 8–24.

Noble, B., Metz, K., Pandolf, K. B., Bell, C. W., Cafarelli, E., and Sime, W. E. (1973). Perceived exertion during walking and running. II. *Med. Sci. Sports*, **5**, 116–120.

Pandy, M. G. and Berme, N. (1988a). A numerical method for simulating the dynamics of human walking. *J. Biomech.*, **21**, 1043–1051.

(1988b). Synthesis of human walking: a planar model for single support. *J. Biomech.*, **21**, 1053–1060.

Reynolds, T. R. (1987). Stride length and its determinants in humans, early hominids, primates and mammals. *Amer. J. Phys. Anthropol.*, **72**, 101–115.

Saini, M., Kerrigan, D. C., Thirunarayan, M. A., and Duff-Raffaele, M. (1998). The vertical displacement of the center of mass during walking: a comparison of four measurement methods. *J. Biomech. Eng.*, **120**, 133–139.

Saunders, J. B. deC., Inman, V. T., and Eberhart, H. D. (1953). The major determinants in normal and pathological gait. *J. Bone Joint Surg.*, **38A**, 543–558.

Shimba, T. (1984). An estimation of center of gravity from force platform data. *J. Biomech.*, **17**, 53–60.

Siegler, S., Seliktar, R., and Hyman, W. (1982). Simulation of human gait with the aid of a simple mechanical model. *J. Biomech.*, **15**, 415–425.

Simon, S. R., Knirk, J. K., Mansour, J. M., and Koskinen, M. F. (1977). The dynamics of the center of mass during walking and its clinical applicability. *Bull. Hosp. Joint Dis.*, **38**, 112–116.

Thirunarayan, M. A., Kerrigan, D. C., Rabuffetti, M., Della Croce, U., and Saini, M. (1996). Comparison of three methods for estimating vertical displacement of center of mass during level walking in patients. *Gait and Posture*, **4**, 306–314.

Thorstensson, A. and Robertson, H. (1987). Adaptations to changing speed in human locomotion: speed of transition between walking and running. *Acta Physiol. Scand.*, **131**, 211–214.

Thorstensson, A., Nilsson, J., Carlson, H., and Zomlefer, R. (1984). Trunk movements in human locomotion. *Acta Physiol. Scand.*, **121**, 9–22.

Waters, R. L., Morris, J., and Perry, J. (1973). Translational motion of the head and trunk during normal walking. *J. Biomech.*, **6**, 167–172.

Whittle, M. W. (1997). Three-dimensional motion of the center of gravity of the body during walking. *Hum. Move. Sci.*, **16**, 347–355.

Appendix

The computer program performs calculations designed to simulate motion of an inverted pendulum that changes its length during the swing so that the height of the point mass follows a sinusoidal function of time, minimum at the beginning and end of the swing (MDS1 and MDS2) and maximum in the middle of the swing (MSS). The motion of the mass is determined by numerical integration performed at Δt intervals of 0.1 msec. For every interval beginning at time t_i and ending at t_{i+1}, the program calculates

(1) $\theta_{i+1} = \theta_i - \omega_i \, \Delta t$

(2) orbital velocity $V_{o,i+1} = V_{o,i} - (W_{o,i} / M)\Delta t$, which is equivalent to calculating $\omega_{i+1} = \omega_i - (W_{o,i}\, R_i / I_i)\Delta t$

(3) $W_{o,i+1}$ and $W_{c,i+1}$, which are simple trigonometric functions of θ_{i+1}

(4) At this point, the mass will have a calculated height (H_{i+1}) determined solely by R_i and θ_{i+1}; however, a new height (H^*_{i+1}) must be determined because of the constraint that the height of the mass follows a sinusoidal path; thus

$$H^*_{i+1} = H_{MDS} + \Delta H \cdot [\sin(\pi t_{i+1}/t_{MSS} - \pi/2) + 1]/2 \qquad (14.11)$$

(5) this new height is accomplished by a change in VSL length; the new VSL length (R_{i+1}) is calculated

$$R_{i+1} = H^*_{i+1} / \cos \theta_{i+1} \quad (14.12)$$

(6) the change in VSL length occurs without any change in angular momentum ($\omega R^2 M$) because no torque has been applied; however, since R has changed, ω can no longer have the value (ω_{i+1}) calculated in step (2); the new value of ω is

$$\omega^*_{i+1} = (R_i / R_{i+1})^2 \quad (14.13)$$

which is equivalent to calculating a new orbital velocity

$$V^*_{o,i+1} = V_{o,i+1}(R_i / R_{i+1}) \quad (14.14)$$

(7) because the VSL length has changed from R_i to R_{i+1}, the mass has an axial velocity (V_c) at t_{i+1}

$$V_{c,i+1} = (R_{i+1} - R_i)/\Delta t \quad (14.15)$$

(8) the difference between the axial velocity at t_i and t_{i+1} allows calculation of the axial acceleration (A_c) of the mass during the interval

$$A_c = (V_{c,\,i+1} - V_{c,\,i})/\Delta t \quad (14.16)$$

(9) the centripetal force (F_c) that would be required to maintain the mass on a circular path with radius R_{i+1} and angular velocity $\omega^*{}_{i+1}$ is

$$F_{c,\,i+1} = M(\omega^*_{i+1})^2 R_{i+1} \quad (14.17)$$

(10) the outward force ($-L_c$) that the VSL would exert on the mass if it were following a circular path with radius R_{i+1} and angular velocity $\omega^*{}_{i+1}$ is calculated as

$$-L_{c,\,i+1} = F_{c,\,i+1} - W_{c,\,i+1} \quad (14.18)$$

(11) an adjustment must then be made to the calculated force that the VSL exerts on the mass to account for the fact that mass has either more or less axial acceleration than required to keep it on a circular path

$$-L^*_{c,\,i+1} = -L_{c,\,i+1} \pm MA_c \quad (14.19)$$

(12) finally, the force that substrate exerts on the VSL also equals $-L^*{}_{c,\,i+1}$

The value of the mass is irrelevant since all calculations of force are expressed relative to body weight.

Each simulation begins with the operator specifying the desired average horizontal velocity ($\overline{V}_h$) during the simulated step. The pendular swing begins at MDS1 at $t = 0$. The program assigns to the mass a high initial angular velocity that knowingly will produce an average horizontal velocity greater than desired. Furthermore, during this invalid swing of the collapsible/extensible inverted pendulum, the mass will attain a maximum vertical height at some point that does *not* correspond to MSS, i.e., at some value of θ other than 0°. The average horizontal velocity produced by this first iteration is used to determine a different initial angular velocity for a second iteration that yields a result closer to $\overline{V}_{\mathrm{h}}$. Further iterations are performed until an initial angular velocity is identified that produces the desired average horizontal velocity, which is the one that also leads to a *H–t* curve that meets the requirement of shape and location of maxima and minima. For this final simulation, the program outputs values of all relevant parameters at intervals of 0.1 msec.

15 *Estimating the line of action of posteriorly inclined resultant jaw muscle forces in mammals using a model that minimizes functionally important distances in the skull*

WALTER STALKER GREAVES
University of Illinois at Chicago

Introduction

The inherent complexity of the masticatory apparatus is frequently simplified by resolving the jaw muscle forces into a single resultant vector (e.g., Weijs and Dantuma, 1981). This approach is reasonable because there is some evidence that essentially all the major jaw adductors are active at virtually the same time, at the point in the power stroke of the chewing cycle where the muscle forces are highest (e.g., Møller, 1966; de Vree and Gans, 1976; Weijs and Dantuma, 1981). The resultant force is therefore an appropriate consideration when constructing a static masticatory model of the chewing apparatus that deals with this critical point in the chewing cycle (e.g., Turnbull, 1970; Greaves, 1995). Moreover, this approach greatly simplifies the analysis of this complicated system.

While the magnitudes of the muscle forces are often of interest, two other characteristics of the resultant force vector (i.e., its orientation and position in space) are perhaps even more critical. These latter two features determine the vector's moment arm and thus the moment of the jaw muscle force. While the masticatory apparatus is a three-dimensional structure, the orientation of the vector of muscle force can sometimes be studied profitably in two dimensions in lateral view. From this vantage point, the vector points dorsally, from the lower jaw up to the skull, and can be inclined either anteriorly or posteriorly. The actual inclination of the vector of muscle force, in lateral view, depends on the relative sizes of the three great jaw-closing muscles – temporalis, masseter, and

Shaping Primate Evolution, ed. F. Anapol, R. Z. German, and N. G. Jablonski. Published by Cambridge University Press.

medial pterygoid. The temporalis slopes up from the lower jaw to attach in the temporal fossa on the lateral side of the brain case. Thus it slopes toward the rear. The other two muscles slope up and toward the front because they attach more anteriorly on the skull. When the temporalis is larger than the other two muscles taken together, the resultant force of the jaw muscles has a posterior inclination. It has an anterior inclination when masseter and pterygoid form the dominant muscle group. Thus, the inclination varies because there is variation in the size of the individual muscles that close the jaw. To date, the actual inclination of a vector has almost always been estimated only from dissections of the jaw-closing muscles. But given the complexity of these structures, its orientation is not known exactly (Weijs and Dantuma, 1981; Greaves, 1988, 1995). Aside from the many difficulties in obtaining a reasonable estimate of the inclination of the vector, an additional problem is the limited amount of anatomical material from modern mammals from which estimates can be made. Estimating the vector in fossil forms is even more problematic because the size, architecture, and other features of the muscles themselves have to be estimated.

An estimate of the orientation of the resultant vector of jaw muscle force and its location in both fossil and extant mammals can be determined using a previous theoretical analysis (Greaves, 1988), which was mainly concerned with determining the maximum average bite force along the jaw. It concluded that the resultant muscle force intersects the tooth row just behind the third molar at 30% of jaw length from the jaw joint (cf. Gans, 1988; Greaves, 2000). In that analysis, distances were measured perpendicular to a vertical resultant vector.

Theoretical background

The inclination of the resultant vector of jaw muscle force will change as muscle activity is varied and when the animal uses the jaw in different ways. The current study accepts the idea that during strong jaw muscle activity at the high point in the power stroke, muscle forces in any given animal can be resolved as a vector with a single line of action. The present study carries forward two previous analyses of the masticatory apparatus (Greaves, 1991, 1998). This new analysis estimates a similar line of action for the vector that was alluded to above, but approaches the problem of the orientation of the vector in a different way than did the 1988 study and relies solely on skull geometry. Moreover, in contrast to the theoretical estimate of 1988, the result of the new study is quite easy to apply in actual cases. The presence of two independent theoretical estimates that agree with available estimates from dissections is encouraging.

The first analysis of the masticatory apparatus (Greaves, 1991) began with the well-known observation that the jaw joint in extant mammals is virtually always located above the level of the upper tooth row. That study suggested that the distance from the jaw joint to the third molar (JM) described an important structural element in both the skull and the lower jaw because these regions connect the jaw joint to the tooth rows (cf. DeMar and Barghusen, 1972). A number of factors may influence the length of this distance, and efficiency is likely to be important. Thus, this distance was expected to be as short as possible in order to minimize the amount of metabolically expensive bone, including a safety factor, used in the construction of both the skull and the lower jaw (Alexander, 1981; Currey, 1984; Wainwright *et al.*, 1976). Bone is metabolically expensive because bone tissue is constantly being added and removed. This remodeling of bones is an important reason for the distance between the joint and the tooth row to be as short as possible. Less bone tissue is used in the construction of parts of the jaw mechanism and less energy is required for its upkeep.

The second study (Greaves, 1998) suggested that there was a second structurally important distance whose length might be minimized. This distance extended from the jaw joint to a point (S) in the temporal fossa that lay on the vector representing the resultant muscle force. The jaw muscles attach to the skull at different regions. A single point (S) that represents in some sense the average of these regions was considered. The idea of an average muscle attachment site is broadly analogous to the idea of using a single vector to represent all the forces of the three major muscles that close the jaw. The distance from the jaw joint up to some point S, taken to be the average muscle attachment site in the temporal fossa, represents another functionally important bony strut in the skull.

As the muscles pull down, the condyle of the lower jaw pushes up and the skull resists this compression. A thick bar of bone extends from the joint to the molar (line JM) in the lower jaw and the equivalent distance is spanned by arches of bone in the skull. Line JS is a simplified representation of that part of the compression-resisting shell of the skull in the region of the temporal fossa that lies above the jaw joint and extends from the jaw joint to point S.

The 1998 analysis considered a number of theoretically possible locations for the jaw joint. That study found that as one of these distances (JM and JS) increased, the other decreased. Therefore, both of these distances could not be a theoretical minimum. However, the study did demonstrate that the *sum* of the two distances could be minimized even though this was not possible for each of the individual distances. Elementary plane geometry indicates that this sum is minimized when these two distances are equal. A third conclusion of this second analysis was that minimizing the sum of the two distances, that is, for

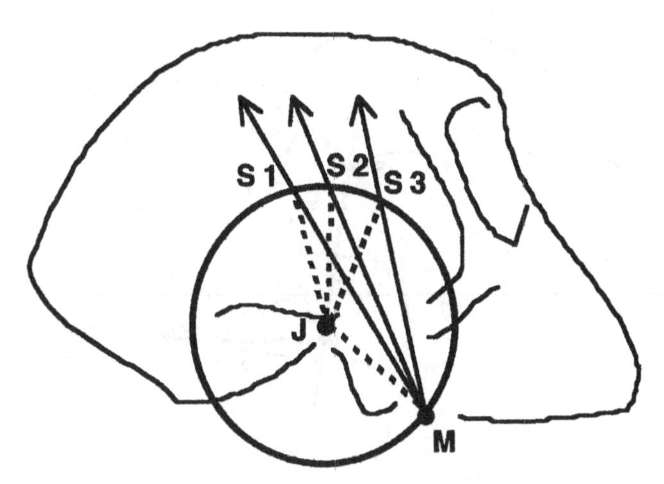

Figure 15.1. The jaw joint (J) is located at the center of a circle that intersects a point just behind the third molar (M). The solid lines radiating from M to S1, S2, and S3 are representative chords of the circle. These chords are *segments* of lines that represent *potential* muscle vectors (arrows). The dashed line from the joint to the third molar (JM) is equal to all of the other dashed lines (JS1, JS2, and JS3) because they are all radii of the same circle. An outline of a generalized skull is included for orientation. See text for further discussion.

them to be equal, required that the jaw joint be located above the level of the tooth row as is the case in the majority of mammals (Greaves, 1998).

The model

On a lateral view of a skull, a circle can be imagined with its center at the jaw joint and its circumference intersecting the rear edge of the third molar (Fig. 15.1). That is, a radius of the circle equals the distance from the joint (J) to the third molar (M). According to the 1998 study noted above, the second distance (JS), from the jaw joint to a point (S) in the temporal fossa, intersects the line of action of the vector and should be equal to the first distance (JM). Thus, part of the line representing a resultant vector of jaw muscle force is a chord of the circle; it begins at the point on the circumference just behind the third molar (M) and passes through point S on a different part of the same circumference. Any number of chords (MS1, MS2, etc.) can be represented in this fashion and each one can be used to indicate the line of action of its vector (Fig. 15.1). In all of these examples, the two distances, one from the joint to the molar (JM) and the other from the joint to a point S (JS1 or JS2, etc.) on a vector, are radii of the same circle and are thus equal. In actual cases there will be only

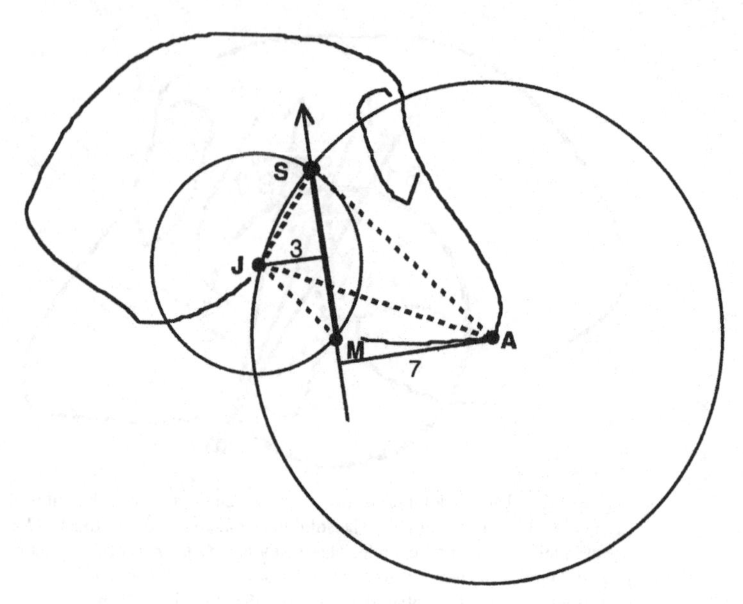

Figure 15.2. The basic lever system of the jaw is represented by the four points J, M, A, and S. The two circles that are centered at J and A intersect at S. Thus, JS = JM and AS = AJ. The solid line SM represents the line of action of the resultant muscle force. The perpendicular distances from SM to J and from SM to A are three and seven units respectively.

one vector with a single line of action. However, the actual location of point S on the circumference of the circle cannot be determined with the procedure just outlined. Point S could be anywhere on that part of the circumference that lies within the temporal fossa.

The search for other possible minimum distances suggests a way to model the jaw mechanism using only four points that are located on the skull (Fig. 15.2). Three of these points are obvious landmarks on the skull of a generalized mammal and roughly define the jaw mechanism in the skull. They are (1) the upper surface of the jaw joint (glenoid cavity), (2) the posterior edge of the third upper molar, and (3) the anterior end of the upper jaw. These three features can be represented as points J, M, and A, respectively. These three points indicate the total length of the jaw (AJ), approximate tooth row length (AM), and the distance from the joint to the molar (JM).

What these three points do not do is define the inclination of the resultant vector of the jaw muscle force that is a critical element of the jaw mechanism. Without knowledge of the muscle vector, relative bite forces that are reasonably accurate cannot be calculated. The resultant extends up and back from behind the third molar. It then intersects point S which is located in the temporal fossa. The three original points J, M, and A can be used to construct this fourth point

(S). These four points then define most of the more basic features of the jaw mechanism.

The jaw joint (J) is both the posterior end of the jaw lever and the fulcrum. The lever is usually a bent lever in mammals (JMA) where the point just behind the third molar (M) is the intersection of the line of action of the muscle vector (Greaves, 1988; cf. Greaves, 2000). The line (MS) from point M to point S defines the inclination of the vector.

Accepting that the skull will be constructed with a minimum amount of bone tissue suggests that some distances between these four points should be as short as possible. Setting JS equal to JM minimizes the sum of these two distances as was described above. Another obviously important distance, because it represents the length of the entire jaw, is AJ. This distance can be paired with the distance from A to S. Distance AS can be taken to be functionally important because it can be thought of as representing a strut that braces the tooth row and upper jaw against the back part of the skull and is usually not a simple bar of bone, but rather a complicated collection of bars, sheets, or arches. Setting AS equal to AJ minimizes their sum.

In practice, the location of point S can be found by constructing two circles with centers at A and J and radii AJ and JM, respectively (Fig. 15.2). The intersection of these two circles defines point S. Thus, JS = JM and AS = AJ. Point S is thus found by minimizing the two compound distances MJS and JAS. (Note that it would be possible to set AS equal to AM. However, the end of the jaw (A) is clearly farther away from a point in the temporal fossa than it is from the third molar. Therefore, if one is confined to a lever system defined solely by the four points mentioned above, AS cannot be equal to AM.)

Note also that there are actually two intersections, but one is not meaningful because it lies outside and below the skull (Fig. 15.2). Point S is expected to lie on the skull in the neighborhood of the temporal fossa. More importantly, AM represents the approximate length of the tooth row. This is not a distance that should be minimized. Rather, tooth row length is basic for a given jaw mechanism. Other features would presumably complement the length of the tooth row. Once point S is determined, a straight line connecting M and S then defines the line of action of the resultant muscle force, although it does not define the length, and therefore the magnitude, of the resultant vector of jaw muscle force.

Given that the joint, the molar, and the anterior end of the jaw form a triangle JMA, three lengths (JM, JA, and AM) are potentially available for use in determining point S. As just mentioned, distance AM is not available because it represents the approximate length of the tooth row. That leaves only JM and JA.

With two sides of the triangle available, where either end can be the center of a circle, there are only four possible pairs of circles. In two of these pairs, the circles do not intersect and thus do not define a point S according to the above criteria. The third pair of circles intersects at point J and a point outside and below the skull. Again, a meaningful point S is not defined. The last pair of circles intersects at two points. One is not meaningful because it lies below and outside the skull. The final point lies within the temporal fossa. Thus, only a single appropriate point on the skull is defined by the various possible constructions.

First, the distance JM was measured from J to S rather than from M to S. That is, point J is taken to be the center of the constructed circle with JM as the radius. If point M is taken to be the center of the circle it intersects the other circle, which has its center at A, at point J. A fourth point S is thus not defined. These two circles also intersect at a second, non-meaningful, point below the skull.

Second, line AJ was measured from A to S and not from J to S (the center of the circle is at A rather than J). If both AJ and JM were measured from J (both circles centered at J), two concentric circles are formed that do not intersect and so a point S is not defined. If the center of the larger circle is at point J and the center of the smaller is at point M, the circles are not concentric but they nevertheless do not intersect and so S is again not defined.

Once the location of point S is determined, SM can be drawn to represent the line of action of the resultant muscle force in lateral view. While the above analysis is two-dimensional, recall that the *entire* masticatory apparatus is a three-dimensional structure and is best considered so in most cases.

At this point, the idea that the line of action of the muscle force should be located 30% of the way along the jaw, so as to maximize the average bite force, can be added to the analysis (Greaves, 1988; cf. Greaves, 2000). This simply means that the joint (J) should be located three units from the line of action while the anterior end of the jaw (A) should be seven units away (Fig. 15.2). This relationship, as well as the equal distances between points J and M and J and S, and also between A and J and A and S, constrains the locations, relative to one another, of the four points (J, M, A, and S) that are used to define the basic jaw lever system. While this range is infinite in a geometric sense, at this point the model is only concerned with that part of the range where (1) the jaw joint lies above the level of the tooth row and (2) the line of action is posteriorly inclined. An anteriorly inclined line of action will be considered below.

The constraints described above severely limit the number of possible configurations of the four points. The relevant range for the groups of points was determined by constructing two lines each parallel to a line SM where the intervening distances were in the ratio of 3 : 7 (Fig. 15.2). Therefore, point M was positioned on the line located between the other two parallel lines, point J on the parallel line three units from line SM, and point A on the remaining

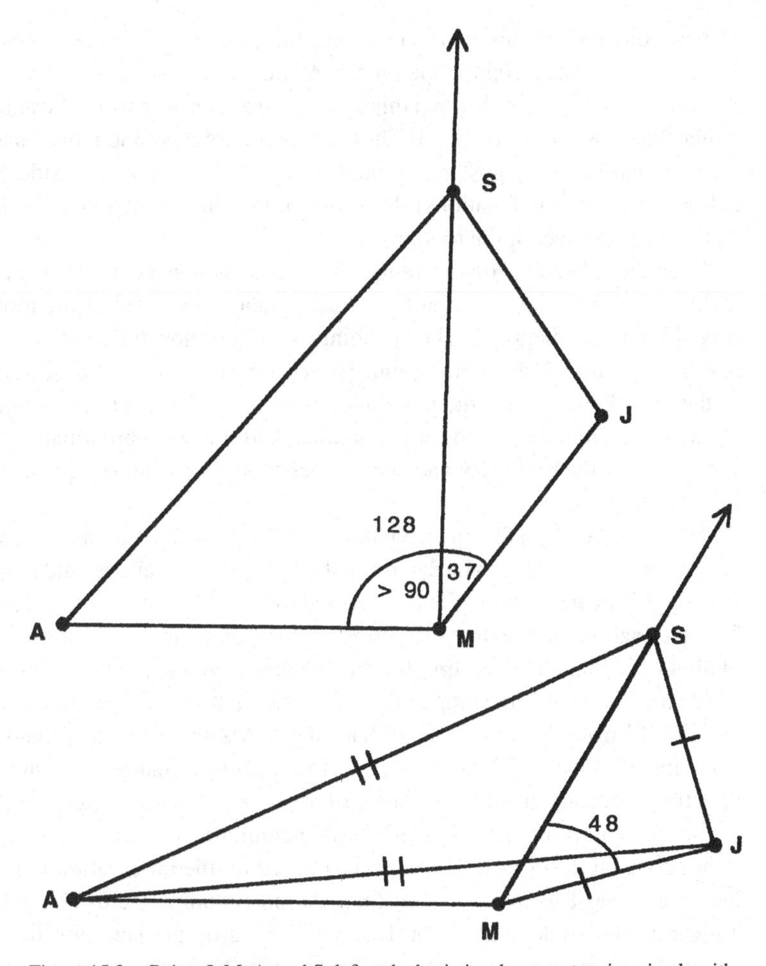

Figure 15.3. Points J, M, A, and S define the basic jaw lever system in animals with a posteriorly inclined muscle vector. An outline of a skull has been omitted. The two drawings represent the relationships between these four points at each end of the range of relevant possibilities. Angle JMS varies from about 37° in the upper drawing to about 48° in the lower. The minimum value of angle JMA is approximately 128°. The jaw joint (J) lies above the level of the tooth row in both cases, but just barely so in the lower drawing. The lines of action (SM) are both posteriorly inclined, but just barely so in the upper drawing.

parallel line seven units in the other direction. This insured that point M was 30% of the way from J to A. Given that point S must be on the same line as point M, the next step was to find locations for A, S, and J, on their respective lines, such that JS = JM and AS = AJ. The range of these appropriate groupings was determined by trial and error (Figs. 15.2 and 15.3).

Near one end of this pertinent range, the estimate of the tooth row (line MA) almost forms a right angle with SM, the line of action (Fig. 15.3, upper diagram). Thus, in cases beyond this point the line of action would be anteriorly rather than posteriorly inclined. These cases do not meet the requirements of the model where only posteriorly inclined lines of action are considered (see below for anterior inclinations). Note that at this end of the range the joint is well above the level of the tooth row.

Near the other end of the range the line of action is clearly posteriorly inclined, but here the jaw joint is just barely above the level of the tooth row (Fig. 15.3, lower diagram). The jaw joint would lie below the level of the tooth row in cases much beyond this point. These cases do not meet the requirements of the model either since the jaw joint lies above the level of the tooth row in almost all mammals; the fossil entelodonts had a joint approximately on the same level as the tooth row and some saber-toothed cats appear to have had a joint below the row.

The angle between the line from the jaw joint out to the third molar (JM) and the line of action (SM) can be determined graphically (see above) and is approximately 37° near one end of the range and 48° near the other end (Fig. 15.3). The inclination of the line of action is thus restricted to a narrow range of about 12° relative to the line JM that extends from the joint to the molar.

On the other hand, the range of angles between the tooth row (line MA) and the line JM from the joint to the molar (angle JMA) is much larger and varies from approximately 128° to 178° (Fig. 15.3). Thus, a change of 1° in the line of action is accompanied by a change of approximately 4.5° in angle JMA.

Note that limiting the range of possible configurations to those with a jaw joint above the level of the tooth row is based on the observation that this is the case in most mammals. Therefore, the maximum angle of about 178° is a given in the model and is not derived. The reason the jaw joint lies above the tooth row in mammals may perhaps be related to the relationship between the muscle vector and the tooth row. As the joint gets lower and lower the vertical component of the muscle force that is normal to the tooth row, and is actually producing the bite force, decreases while the component pulling the jaw posteriorly, and which produces no bite force, increases. At one end of the range, where the vector is almost perpendicular to the tooth row and the jaw joint is high above the teeth, the component of the muscle force that closes the jaw and produces bite force is very large. The component that pulls the jaw to the rear is very small. If the muscle force is one unit, almost one unit of force pulls the jaws together.

Even in the middle of the range, the normal component producing bite force is still large. Only at the opposite end of the range where the joint is low and the vector has a strong rearward inclination does the normal component become much smaller. In this case, the normal component is about the same size as the posterior component. To produce approximately one unit of force perpendicular to the tooth row, almost 1.4 units of muscle force is required.

This latter structure of the jaw mechanism is reasonable in animals like carnivores where large posterior force components are required to deal with struggling prey. Extending the range beyond this point, with the joint at the same level as the tooth row or below, the posterior component is larger than the normal component. When the jaw joint is limited to a location near the level of the tooth row the muscle force component pulling the jaw up is always larger than the component pulling the jaw to the rear. Accepting the importance of bite force, smaller bite force components are unlikely to be selected for.

Animals that need less backward force are expected to have muscle vectors that are more vertically inclined. Animals that need a small posterior component will have vectors that approach perpendicularity with the tooth row. Then most of the muscle force pulls the jaws together and very little pulls the lower jaw to the rear.

The minimum value of angle JMA (128°) is derived from the model and depends upon the minimum value of angle JMS. This angle can be no smaller than about 128° because it is the sum of 37° and an angle slightly larger than 90° (Fig. 15.3). The actual minimum value of angle JMA in a reasonably large sample of primates and pigs (animals with both a posteriorly inclined vector and usually a third molar) is close to 128°, suggesting that the model has correctly predicted the posterior vector inclination in these groups of animals.

Measurements

One hundred and twenty-seven specimens representing 38 species of primates and 6 species of suids from the collections of the Field Museum of Natural History were examined for this study (Table 15.1). Angle JMA was measured in lateral view using the midpoint of the glenoid (mandibular) fossa (as an estimate of point J), the posterior alveolar border of the third molar (as an estimate of point M), and the end of the first incisor (as an estimate of point A). Angle JMA varied from 135° to 174°. Using data in Spencer (1995) that were taken from almost 850 anthropoid primates, a similar angle was calculated. In this dataset, angle JMA varied from 129° to 176°. These minimum values for angle JMA are virtually the same as those estimated by the model (128°).

Table 15.1. *Specimens measured for this study, with their JMA measurements*

Species	Sex	Number	Angle JMA
Lemur catta	M	85134	165
	F	157993	166
	M	89766	165
L. fulvus	F	85136	165
	?	8337	165
	M	53065	166
Hapalemur griseus	F	57631	164
	?	5652	165
Varecia variegata	?	8347	172
	?	81541	172
	M	129349	168
	?	134468	168
Propithecus verreauxi	?	8346	160
	F	89204	158
	?	8341	158
	?	8342	158
Galago senegalensis	M	74539	158
	M	67151	158
	F	67149	158
Otolemur crassicuadatus	M	96274	165
	M	85957	172
	M	83614	170
Nycticebus coucang	F	99616	166
	M	37988	165
	M	37987	165
Perodicticus potto	M	148987	173
	F	94239	172
Daubentonia madagascariensis	?	15529	169
Saguinus fuscicollis	F	71011	165
	F	71013	165
	M	71006	165
Saimiri sciureus	M	87823	165
	M	87824	164
	M	88240	165
Aotus lemurinus	M	68860	163
	F	68858	163
	F	68850	160
Callicebus brunneus	M	21541	160
	M	52489	153
	M	78673	160
Alouatta belzebul	F	92089	150
	M	35078	145
	M	92090	145
Cacajao calvus	M	88814	155
	M	88817	155
	M	88818	155

Table 15.1 *(cont.)*

Species	Sex	Number	Angle JMA
Chiropotes albinasus	F	94926	158
	F	94927	158
	F	94928	158
Pithecia aequatoralis	M	86993	158
	F	86994	159
Cebus albifrons	F	16564	163
	F	16565	163
	F	16566	161
Ateles belzebuth	F	92124	161
	M	92125	160
	F	92123	160
Lagothrix lagothricha	M	98049	151
	M	98050	157
	F	98051	158
Cercocebus albigena	M	24303	158
	F	24298	150
	F	24312	157
Cercopithecus aethiops	M	27176	161
	F	27177	162
	M	27278	155
Erythrocebus patas	F	60677	167
Macaca arctoides	M	105682	160
	M	105683	160
	M	99617	160
M. fascicularis	M	99651	160
	F	87428	160
	?	87431	160
M. nemestrina	M	99691	160
	M	99673	152
	F	99685	156
Miopithecus talapoin	F	121359	160
	?	129458	165
	M	95273	165
Papio anubis	F	27043	165
	F	27042	165
	?	104963	165
Theropithecus gelada	M	27184	148
	M	27039	148
	M	27187	150
Colobus guereza	M	8175	161
	M	17697	161
	M	17696	155
Presbytis entellus	F	82811	174
	M	82810	160
	F	35816	161

Table 15.1 *(cont.)*

Species	Sex	Number	Angle JMA
Pygathrix roxellana	M	31143	158
	F	31140	158
	F	31142	160
Hylobates concolor	M	46496	165
	F	38017	155
	F	38016	155
H. lar	M	99739	147
	M	99740	147
	F	99762	169
Pongo pygmaeus	M	91723	135
	M	19023	135
	?	19025	153
Pan troglodytes	F	29823	154
	F	18403	151
	?	156701	146
Gorilla gorilla	M	134482	135
	M	126045	138
	?	99092	136
Hylochoerus meinertzhageni	F	49148	155
	M	49147	160
	M	27530	151
Phacchoerus aethiopicus	F	26988	158
	F	27130	158
	M	26989	145
Sus barbatus	M	62825	156
	M	62830	155
	?	85916	160
Tayassu pecari	M	62080	165
	M	79930	165
	M	79927	170
Choeropsis liberiensis	?	125150	170
	F	140919	170
	M	135777	170
Hippopotamus amphibius	M	34929	165
	?	34927	165
	F	34928	165

Discussion

The resultant vector's orientation can be estimated easily on a skull because the jaw joint, the third molar, and the anterior end of the jaw are reasonably obvious landmarks although the exact points to be used have to be defined. The graphical analysis above indicated that angle JMA changes by a little more than 4.5° for every change of 1° in angle JMS. Thus, after constructing a table of values, the inclination of the line of action (angle JMS) can be read off after measuring angle JMA. As an example, the following is a reduced version of such a table (JMA/JMS): 128/37; 137/39; 146/41; 160/44; 168/46; 178/48. Estimating angle JMS with such a table is perhaps easier than using pairs of dividers (or a compass) to find point S, determine line MS, and finally angle JMS (cf. Fig. 15.2).

Note that in some animals the third molar has been lost during the evolution of the group. In these cases the line of action of the muscle vector does not intersect the jaw behind the remaining most posterior tooth, but rather some distance behind it. According to previous analyses (Greaves, 1988; cf. Greaves, 2000), the vector intersects the jaw 30% of the way along the jaw, regardless of whether there is a third molar present or not, because the loss of a single tooth does not modify the basic plan of the masticatory apparatus. The difficulty in these instances is that estimating the inclination of the line of action requires knowledge of the length of the line from the joint to the third molar. If this tooth is not present in a species, this line (JM), and therefore the line of action, cannot be determined unambiguously. In these cases, the anteroposterior length of the missing tooth can be estimated and the estimate of the resulting line of action compared with that determined from an application of the 1988 analysis; the vector that divides the jaw in a 3 : 7 ratio is to be determined. Unfortunately, the above caveat applies to many modern and fossil Carnivora.

Nevertheless, even in these cases, reasonably good estimates of lines of action are possible without the use of the 1988 analysis. Generally in carnivores, the location of J, the jaw joint (glenoid), is low relative to line MA, the tooth row, and this relationship changes little when estimated molars of varying sizes and locations are added to the rear of the tooth row. Because the joint is low in such cases, the angle between JM and the line of action will be closer to 48°. It will not be close to 37° which is typical of an animal with a very high joint (Fig. 15.3).

Posteriorly directed lines of action have been emphasized in this chapter. Yet many mammals, such as antelopes and rodents, have resultant muscle forces that are anteriorly, rather than posteriorly, inclined relative to the tooth row because the forwardly pulling masseter and pterygoid, taken together, form the dominant jaw muscle group (e.g., Becht, 1953; Maynard Smith and Savage, 1959). In these cases, the point S is not located within the temporal fossa, but rather is positioned out on the face, for example on the maxilla. As previously mentioned, the three points defining the jaw (A, M, and J) define in turn three sides of a triangle (AM, MJ, and JA). Of these three sides only two were used to define point S located within the temporal fossa because the length of the tooth row, AM, was not used. Distances JM and AJ were set equal to JS and AS, respectively, to define point S. In these cases line JM and line AJ were rotated around points J and A, respectively, until they intersected at their end points at

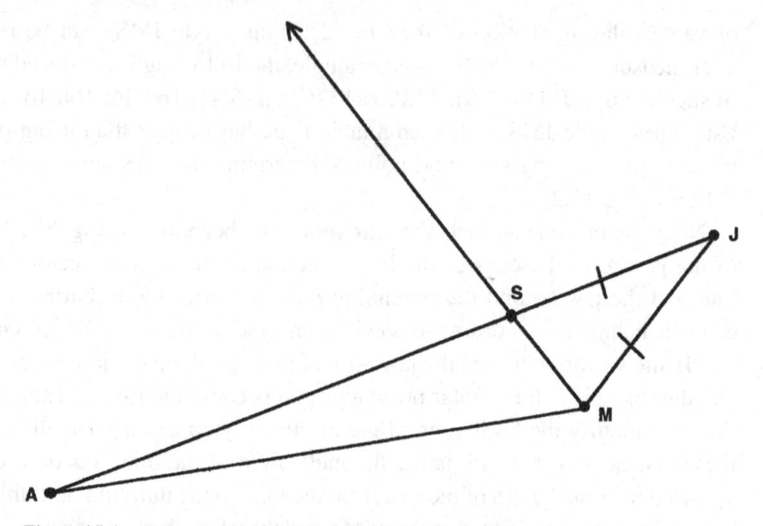

Figure 15.4. Points J, M, A, and S define the basic jaw lever system in animals with an anteriorly inclined muscle vector. Line JS = JM. Point S is located on line AJ at a distance JS (= JM) from point J. See text for further discussion.

point S. In animals with a posteriorly inclined vector, a tract of bone from A to S seems reasonable. This element buttresses the tooth row against the high brain case.

In animals with an anterior vector such a buttress is often not required. For example, selenodont artiodactyls (antelope, sheep, etc.) are animals with relatively long faces that are not very deep and so line AJ approximates the main structural element of the skull. An element like AS in such animals is not expected; it lies so close to line AJ that it is redundant.

For animals with an anteriorly inclined vector, the line AM, representing the tooth row, is again not available as a distance to be paired with a second distance. A distance AS is not constructed from the distance AJ because line AJ already approximates the main anteroposterior structural support of the skull and a strut representing distance AS seems redundant. Therefore, distance JM is the only span available that can be used to construct a second distance of equal length (JS) to be used to determine a point S. Moreover, since line AM is not available, line JS interacts with the only remaining side, which is AJ. The end point of line JS, which is equal to line JM, intersects line AJ at point S only when these two lines are coincident (Fig. 15.4). Thus, point S lies on line AJ at a distance JM (= JS) from the jaw joint and determines the line of action (MS) when the muscle vector is anteriorly oriented. The angle between the vector and line AM is always less than 90°.

Angle JMA is observed to be less than 180° in most mammals because the jaw joint is typically above the level of the tooth row. In hypothetical cases, with jaw joints located very far above the tooth row, this angle approaches 115° and the anterior component of muscle force is reasonably large. Since in most cases there appears to be little need for very strong anterior force components, extremely high jaw joints are not expected. For many selenodont artiodactyls with high, but not excessively high joints, angle JMA approaches 130–140° and the anterior component is not excessive. It follows that the angle between the vector and line JM (angle JMS) is often near 70° in these animals.

Comparing anterior and posterior lines of action, with the distance from the jaw joint to the third molar (JM) held constant, the lengths of the tooth row (MA) and line JA are significantly longer in the former case (Figs. 15.2 and 15.4). This is true because the distances from the line of action to J and A are maintained at a 3 : 7 ratio. If line JA is held constant, then line JM differs in these two cases. This result is consistent with the observation that animals with posteriorly aligned vectors and very large temporalis muscles usually have a relatively long distance between the jaw joint and the third molar. Animals with anteriorly inclined vectors and large masseter/pterygoid complexes have relatively short distances between the joint and this tooth (Greaves, 1991).

The major result of this analysis suggests that relative to the tooth row, a clear relationship exists between the inclination of the resultant vector of jaw muscle force and the height of the glenoid in a majority of mammalian jaw mechanisms. While this is perhaps surprising, it is generally consistent with available estimates and with a series of earlier results. Taken together, they tend to bolster the view that the structural plan of the jaw mechanism of most mammals is remarkably similar when considered at its most basic level.

Acknowledgments

I thank W. Stanley, L. Heaney, and B. Patterson of the Field Museum of Natural History, Chicago, for permission to study specimens in their charge, and M. Greaves for discussion and editorial assistance.

References

Alexander, R. McN. (1981). Factors of safety in the structure of animals. *Sci. Prog. Oxford*, **67**, 109–130.

Becht, G. (1953). Comparative biologic-anatomical researches on mastication in some mammals. I and II. *Proc. Kon. Akad. Wetensch. Amsterdam*, Series C, **56**, 508–527.

Currey, J. (1984). *The Mechanical Adaptations of Bones*. Princeton, NJ: Princeton University Press.

DeMar, R. and Barghusen, H. R. (1972). Mechanics and the evolution of the synapsid jaw. *Evolution*, **26**, 622–637.

de Vree, F. and Gans, C. (1976). Mastication in pygmy goats (*Capra hircus*). *Ann. Soc. Roy. Zool. Belg.*, **105**, 255–306.

Gans, C. (1988). Muscle insertions do not incur mechanical advantage. *Acta. Zool. Cracoviensia*, **31**, 615–624.

Greaves, W. S. (1988). The maximum average bite force for a given jaw length. *J. Zool.*, **214**, 295–306.

(1991). The orientation of the force of the jaw muscles and the length of the mandible in mammals. *Zool. J. Linn. Soc.*, **102**, 367–374.

(1995). Functional predictions from theoretical models of the skull and jaws in reptiles and mammals. In: *Functional Morphology in Vertebrate Paleontology*, ed. J. J. Thomason. Cambridge: Cambridge University Press. pp. 99–115.

(1998). The relative positions of the jaw joint and the tooth row in mammals. *Can. J. Zool.*, **76**, 1203–1208.

(2000). Location of the vector of jaw muscle force in mammals. *J. Morphol.*, **243**, 293–299.

Maynard Smith, J. and Savage, R. J. G. (1959). The mechanics of mammalian jaws. *School Sci. Rev.*, **40**, 289–301.

Møller, E. (1966). The chewing apparatus. *Acta Physiol. Scand.*, **69** (Suppl.), 1–229.

Spencer, M. A. (1995). *Masticatory System Configuration and Diet in Anthropoid Primates.* Ph.D. thesis, State University of New York, Stony Brook.

Turnbull, W. D. (1970). Mammalian masticatory apparatus. *Fieldiana: Geology*, **18**, 147–356.

Wainwright, S. A., Biggs, W. D., Currey, J. D., and Gosline, J. M. (1976). *Mechanical Design in Organisms*. London: Edward Arnold.

Weijs, W. A. and Dantuma, R. (1981). Functional anatomy of the masticatory apparatus in the rabbit (*Oryctolagus cuniculus* L). *Neth. J. Zool.*, **31**, 99–147.

Part V
Primate diversity and evolution

Once again, though my earlier work was clearly aimed at understanding functional adaptations in specific bone–joint–muscle units, it also equally clearly led into interests in primate diversity and evolution. I was not, of course, a primate taxonomist, never having had the requisite training. I never worked in field situations (being allergic to high temperatures, heavy rainfall, high altitudes, mosquitoes, leeches, etc). I therefore never participated in field observations of living species or in field discoveries of fossils. And though I was never formally educated in mathematics and statistics, and never capable myself of making advances in them, I was always a user who was interested in how such methods could be applied to data. Especially was I interested in the kinds of questions that the above methods and data might answer.

I have thus remained enormously interested in all such studies carried out by others – indeed, the data of others were essential to some of the investigations that I myself made. Mainly working with colleagues, however, I may have been perhaps the first to apply full multivariate statistical analyses to the data of field observation, of the niche. Likewise, through colleagues and students, I may have been amongst the first to use morphometric methods as tools to go beyond the data themselves, to seek correlations with behavior, with the niche, with development, and with evolution. The following chapters on the niche, on cladistics, and on the development of morphometrics itself carry all this so much further. These studies (and many others not represented in this book) go so very far beyond what I originally envisaged.

Charles Oxnard

16 *The evolution of primate ecology: patterns of geography and phylogeny*

JOHN G. FLEAGLE
Stony Brook University

KAYE E. REED
Arizona State University

Introduction

Over four decades, Charles Oxnard has been a relentless pioneer in expanding the quantitative horizons of research in primate and human evolution. His many works using multivariate analyses to elucidate and amplify our understanding of the primate shoulder, the primate foot, primate locomotion, prosimians, primate limb proportions, and the relationships of early hominids are well known and widely cited (Ashton *et al.*, 1965, 1975, 1976; Oxnard, 1981, 1984). Less widely cited are his efforts with Robin Crompton and Susan Lieberman to use many of the same quantitative techniques to examine broad patterns in primate behavior and ecology (Crompton *et al.*, 1987; Oxnard *et al.*, 1990). In recent years we have made several efforts to redress this oversight (Fleagle and Reed, 1996, 1999a; Reed, 1999), and it seems particularly appropriate to provide here a general summary of that work. Charles Oxnard is more than a gifted quantitative biologist; he is also a person who delights in reducing the seemingly insurmountable complexity of nature to simple and often esthetically pleasing patterns. Yet, at the same time, he has always been keen to push his analyses one more step and demonstrate that a dataset may yield very different patterns when viewed from a slightly different perspective. Accordingly, in the spirit of Charles's work we will concentrate on some of the broader patterns that emerge from our studies of primate communities when we look at the same dataset from a slightly different perspective.

Studies of primate behavior and ecology are well suited for multivariate treatments. Since the late 1950s hundreds of primatologists have spent many thousands of hours recording detailed quantitative data about the diet, locomotion,

Shaping Primate Evolution, ed. F. Anapol, R. Z. German, and N. G. Jablonski. Published by Cambridge University Press.

Table 16.1. *Ecological variables used in multivariate comparisons (see Fleagle and Reed, 1996)*

Body size
Activity pattern
Fruit in diet
Leaves in diet
Fauna in diet
Leaping
Arboreal quadrupedalism
Terrestrial quadrupedalism
Suspensory behavior
Bipedalism

and activity patterns of hundreds of individual populations. Most of these data have appeared in autecological studies of the behavior and ecology of a single species. In a few instances, comparative studies of several sympatric taxa have vastly expanded our understanding of primate socioecology (e.g., Struhsaker, 1975; Chivers, 1980; Fleagle and Mittermeier, 1980; Mittermeier and van Roosmalen, 1981; Terborgh, 1983). Even broader studies, addressing the similarities and difference among the primates of different sites and different continents are rarer still, and have been largely qualitative (e.g., Bourliere, 1985; Terborgh and van Schaik, 1987; but see Kappeler and Heymann, 1996). Our efforts have been aimed at remedying this gap in our understanding of ecological similarities and differences of primate communities by undertaking quantitative analyses of primate communities at various levels. These studies have been focused both within and between major biogeographic areas to understand not only the major factors that influence patterns of diversity (Reed and Fleagle, 1995) but also to compare patterns of "niche use" in different regions (Fleagle and Reed, 1996). Subsequently, we have incorporated information about patterns of phylogenetic branching into our studies to examine more broadly the relationship between phylogenetic divergence time and ecological diversity (Fleagle and Reed, 1999a). In this paper we examine more closely the relative roles of geography and phylogeny in shaping patterns of primate ecological diversity.

Methods

Our analyses (see Fleagle and Reed, 1996 for details) are based on quantitative measurements of ten ecological variables (Table 16.1) for 70 different populations from eight extant primate communities (or assemblages). In order to sample the diversity of living primates as broadly as possible we chose data from two well-studied communities from each of four major biogeographic regions. These communities include Tai and Kibale Forests from Africa, Kuala Lompat and Ketambe from Asia, Ranomafana and Morondava (Marasolaza) from Madagascar, and Suriname (Raleighvallen-Voltzberg) and Manu (Cosha Cachu) from South America. Our choice of communities was largely dictated by the fact that at these sites there have been broad studies of many aspects of the behavior of most of the species, so it was possible to get a nearly complete dataset. More than 90% of the data were available from studies of the species at these sites. In cases where a species, or some aspects of a species, had not been studied at the site we used data from the same species at a nearby site. Certainly two sites from each region is not sufficient to provide a broad documentation of ecological diversity within a continent. However, these sites have some of the most diverse communities in their respective regions, in part because primatologists have often chosen diverse communities as sites for long-term studies. It is also worth noting that the two sites from each region encompass different biogeographical provinces.

We used principal coordinates analysis (PCO) to reduce this 70×10 matrix to a series of composite factors. A Pearson correlation matrix was calculated and this matrix was then used in the creation of factor scores using principal coordinates analysis (Gower, 1966; Pimentel, 1979) to summarize the ecological space along a new series of orthogonal vectors that maintain the original relationships among the taxa in total space.

Results

Because of intercorrelations among the variables, much of the ecological diversity could be encompassed in a few composite factors generated by the PCO. In this study, the first two factors produced by the analysis contained 28% and 25% of the information, respectively, so that a bivariate plot of these first two principal coordinates displayed over 50% of the variation contributing to the differences among the taxa. Subsequent factors contain much less information and are less amenable to interpretation (see Fleagle and Reed, 1996). Despite their composite nature, these first two variables were readily interpretable in terms of the general aspects of behavior and ecology that they represented (Fig. 16.1). The first factor is negatively correlated with nocturnality and leaping, while positively correlated with body size, diurnality, frugivory, climbing, and terrestrial quadrupedalism. The second factor is most highly correlated with folivory in a positive direction and with arboreal quadrupedalism and faunivory in a negative direction. Thus the 70 primate populations are spread out

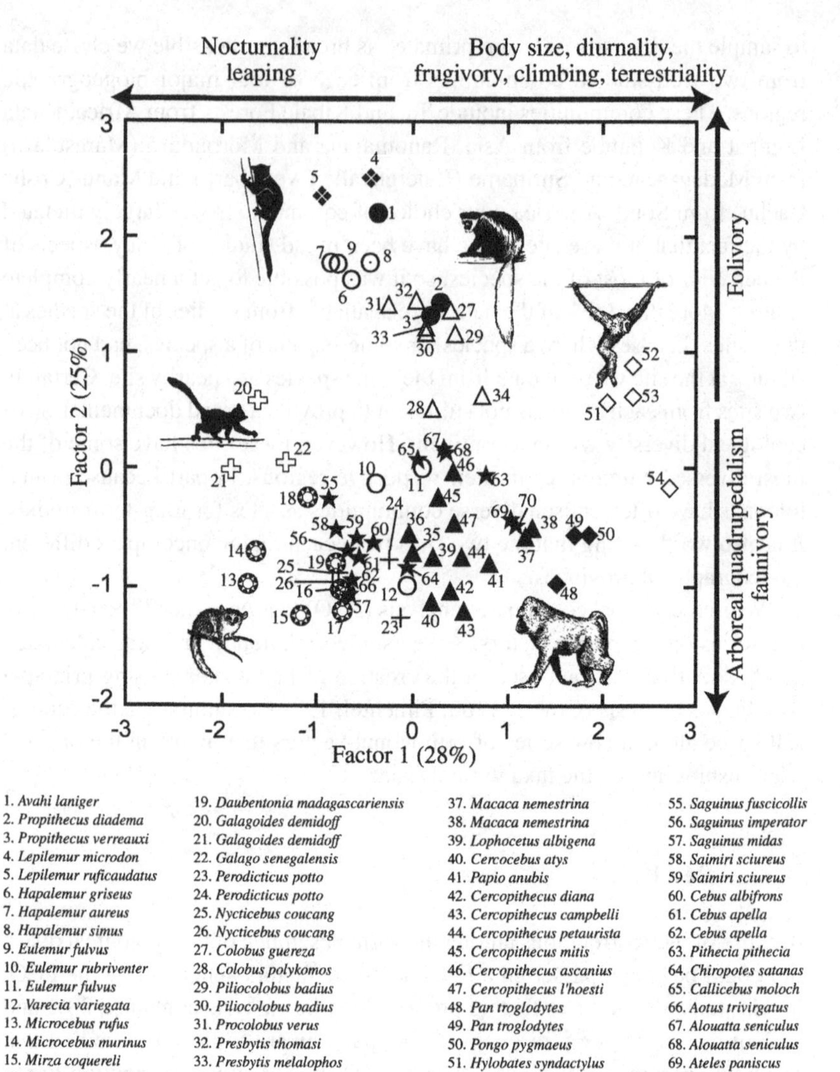

Figure 16.1. Factors 1 and 2 of the principal coordinates analysis of 70 primate populations using ten original ecological variables. Symbols indicate taxonomic groups. For details of the analysis see Fleagle and Reed (1996). Note that the species cluster according to phylogeny.

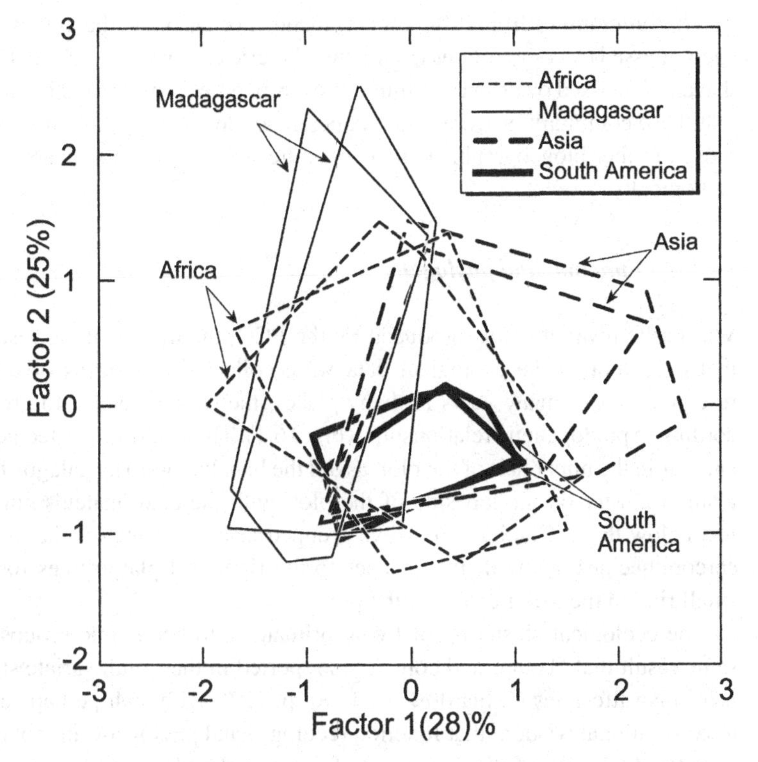

Figure 16.2. Polygons outlining the ecospace occupied by species comprising eight individual primate communities from four major biogeographical regions. Note that communities within each region occupy similar ecospace, and communities from different regions occupy distinct and characteristic ecospaces. For details see Fleagle and Reed (1996).

across the two-dimensional plot according to their ecological similarities and differences.

Because our initial goal was to compare the primate communities within and between continents, we plotted the species of each community and compared the ecospace occupied by these communities, both within and between continents (Fig. 16.2). There were three main results from this exercise. First, communities within continents were remarkably similar, despite considerable differences in species diversity in some. Thus, communities within a continent tended to occupy an ecospace of the same size and shape. Second, communities from different continents showed consistent differences. For example, Malagasy communities have more noctural and folivorous taxa, Asian communities have more suspensory taxa, and African communities have more terrestrial, frugivorous taxa. Finally, there were differences in the size of the ecospace occupied by the communities from different continents. For example, the graphical area encompassed by communities from South America is considerably smaller than that encompassed by primate communities from other continents. This suggests that these communities from South America are less ecologically diverse than those of other biogeographical areas, i.e., the species have not diverged as far ecologically.

Function and phylogeny

Within the bivariate plots produced by the PCO, the most striking result was that even though the ecological data were collected by dozens of different researchers over many years in many places, the species tended to group according to phylogenetic relationships (Fig. 16.1). Thus indriids plotted near one another in the upper half of the plot, as did the hapalemurs. The galagos formed a small cluster on the left side of the plot, with the cheirogaleids clustering just below them. Colobine monkeys grouped near the center of the plot with cercopithecines below them and apes to the right. All platyrrhines formed a small ring in the lower center of the plot.

The ecological clustering of living primates into taxonomic groups is the same result that Oxnard and colleagues reported in their multivariate study of prosimian lifestyles (Oxnard *et al.*, 1990, p. 127). Although perhaps counter to conventional wisdom that function (ecology) and phylogeny are not closely correlated because of the frequency of functional and ecological homoplasy, in fact this should not be a surprising result. Most groups of living primates are characterized by distinctive ways of life. Thus, gibbons are medium-sized, diurnal, very suspensory frugivores and folivores. Indriids are small to medium-sized, leaping frugivores and folivores. Galagos are small, leaping, nocturnal faunivores and frugivores, and so forth. Although there is certainly evidence for ecological convergence among primates – for example there is overlap between some larger indriids and colobines, and between lorises and cheirogaleids – it is often limited. For example, *Ateles* plots closer to gibbons than does any other platyrrhine because of its suspensory habits, and is certainly the closest platyrrhine equivalent to the lesser apes. Nevertheless, spider monkeys are still more quadrupedal than any of the gibbons and lie about midway between lesser apes and other platyrrhines.

Discussion

The view that function and phylogeny are not dichotomous ways of grouping primates, but rather flip sides of the same coin, has been a common theme in

Charles Oxnard's work for decades. Early on, in the midst of the classic work by Ashton and others on the primate shoulder, they noted that while the scapula grouped taxa according to forelimb use during locomotion, there was also a strong phylogenetic signal in the shoulder data (Ashton *et al.*, 1965). Oxnard returned to this theme again in the late 1970s and early 1980s in his analyses of Adolph Schultz's data on primate limb skeletons (Ashton *et al.*, 1975, 1976; Oxnard, 1978, 1981). They showed that separate analyses of the forelimb elements or the hind-limb elements grouped according to limb function. However, when the forelimb and hind-limb data were combined into a single analysis, primates tended to cluster according to phylogenetic groups. Obviously function has evolved along phylogenetic lines. Indeed Oxnard and colleagues repeatedly pointed out how the phylogenetic groupings in their functional data were congruent with the molecular phylogenies of primates that were just becoming common in the literature at the time (Oxnard, 1978, 1981).

Ecology and evolution

In view of the striking association between ecology and phylogeny in our initial analyses, we decided to look more closely at this relationship (Fleagle and Reed, 1999a). In particular we looked at the relationship between ecological distance and phylogenetic distance using a multivariate measure of ecological distance (D^2) based on our ecological matrix and using the divergence time between pairs of primate taxa generated by recent studies in molecular systematics. We chose the divergence dates from a single broad comparative study (Porter *et al.*, 1997) to avoid mixing dates from different studies based on different molecules.

When our matrix of ecological distances is compared with a comparable matrix of divergence times, there is a significant correlation for our entire sample of primate taxa drawn from eight communities in four biogeographic areas. Moreover, when we regressed four different measurements of the ecological diversity within the eight communities against the average divergence time among the taxa within those communities, there was always a positive correlation (Fleagle and Reed, 1999a, Fig. 6.15). In order to investigate the relationship between ecological distance and phylogenetic distance further, we looked at the role of geography, as the primate communities in different regions have very disparate histories.

The role of geography

When we looked at the relationship between a multivariate measure of ecological distance (D^2) and calculated divergence time for individual pairs of species

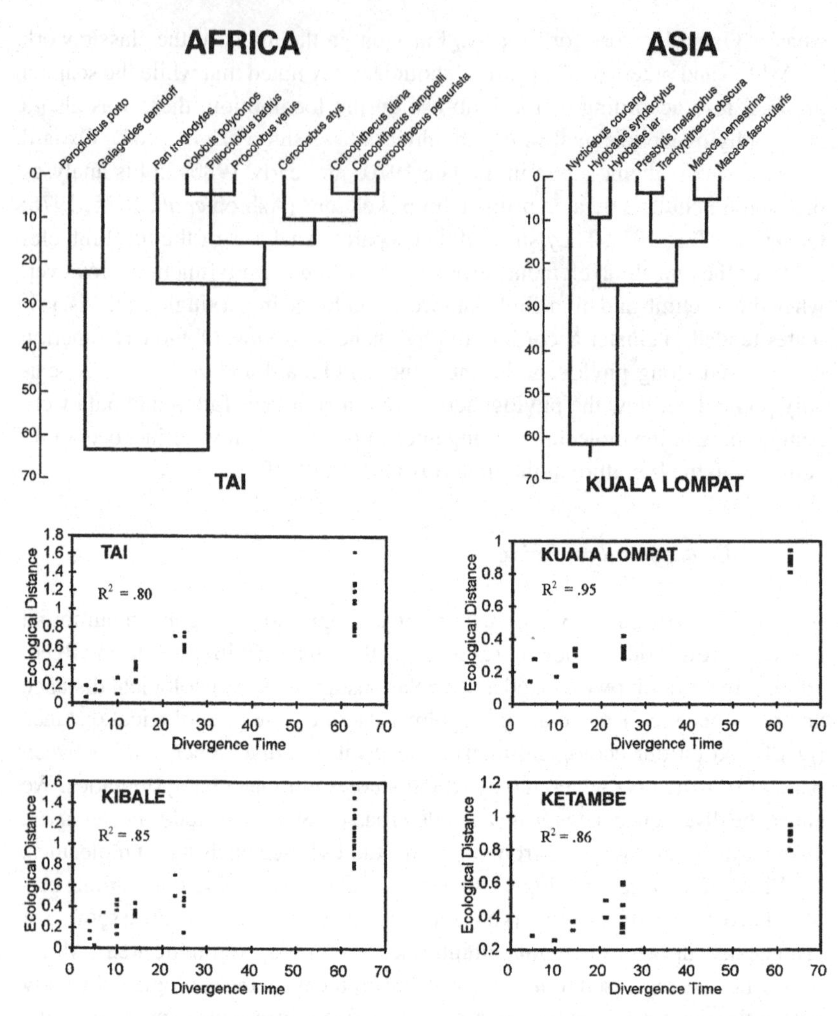

Figure 16.3. Above: molecular phylogeny for the primate species from one community in Africa and one community in Asia showing estimated divergence times for major branches. Below: regressions of ecological distance between pairs of species and reconstructed divergence times based on molecular phylogeny. Note that for the African and Asian communities the correlation between ecological distance and phylogenetic distance (divergence time) is very high.

within communities, we found two very different types of relationships. At the two Asian localities (Kuala Lompat and Ketambe) and at the two African localities (Tai and Kibale) there is a very high correlation between ecological distance and divergence time for pairs of individual taxa (Fig. 16.3). Recently divergent taxa, for example, pairs of colobines or pairs of cercopithecines, are

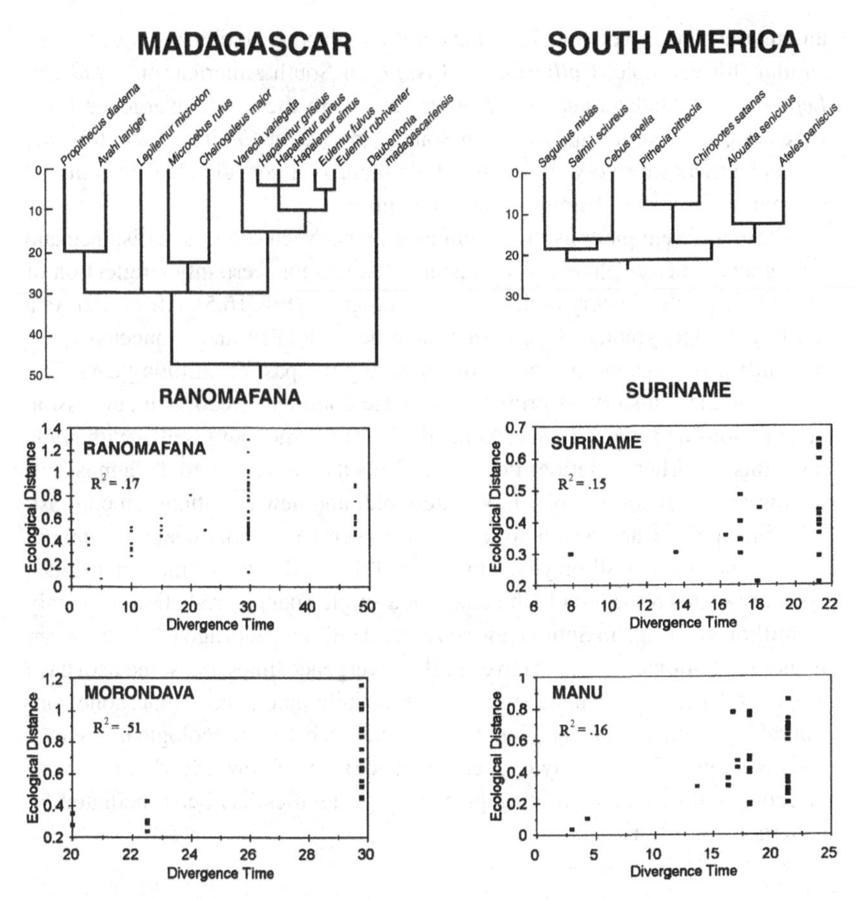

Figure 16.4. Above: molecular phylogeny for the primate species from one community in Madagscar and one community in South America showing estimated divergence times for major branches. Below: regressions of ecological distance between pairs of species and reconstructed divergence times based on molecular phylogeny. Note that for the Malagasy and South American communities the correlation between ecological distance and phylogenetic distance (divergence time) is very low.

very similar ecologically; and distantly divergent taxa, for example, lorises and gibbons, or galagos and chimpanzees, are very different.

However, for the localities in Madagascar (Ranomafana and Morondava) and South America (Suriname and Manu) there is a very low correlation between ecological distance and divergence time for pairs of taxa (Fig. 16.4). This is because the divergence times in these regions tend to be clustered around single time periods 21 million years ago for South America and 30 million years ago for Madagascar when many modern families and subfamilies split. Moreover, among the pairs of taxa that have the common divergence times, some are very similar, for example, *Callicebus* and *Aotus* in South America, or *Avahi* and *Lepilemur* in Madagascar; while other pairs with the same divergence time, for example, *Saguinus* and *Ateles* in South America, or *Cheirogaleus* and *Indri* in Madagascar, are very different. Thus, there is a considerable diversity of ecological distances at single divergence times.

These different patterns in the relationship between ecological distance and divergence time (or phylogenetic distance) are almost certainly a reflection of the biogeographic history of these different regions (Fig. 16.5). Africa and Asia are large biogeographic regions that have been isolated and connected intermittently with one another many times during the past 65 million years. The paleontological history of primates on these continents records a succession of radiations and major replacements during this time, but usually with some remnants of earlier radiations persisting. Thus they have "layered" faunas containing a combination of old, moderately old, and new radiations. In contrast, both Madagascar and South America have been island continents throughout most of the last 65 million years. On both of these islands, the modern primate diversity seems to be largely the result of a single adaptive radiation – roughly 20 million years ago in South America and 30 million years ago in Madagascar. In these radiations, most taxa have similar divergence times, but some taxa have remained largely unchanged from that time while others have undergone considerable adaptive change. Thus the correlation between ecological distance and divergence time (or phylogenetic distance) is very low. Clearly the pattern of ecological diversity in living primate communities has been mediated by geographical history.

An ecological clock?

One of the many lessons to be learned from Charles Oxnard's work is that there is always another story and another hidden pattern somewhere in any dataset if you look hard enough. Accordingly, we decided to see if, despite the very different patterns in the correlation between ecological distance and phylogenetic distance, there still might be some regularity in the relationship. And indeed there is. Despite the fact that the correlation between ecological distance and phylogenetic distance is very high in the African and Asian localities and very low in the Malagasy and South American localities, in all cases the relationship is positive and significant. Moreover, if we combine the taxa from the different localities within each biogeographical area, and examine the relationship between ecological divergence and phylogenetic distance, we find that there is

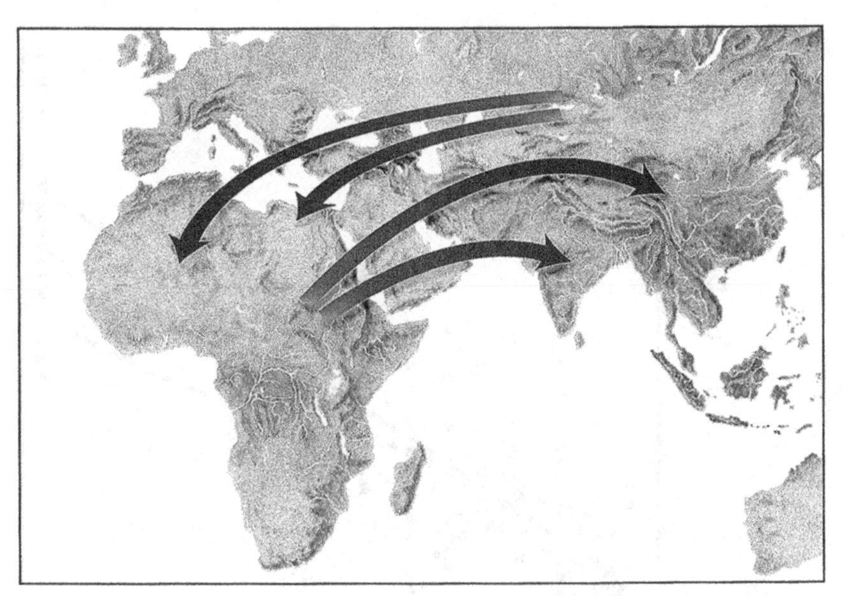

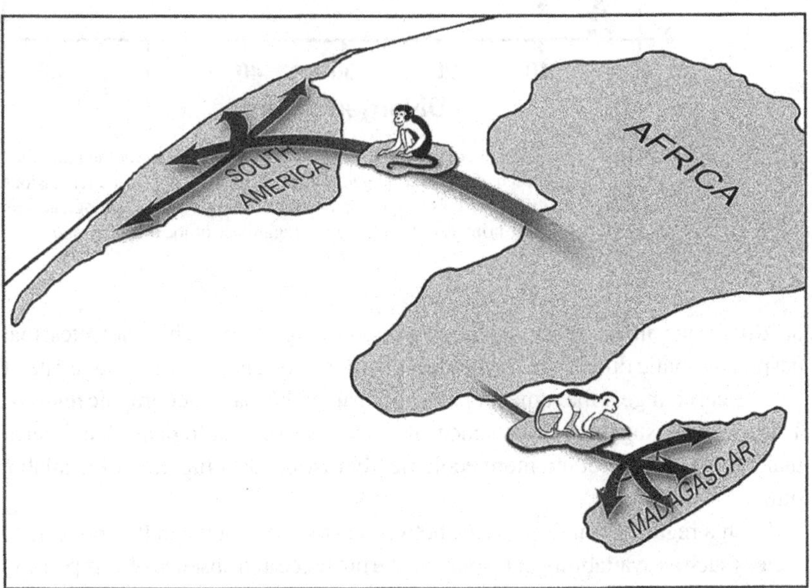

Figure 16.5. Above: the primates of Africa and Asia have layered composition resulting from a long history of faunal interchange among continents and a series of successive radiations that partly replaced older radiations. Below: in contrast, the primate faunas of Madagascar and South America are largely the result of explosive adaptive radiations that took place soon after they were initially colonized approximately 20–30 million years ago.

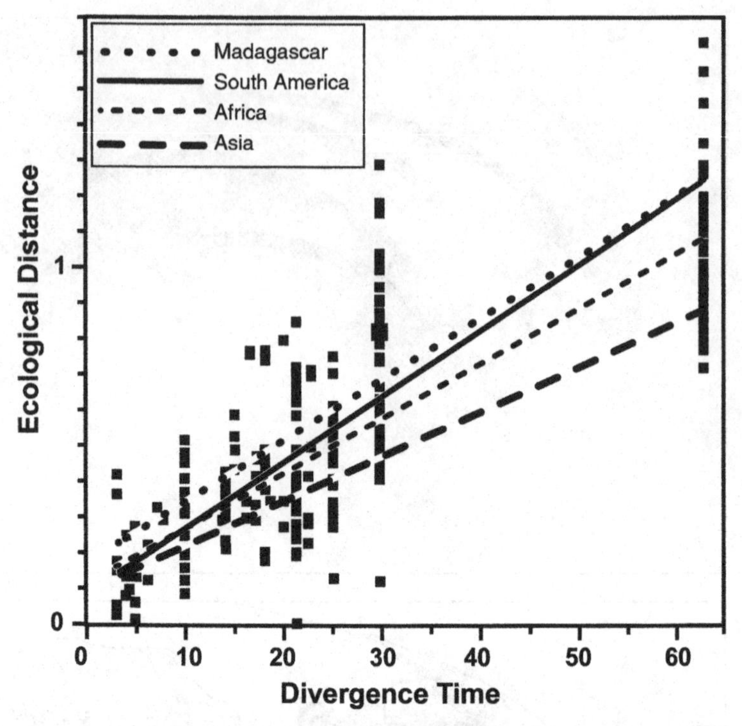

Figure 16.6. Regression of ecological distance and divergence time for the individual pairs of species from each of the four biogeographic regions. Note that the regressions are not signficantly different. This suggests an overall common rate of ecological divergence with time for primates of each region, regardless of local history and environment.

no difference in the slope of the regression (Fig. 16.6). This indicates that despite dramatic differences in the strength of the correlation, the average rate of ecological change is the same for primates in all of the major geographic regions. These results suggest the presence of some sort of "ecological clock" such that primate taxa become more ecologically distinct as a function of available time.

Such a regularity in differences between taxa does not mean that proximate factors such as availability of resources, the presence and absence of competition from other taxa, and the myriad climatic changes throughout the Cenozoic have not played a role in the evolution of species-specific adaptations or the diversity of individual communities, only that at some general level there is a broad constraint on how fast primate adaptations have evolved over the past 60 million years, or that the stochastic interaction of the numerous proximate factors leads to a regular rate of ecological change. Indeed, as the geographic patterns discussed above indicate, there has been considerable variation in the pattern of evolutionary change from continent to continent. Moreover, these results are based on the correlation between ecological distance and divergence time for extant taxa. In many cases the ecological diversity of extant taxa is considerably less than that of extinct communities (Fleagle, 1999; Godfrey *et al.*, 1997), and it is not unlikely that current threats of extinction could lead to even more changes in the near future (Jernvall and Wright, 1998). It is unclear how the inclusion of extinct taxa would change these patterns. Unfortunately we have no divergence times for the extinct species.

Nevertheless, the existence of an "ecological clock" accords well with one of the more striking results of our initial study – the low diversity among platyrrhine primates (Fleagle and Reed, 1996). By any measure of ecological diversity, the most diverse platyrrhine communities are much less diverse than the communities on other continents, and the greatest differences among individual taxa are much less than those seen on other continents (Fig. 16.2). Results of both molecular systematics and the platyrrhine fossil record suggest that the divergence times for living platyrrhine taxa are on the order of about 20 million years, much less than the divergence times for taxa on other continents (e.g., Porter *et al.*, 1997; Fleagle *et al.*, 1997; Fleagle, 1999; Fleagle and Tejedor, 2002). In any case, this intriguing pattern needs to be tested with further analyses of more communities as well as comparative investigations of the role of resources and competitors in primate community evolution (e.g., Fleagle and Reed, 1999b).

Summary

One of Charles Oxnard's many pioneering efforts in biological anthropology involved the use of multivariate analyses in the study of ecological diversity among prosimian primates. We have expanded this approach in a comparison of primate communities in Africa, Asia, Madagascar, and South America. Many of our results concur with those reported by Oxnard and colleagues, including the observation that ecological and phylogenetic groupings are remarkably concordant. There is a positive correlation between the ecological distance between pairs of taxa and the time since their last common ancestor. However, the strength of the correlation between ecological divergence and phylogenetic divergence reflects the history of different geographical regions. It is much higher among the primates in Africa and Asia, in which the faunas show phylogenetic layering, than among the primates from the island continents of Madagascar and South America, which are the result of explosive radiations. Despite the geographical differences, the average rate at which ecological distance between taxa has evolved through time is the same for primates of each region, suggesting the presence of an "ecological clock" underlying the evolution of ecological diversity.

Acknowledgments

We thank Fred Anapol for inviting us to contribute a paper to this volume, and Nina Jablonski and Rebecca German for inviting us to participate in the symposium honoring Charles Oxnard at the Annual Meeting of the American Association of Physical Anthropologists in San Antonio, Texas in 2000. We thank Chris Heesy and Luci Betti-Nash for bibliographic and artistic assistance. Most of all we thank Charles Oxnard for his many stimulating and insightful contributions to biological anthropology over the past four decades. This paper is dedicated to Charles Oxnard and to the late Waylon Jennings, both masters at distilling simple and aesthetically pleasing patterns from the seeming chaos of life.

References

Ashton, E. H., Oxnard, C. E., and Spence, T. F. (1965). Scapular shape and primate classification. *Proc. Zool. Soc. Lond.*, **145**, 125–142.

Ashton, E. H., Flinn, R. M., and Oxnard, C. E. (1975). The taxonomic and functional significance of overall body proportions in Primates. *J. Zool. Lond.*, **175**, 73–105.

Ashton, E. H., Flinn, R. M., Oxnard, C. E., and Spence, T. F. (1976). The adaptive and classificatory significance of certain quantitative features of the forelimb in primates. *J. Zool. Lond.*, **179**, 515–556.

Bourliere, F. (1985). Primate communities: their structure and role in tropical ecosystems. *Int. J. Primatol.*, **6**, 1–26.

Chivers, D. J. (1980). *Malayan Forest Primates*. New York, NY: Plenum Press.

Crompton, R. H., Lieberman, S. S., and Oxnard, C. E. (1987). Morphometrics and niche metrics in prosimian locomotion: an approach to measuring locomotion, habitat, and diet. *Amer. J. Phys. Anthropol.*, **73**, 149–177.

Fleagle, J. G. (1999). *Primate Adaptation and Evolution*. 2nd edn. San Diego, CA: Academic Press.

Fleagle, J. G. and Mittermeier, R. A. (1980). Locomotor behavior, body size, and comparative ecology of seven Surinam monkeys. *Amer. J. Phys. Anthropol.*, **52**, 301–314.

Fleagle, J. G. and Reed, K. E. (1996). Comparing primate communities: a multivariate approach. *J. Hum. Evol.*, **30**, 489–510.

(1999a). Phylogenetic and temporal perspectives on primate ecology. In: *Primate Communities*, ed. J. G. Fleagle, C. Janson, and K. E. Reed. New York, NY: Cambridge University Press. pp. 92–115.

(1999b). Why are platyrrhine monkeys not more diverse? *Congreso Internacional: Evolucion Neotropical del Cenozoico*. La Paz. p. 22.

Fleagle, J. G. and Tejedor, M. A. (2002). Fossil primates of southern South America, In: *The Primate Fossil Record*, ed. W. Hartwig. Cambridge: Cambridge University Press. pp. 161–174.

Fleagle, J. G., Kay, R. F., and Anthony, M. R. L. (1997). Fossil new world monkeys. In: *Vertebrate Paleontology in the Neotropics. The Miocene Fauna of La Venta, Colombia*, ed. R. F. Kay, R. H. Madden, R. L. Cifelli, and J. J. Flynn. Washington, DC: Smithsonian Institution Press. pp. 473–495.

Godfrey, L. R., Jungers, W. L., Reed, K. E., Simons, E. L., and Chatrath, P. S. (1997). Subfossil lemurs: inferences about past and present primate communities in Madagascar. In: *Natural Change and Human Impact in Madagascar*, ed. S. M. Goodman and B. D. Patterson. Washington, DC: Smithsonian Institution Press. pp. 218–256.

Gower, J. C. (1966). Some distance properties of latent root and vector methods used in multivariate analysis. *Biometrica*, **53**, 325–338.

Jernvall, J. and Wright, P. C. (1998). Diversity components of impending primate extinctions. *Proc. Nat. Acad. Sci.*, **95**, 11279–11283.

Kappeler, P. M. and Heymann, E. W. (1996). Nonconvergence in the evolution of primate life history and socio-ecology. *Biol. J. Linn. Soc.*, **59**, 297–326.

Mittermeier, R. A. and van Roosmalen, M. G. M. (1981). Preliminary observations on habitat utilization and diet in eight Surinam monkeys. *Folia Primatol.*, **36**, 1–39.

Oxnard, C. E. (1978). Concordance and disconcordance in primate relationships. *Amer. Zool.*, **18**, 648.

(1981). The place of man among the primates: anatomical, molecular and morphometric evidence. *Homo*, **3**, 23–53.

(1984). *The Order of Man: a Biomathematical Anatomy of the Primates*. New Haven, CT: Yale University Press.

Oxnard, C. E., Crompton, R. H., and Lieberman, S. S. (1990). *Animal Lifestyles and Anatomies: the Case of the Prosimian Primates*. Seattle, WA: University of Washington Press.

Pimentel, R. A. (1979). *Morphometrics: the Multivariate Analysis of Biological Data*. Dubuque, IA: Kendall/Hunt.

Porter, C. A., Page, S. L., Czelusniak, J., *et al.* (1997). Phylogeny and evolution of selected primates as determined by sequences of the e-globin locus and 5' flanking regions. *Int. J. Primatol.*, **18**, 261–295.

Reed, K. E. (1999). Population density of primates in communities: differences in community structure. In: *Primate Communities*, ed. J. G. Fleagle, C. H. Janson, and K. E. Reed. New York, NY: Cambridge University Press. pp. 116–140.

Reed, K. E. and Fleagle, J. G. (1995). Geographic and climatic control of primate diversity. *Proc. Nat. Acad. Sci.*, **92**, 7874–7876.

Struhsaker, T. T. (1975). Food habits of five monkey species in Kibale forest, Uganda. In: *Recent Advances in Primatology. Vol. 1, Behavior*, ed. D. J. Chivers and J. Herbert. New York, NY: Academic Press. pp. 225–248.

Terborgh, J. (1983). *Five New World Primates*. Princeton, NJ: Princeton University Press.

Terborgh, J. and van Schaik, C. P. (1987). Convergence vs. nonconvergence in primate communities. In: *Organization of Communities Past and Present*, ed. J. H. R. Gee and P. S. Giller. London: Blackwell. pp. 205–226.

17 *Charles Oxnard and the aye-aye: morphometrics, cladistics, and two very special primates*

COLIN P. GROVES
Australian National University

The aye-aye and the question of strepsirhine monophyly

The two suborders of primates, Haplorhini and Strepsirhini, are distinguished clearly by four well-studied and anatomically complex characters:

(1) The possession by Strepsirhini of a rhinarium. A rhinarium is widespread among mammals generally, and its possession by Strepsirhini among primates is undoubtedly plesiomorphic.
(2) The presence in Strepsirhini of a tapetum lucidum in the eye. Although a tapetum is likewise widespread among mammals, its structure differs from one group to another, so its polarity among primates is unclear.
(3) The existence of a fovea in the retina in the Haplorhini. A retinal fovea appears to be unique to Haplorhini, except that an approach to it is seen in *Lemur catta* and *Hapalemur griseus* (Pariente, 1970). The most parsimonious interpretation is that it is a haplorhine apomorphy, and that there has been some convergent development of one in the two lemurid taxa cited.
(4) A complex of characters of the fetal membranes, such that the placenta is hemochorial in Haplorhini but epitheliochorial in Strepsirhini. Martin (1990), noting that some trophoblastic invasion occurs in two species of galago, argued that the primitive state is endotheliochorial, and that consequently both epitheliochorial and hemochorial states are derived; but the cladistic analysis of Luckett (1993) concluded that the condition in the Strepsirhini is primitive.

It would appear, therefore, that of these four characters the Strepsirhini are questionably derived in only one, the presence of a tapetum. Otherwise, they

Shaping Primate Evolution, ed. F. Anapol, R. Z. German, and N. G. Jablonski. Published by Cambridge University Press.

are defined only by plesiomorphic retentions. What, then, is the evidence that the Strepsirhini really are monophyletic?

Three undoubtedly derived conditions do occur in almost all the Strepsirhini: the toilet claw, the dental comb, and the ansa coli. A toilet claw also occurs, in *Tarsius* (on toes 2 and 3, rather than merely on toe 2 as in strepsirhines), but no other primate has a dental comb or an ansa coli. The spoiler, however, is the aye-aye.

The aye-aye (*Daubentonia madagascariensis*) is one of the Malagasy lemurs, generally assigned to its own, monotypic family, Daubentoniidae. Unlike any other strepsirhine it has claws on all its digits except the hallux, not merely on the second toe; and it lacks a dental comb, instead having a single pair of incisors, grossly hypertrophied labiolingually, and growing from persistent pulps. The questions which arise are whether the claws are plesiomorphic (so that its ancestors never possessed the nails-plus-toilet-claw combination) or autapomorphic (and perhaps submerging an originally developed toilet claw); and whether the undoubtedly apomorphic condition of the incisors has caused an ancestral dental comb to be lost. The answers to these two questions will help to decide whether the aye-aye is (1) the sister-group of all other primates, (2) a strepsirhine, but the sister-group of all other strepsirhines, (3) a lemuriform, the sister-group of all other lemurs, or (4) lemuriform, and a highly specialized derivative of one of the other lemuriform groups.

The autapomorphic nature of the aye-aye

A paper by Oxnard (1981a) approached this issue from a morphometric perspective. Comparing its body proportions and skeletal measurements with other primates, first system by system and then overall, *Daubentonia* appeared strikingly distinct: it "is more different in its overall bodily morphology from any primate than is any primate from any other" (Oxnard, 1981a, p. 14). Whether the comparison is with all other primates, or restricted to other quadrupeds, or to other strepsirhines, the aye-aye is remarkably different.

Tattersall (1982) took exception to the implications, as he saw them, of this study. If Oxnard really meant to imply that the autapomorphic states of the aye-aye exclude it from a close evolutionary relationship with others, then this is clearly false: they reflect "no more than its adaptive history subsequent to its divergence" (Tattersall, 1982, p. 13). Actually, Tattersall's reading of Oxnard's paper may be in error, as Oxnard (1981b) had already made clear in a companion paper in which he compared the phylogenetic groupings found in most morphometric studies with those found by biomolecular studies and those "inherent in classical morphology." In such cases, he found, morphometric and molecular groupings tend to concur quite closely, whereas "classical morphology" (note that these were from pre-cladistic days) often told quite a different story. Thus, the morphometric uniqueness of the aye-aye would have profound implications for our understanding of its phylogenetic position, and served actually to throw in question earlier assumptions based on morphology.

Is the aye-aye really a lemuriform, or a strepsirhine at all?

That there is a problem with the relationships of the aye-aye was first seriously pointed out by Groves (1974), who questioned whether primatologists might not have assumed that it must be a member of the Lemuriformes simply because it lives on Madagascar. Groves claimed that, unlike all other primates, the aye-aye retains primitive features in several regions:

(1) the orbit, which is dominated by a maxillary plate and has neither an ethmoid exposure nor a frontopalatine suture;
(2) the nasal cavity, which has more turbinals than other primates;
(3) the appendages, which have a deep stratum unlike any other strepsirhine, and the deep stratum is thicker than in any other primate;
(4) the muscular system, which has a two-part M. obturator externus and, in some individuals, two Mm. peroneo-tibiales, unlike any other primate;
(5) inguinal nipples.

Looking over this list, some of the features do not look quite so convincing today. The pattern of bones in the orbit is variable, and frontopalatine contact does occur in the aye-aye (Cartmill, 1978). The turbinal count is partially confirmed by Tattersall and Schwartz (1974), although it is essentially just one of a number of complexities in strepsirhines. It is now known from a recent study by Soligo and Müller (1999) that at least one other strepsirhine (*Lemur catta*) does have two-layered appendages, although their relative thickness was not studied by Soligo and Müller, and the coverage is in great need of being extended. The muscular characters need to be examined in a wider range of specimens, and of taxa.

Groves (1974) noted in addition that, unlike (other?) strepsirhines but like at least some haplorhines, the aye-aye has an occipital insertion of M. trapezius, a coracoid origin for M. coracobrachialis medius, a double origin for M. flexor digitorum sublimis, and a relatively simple carotid pattern. These characters, too, need further study, particularly as Cartmill (1978) noted how small were the sample sizes for carotid artery studies.

Despite these caveats, the question needs to be faced. Does the aye-aye retain primitive features not seen in any other primate? If so, is it the sister-group of

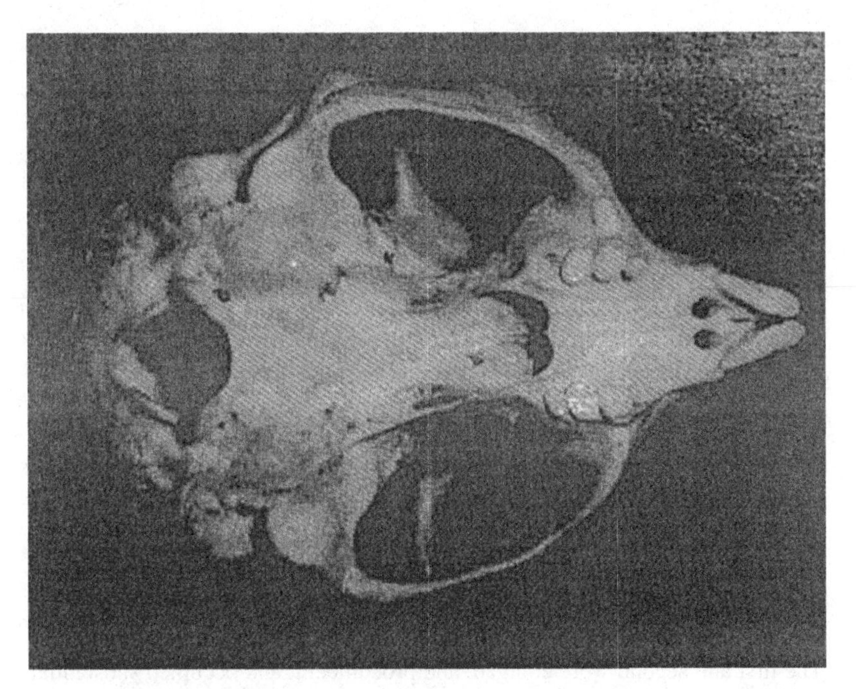

Figure 17.1. Basal view of adult aye-aye cranium.

all other primates, or are its primitive features actually character reversals, or have the haplorhines and the (other?) strepsirhines evolved the derived states in parallel? If it is really and truly a strepsirhine, the same questions: does it retain primitive features which are derived in other strepsirhines, and what is their status, homologies, or homoplasies?

Which brings us back to the three strepsirhine synapomorphies. One is now settled: the aye-aye does have a toilet claw. Soligo and Müller (1999) have shown very neatly that, even though the aye-aye has claws on all its toes except the hallux, the one on the second toe is clearly different from the others. It is probable, too, that it possesses an ansa coli, and although this information comes from the "older literature" there seems no reason to doubt it. There remains one: did its ancestors ever have a dental comb?

Did the proto-*Daubentonia* have a dental comb?

The adult aye-aye has a single pair of anterior teeth in each jaw (Fig. 17.1). These are mesiodistally compressed and sagittally expanded, enamel is confined to the labial surface, and the root is enormously elongated and open throughout life.

Traditionally, these enlarged front teeth are described as permanent incisors, but this has been questioned (see below). The deciduous dentition has been much discussed. In a newborn aye-aye Peters (1866a) described two deciduous molars, and four pairs of front teeth in the upper jaw, three in the lower, which he interpreted as di^1, I^1, di^2, and c'; and di_1, I_1, and di_2 respectively – but he admitted that the di_2 could alternatively be a deciduous canine. The permanent incisors were just the tips poking through. In a suckling aye-aye, the permanent incisors were now fully erupted, and had pushed out the deciduous anteriors and, in the upper jaw, pushed the laterals sideways. Later the same year Peters (1866b) studied another newborn, in which the tips of the permanent incisors were still hidden. The deciduous dental formula was therefore (in modern notation) 212/202 or, possibly, 212/112.

Tattersall and Schwartz (1974) suggested that these hypertrophied anterior teeth are not incisors but canines, but they agreed that they belong to the permanent series. They disagreed with Peters' interpretation of the first deciduous maxillary tooth as a canine; for them, it is a deciduous molar.

Luckett and Maier (1986), on the basis of their examination of the condition in a late fetus, offered yet another interpretation. Three pairs of front teeth were present; all were in the premaxilla hence they must be incisors, not canines. The first and second were elongate and procumbent, and occupied substantial alveoli, the third was small and vertical with only a shallow alveolus. Only the second incisor, the largest, had enamel (restricted to the labial surface). There was no sign that successional teeth were developing. The situation was similar in the mandible. From this, the authors concluded that the front teeth in the adult are persistent deciduous second incisors.

The matter was reconsidered in detail by Ankel-Simons (1996), who reasserted the traditional interpretation of the hypertrophied teeth as permanent incisors and of the first maxillary tooth as a canine. She found that an aye-aye that had died at birth in the Duke University Primate Center had only two pairs of teeth in the premaxilla (di^1 and I^1), and no lateral deciduous incisor was visible even on *x*-ray; so, for her, the correct deciduous dental formula is 112/112, although she allowed that there might be polymorphism in the presence or absence of di^2.

Infant *Daubentonia* specimens are not common in collections, but in Naturalis (formerly Rijksmuseum van Natuurlijke Historie) in Leiden there is a skull (cat.ost.B) of an early juvenile (Fig. 17.2). In the maxilla, the first molar is freshly erupted on one side, and nearly erupted on the other; the second molar is well on its way towards eruption; the third molars are visible in their crypts. All the deciduous molars are shed, and permanent P^4 is in process of eruption. A remnant alveolus is still present somewhat behind the premaxillary suture (Fig. 17.3); in agreement with Ankel-Simons (1996), I interpret the

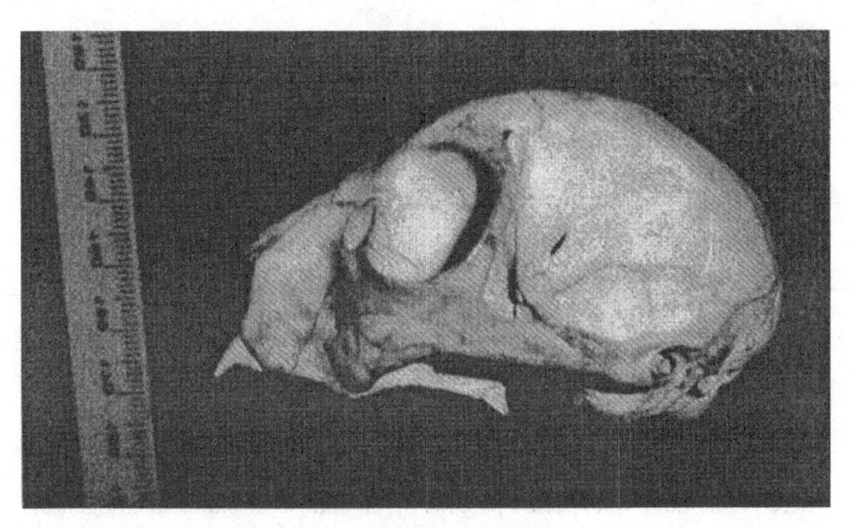

Figure 17.2. Lateral view of juvenile aye-aye cranium.

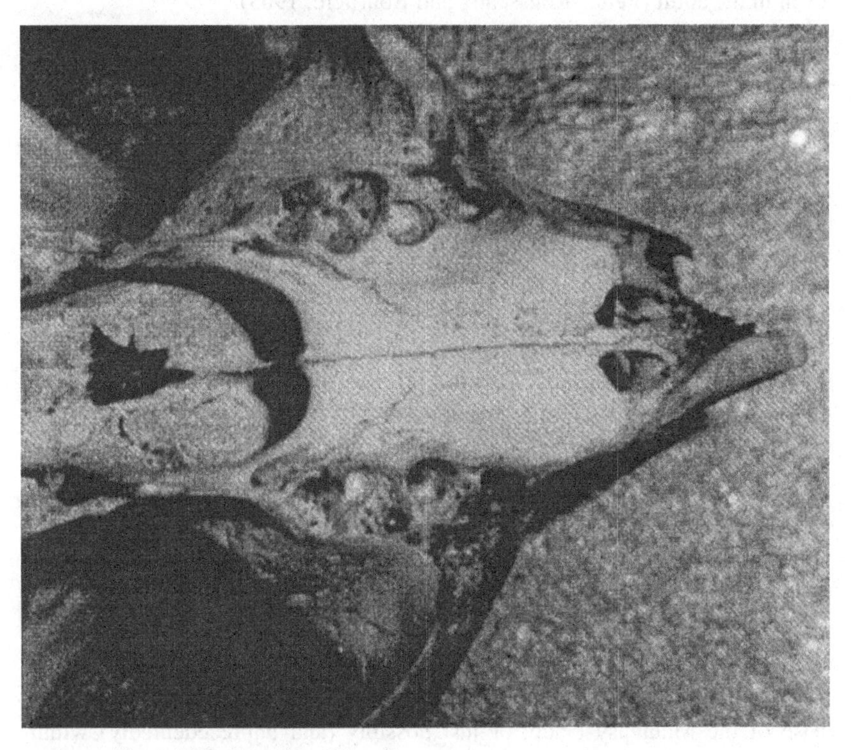

Figure 17.3. Dentition of juvenile aye-aye.

tooth that had occupied this as a deciduous canine. No alveolus remains for di_2/c' (whichever it is). A functional pair of incisors is present in both jaws, but still peg-like, not yet expanded sagittally along the lateral sides of the incisive foramina as in the adult. In the premaxilla there is a pair of alveoli behind them. There is no remaining trace of di^1 or their alveoli, but the presence of alveoli for di^2 suggests that Ankel-Simons' (1996) alternative interpretation is correct: whether these teeth are present or not is subject to polymorphism.

The ancestral aye-aye, therefore, may or may not have had a dental comb, but its deciduous upper incisor pairs are, polymorphically, either one or two in number.

The aye-aye and its eggs

There is one quite unexpected peculiarity. In almost all mammals, oogenesis has ceased by the time of birth: by that time, they have all the ova they are ever going to have. In the Lorisiformes and the aye-aye alone, oogenesis persists even in the adult (Petter-Rousseaux and Bourlière, 1965).

Indriid affinities for the aye-aye?

Tattersall and Schwartz (1974) noted some resemblance in the deciduous mandibular premolars between *Daubentonia* and *Indri*, and later (Schwartz and Tattersall, 1985) definitively associated the aye-aye with the Indriidae in a clade. They cited as cranial similarities the following characters:

(1) a rounded, globular cranium
(2) deepened facial skeleton and mandible
(3) expanded jaw angles
(4) rounded mandibular condyles

There are some similarities in the globular neurocranium, deepened splanchnocranium, and rounded condyles to *Propithecus*, but these apply equally well to some non-indriids, especially *Prolemur simus*, and conversely they do not apply at all to *Indri*. The mandible deepens markedly posteriorly in the Indriidae, but is uniformly deep in the aye-aye. The jaw angles are expanded noticeably in the Indriidae, but not at all in the aye-aye.

A combination dataset including some behavioral characters (Stanger-Hall, 1997) definitely excluded the aye-aye from the Indriidae, and placed it at the base of the Malagasy clade, or just possibly (and unprecedentedly) within the cheirogalid clade. According to Gebo and Dagosto (1988), the Indriidae and Lemuridae share derived conditions of the pedal skeleton; these are absent in *Daubentonia*, as in the Cheirogaleidae.

There are two only possible pieces of evidence that the aye-aye is a lemuriform at all:

(1) The structure of the auditory bulla, with the tympanic ring inside it. The polarity of this character is not clear: it may be a synapomorphy with other Malagasy lemurs, or it may be a symplesiomorphy of all strepsirhines (or even of all primates).
(2) The pattern of the palmar and plantar pads. In Lemuriformes, the first interdigital pad is fused with the thenar on both palm and sole (Rumpler and Rakotosamimanana, 1972), an obviously derived condition. But it is unclear, from these authors' diagram, whether the aye-aye possesses this fusion or whether the thenar pad has simply disappeared.

The molecular evidence (Dutrillaux and Rumpler, 1995; Yoder, 1997) is that the aye-aye is certainly not an indriid. It is at least the sister-group of all other Malagasy lemurs (Porter *et al.*, 1995, ε-globin; Stanger-Hall and Cunningham, 1998, 16S); possibly as divergent from these as they in turn are from the Lorisiformes (Adkins and Honeycutt, 1994, COII; but see Stanger-Hall and Cunningham, 1998). There is no other indication that it might be on the lorisiform stem, so the meaning of their persistent oogenesis remains a mystery.

Conclusions

The aye-aye is not an indriid or indrioid, indeed is not closely related to any other lemuriform at all. At present it is impossible to determine whether it is: (1) the sister-group to all the other Malagasy lemurs, (2) the sister-group to all the other strepsirhines, or (3) just possibly a sister-group to the Lorisiformes. Given this uncertainty, it is advisable to allocate it to a third infraorder of strepsirhines. Just as I proposed a quarter of a century ago (Groves, 1974), the Strepsirhini should be divided into Lorisiformes, Lemuriformes, and Chiromyiformes.

Although a distance method and avowedly non-cladistic, multivariate morphometrics does appear in the aye-aye case (and probably in other cases too: see the discussion on the tarsier in Shoshani *et al.*, 1996) to offer insights into phylogenetic affinities. The mechanism is probably somewhat as follows. Evolutionary rates are strongly positively skewed: the slowest possible evolutionary rate, complete stasis, is a pervasive reality in lineage after lineage, whereas the more rapid the rate of change the less probable it is. Thus the large genetic changes reflected in a large morphometric distance are more likely to be due to time than to high evolutionary rates.

Strongly autapomorphic organisms, like the aye-aye or Charles Oxnard, may therefore offer unexpected but satisfying insights into the evolutionary process.

References

Adkins, R. M. and Honeycutt, R. L. (1994). Evolution of the primate cytochrome c oxidase subunit II gene. *J. Mol. Evol.*, **38**, 215–231.

Ankel-Simons, F. (1996). Deciduous dentition of the aye aye, *Daubentonia madagascariensis. Amer. J. Primatol.*, **39**, 87–97.

Cartmill, M. (1978). The orbital mosaic in prosimians and the use of variable traits in systematics. *Folia Primatol.*, **30**, 89–114.

Dutrillaux, B. and Rumpler, Y. (1995). Phylogenetic relations among Prosimii with special reference to Lemuriformes and Malagasy nocturnals. In: *Creatures of the Dark: the Nocturnal Prosimians*, ed. L. Alterman, G. A. Doyle, and M. K. Izard. New York, NY: Plenum Press. pp. 141–150.

Gebo, D. L. and Dagosto, N. (1988). Foot anatomy, climbing, and the origin of the Indriidae. *J. Hum. Evol.*, **17**, 135–154.

Groves, C. P. (1974). Taxonomy and phylogeny of prosimians. In: *Prosimian Biology*, ed. R. D. Martin, G. A. Doyle, and A. C. Walker. London: Duckworth. pp. 449–473.

Luckett, W. P. (1993). Developmental evidence from the fetal membranes for assessing archontan relationships. In: *Primates and their Relatives in Phylogenetic Perspective*, ed. R. D. E. MacPhee. New York, NY: Plenum Press. pp. 149–186.

Luckett, W. P. and Maier, W. (1986). Developmental evidence for anterior tooth homologies in the aye-aye, *Daubentonia. Amer. J. Phys. Anthropol.*, **69**, 233.

Martin, R. D. (1990). *Primate Origins and Evolution.* London: Chapman and Hall.

Oxnard, C. E. (1981a). The uniqueness of *Daubentonia. Amer. J. Phys. Anthropol.*, **54**, 1–21.

(1981b). The place of man among the primates: anatomical, molecular and morphometric evidence. *Homo*, **32**, 149–176.

Pariente, G. (1970). Rétinographies comparées des Lémuriens malgaches. *C. R. Acad. Sc. Paris*, **270D**, 1404–1407.

Peters, W. (1866a). Ueber die Säugethiergattung *Chiromys. Abh. Kön. Preuss. Akad. Wissensch. Berlin*, **1886**, 78–100.

(1866b). Nachtrag zu seiner Abhandlung über *Chiromys. Abh. Kön. Preuss. Akad. Wissensch. Berlin*, **1886**, 221–222.

Petter-Rousseaux, A. and Bourlière, F. (1965). Persistence du phénomène d'ovogénèse chez l'adulte de *Daubentonia madagascariensis. Folia Primatol.*, **3**, 241–244.

Porter, C. A., Sampaio, I., Schneider, H., Schneider, M. P. C., Czelusniak, J., and Goodman, M. (1995). Evidence on primate phylogeny from ε-globin gene sequences and flanking regions. *J. Mol. Evol.*, **50**, 30–55.

Rumpler, Y. and Rakotosamimanana, B. R. (1972). Coussinets palmo-plantaires et dermatoglyphes des représentants des lémuriformes malgaches. *Bull. Assoc. Anat.*, **154**, 1127–1143.

Schwartz, J. H. and Tattersall, I. (1985). Evolutionary relationships of living lemurs and lorises (Mammalia, Primates) and their potential affinities with European Eocene Adapidae. *Anthropol. Pap. Amer. Mus. Nat. Hist.*, **60**, 1–100.

Shoshani, J., Groves, C. P., Simons, E. L., and Gunnell, G. F. (1996). Primate phylogeny: morphological vs molecular results. *Mol. Phylogenet. Evol.*, **5**, 102–154.

Soligo, C. and Müller, A. E. (1999). Nails and claws in primate evolution. *J. Hum. Evol.*, **36**, 97–114.

Stanger-Hall, K. F. (1997). Phylogenetic affinities among the extant Malagasy lemurs (Lemuriformes) based on morphology and behavior. *J. Mamm. Evol.*, **4**, 163–194.

Stanger-Hall, K. F. and Cunningham, C. W. (1998). Support for a monophyletic Lemuriformes: overcoming incongruence between data partitions. *Mol. Biol. Evol.*, **15**, 1572–1577.

Tattersall, I. (1982). Two misconceptions of phylogeny and classification. *Amer. J. Phys. Anthropol.*, **57**, 13.

Tattersall, I. and Schwartz, J. H. (1974). Craniodental morphology and the systematics of the Malagasy lemurs (Primates, Prosimii). *Anthropol. Pap. Amer. Mus. Nat. Hist.*, **52**, 139–192.

Yoder, A. (1997). Back to the future: a synthesis of strepsirhine systematics. *Evol. Anthropol.*, **6**, 11–22.

18 *From "mathematical dissection of anatomies" to morphometrics: a twenty-first-century appreciation of Charles Oxnard*

FRED L. BOOKSTEIN
University of Michigan

F. JAMES ROHLF
Stony Brook University

Introduction

In a series of books in the 1970s and 1980s, Charles Oxnard, with the help of colleagues and students, propounded a nearly complete biometric methodology for "biomathematical anatomy" in the comparative context, and used it to explore the functional diversity of the primates, including the prosimians, in extraordinary anatomical detail. Today, twenty years later, is a good time to look back at this accomplishment. In retrospect, Oxnard captured the methodological challenges of this domain quite precisely: problems of statistical representation, sample summary, and correlations among structural, functional, and phylogenetic information. To proceed, he selected from a wide range of toolkits that had already been set on the table by others: tools of multivariate ordination, graphical data display, biomechanical engineering, and phenetics. There resulted masterful book-length expositions of methods and comparative findings all across the order Primates, from prosimian to human, from extinct to extant.

One of us (FLB) first encountered Oxnard's work in reviewing the literature for his thesis (Bookstein, 1978) on the measurement of biological shape – the work that, after some vicissitudes, became contemporary morphometrics. Oxnard's 1973 book is subtitled "Some mathematical, physical, and engineering approaches," and Bookstein learned quickly that if Oxnard had not noted an approach here, either it was not worth noting or it had been invented after 1973.

Shaping Primate Evolution, ed. F. Anapol, R. Z. German, and N. G. Jablonski. Published by Cambridge University Press.

Multivariate morphometrics, Blum's medial (symmetric) axis, cluster analysis, analysis of outlines by tangent lines, optical Fourier analysis, optical stress analysis, Cartesian transformation grids – all were recapitulated in that single short book, as a potential toolkit for graduate students in biomathematics (like Bookstein), in comparative anatomy, or in applications of computer graphics in biology, or for anybody else interested.

It was with great pleasure that Bookstein noted a first example of the reverse diffusion, when a figure out of his dissertation made an appearance on page 74 of Oxnard (1984): comparisons among *Homo sapiens*, Neanderthal, australopithecine, and chimpanzee skulls (in midsagittal section) by principal strains of transformation grids from *H. sapiens* onto each of the others. Still, in the figure caption there, Oxnard commented that "the method is not quite so easy to comprehend and, until comparisons are examined between each of the other pairs . . . it is less easy to see what is the overall meaning of the comparison." In other words, there was not yet a multivariate analysis protocol to go with the grids of Bookstein (1978), and Oxnard was waiting patiently until one might be supplied. As it happened, there never would be such a multivariate analysis. The method Oxnard cited, the biorthogonal grid, was retracted the very next year (though this is not cause and effect!) and replaced in full by the combination of shape coordinates and the thin-plate spline, topics of the next two sections of this essay. Both of these components of the replacement methodology are themselves borrowed, and it is purely a matter of luck that Bookstein, rather than Oxnard, was the researcher to notice that they were worth borrowing. But the idea of borrowing – the idea that comparative and functional anatomy should enthusiastically share toolkits with quantifications from other branches of natural science – and the specific willingness to borrow from applied mathematics as well as from bioengineering, was Oxnard's particular style.

The other of us (FJR) first encountered Oxnard's work a little earlier, in the context of the optical Fourier analysis Oxnard was pioneering in the 1970s. It is no accident that these were optical analyses, as statistical computing of the period was wholly incapable of coping with the bandwidths involved. An actual statistical method did not follow until the middle 1980s (Rohlf and Archie, 1984), and then only for outlines. Oxnard himself seems not to have converted these optical analyses to fully multivariate form until the middle 1990s (e.g., Oxnard, 1997).

To both of us, it was clear that we had encountered a bold biomathematical and biostatistical imagination indeed. His alternative mathematical treatments were all in the pursuit of our common goal of animal ordinations that were sensible on many different criteria simultaneously: sensible as descriptions of metric similarity, sensible as correlates of similar function, and sensible as reflections of evolutionary history.

Oxnard's methodology: a précis

A good guided tour through Oxnard's toolkit is set out in the third chapter of his 1984 book, entitled "Mathematical 'dissection' of anatomies." Chapter 2 of this book has already emphasized "linear relationships," one-dimensional orderings on single variables or composite axes, and Chapter 3 begins with a survey of the existing multivariate morphometric toolkit (cf. Blackith and Reyment, 1971) with its devices such as clusters, separatrices, discriminant functions, canonical variates, principal components, minimum spanning trees, and the "Andrews plot" (conversion of a set of vectors to an interwoven set of plane curves). A second list of tools, the "new methods," includes analysis of outlines (tangent angle, the medial axis), optical Fourier analysis of textures, and transformation grids. This singularly magisterial chapter concludes with a series of small discussions of "problems," dilemmas of analytic strategy just as valid today as they were twenty years ago: questions of "few versus many measurements," the role of instrumentation in procuring three-dimensional data, the role of landmark points, and the wide variety (and inconsistency) of the ways in which empirical findings according to any of these approaches may be challenged logically, biologically, or statistically.

As his arguments unfold, it becomes clear that Oxnard is interested simultaneously in three different ways of organizing his empirical analyses. A first method is by their covariance structure (the multivariate ordination of rotations and projected sums of squares that lies at the core of "multivariate morphometrics"). A second approach is by function: the author is constantly asking how animal behaviors or locomotive strategies assort with the multivariate ordinations based on form. The third organization is by actual evolutionary history, in an approach that combines the explicitly phenetic with a mild privileging of orthogenesis, monotone evolution (a preference that was, we think, shared with the entire grand tradition of comparative primatology before him).

"The problem of geometry"

Throughout the books reviewed here Oxnard is aware of the difference between size measures and what he calls "proportions" or "indices" (i.e., shape measures). At many junctures he is concerned with the dependence of findings on choice of the one domain or the other, and on choices within each domain separately, but he does not seem ever to raise the issue of representing these domains themselves as mathematical spaces in their own right.

In this oversight he is, of course, in excellent company, including, early on, both of the authors of this essay. Right through the decades of Oxnard's highest productivity, the role of figures such as on page 168 of Oxnard (1984) was mainly as a guide to the placement of calipers on a form. Several possibilities were thereby overlooked that, taken together, generated the "revolution in morphometrics" within which our retrospective comments here are couched. The only formal geometric space to which Oxnard refers in the course of his analyses is the multivariate morphometric space of linear combinations of the variables supplied. He is aware, of course, of the Cartesian two- and three-dimensional spaces of the original image data, and skillfully manipulates their intrinsic Euclidean geometry in the course of the image-processing procedures he recommends (Fourier analysis, outline analysis, medial-axis analysis), but the possibility that measurement spaces might likewise have a geometry had not yet occurred to him. The missing aperçu came not from our preferred source literatures of bioengineering and applied statistics, but from a totally different source, the pure Riemannian geometry of shape spaces.

Nowadays we might start from the space of "all possible shape measures" on any dataset of homologous points or homologously labeled outlines, and build the multivariate analyses in a form that no longer presumes particular variables and certainly not particular length measures. For "net differences of proportion" in most applications to single rigid components (such as the locomotively crucial scapula and pelvis), we use Procrustes distance, which is an a-priori formula, not dependent on observed sample covariances.

It is understandable that Oxnard might have overlooked this possibility, as it is not analogous at all to any of his other borrowed tools. In the modern presentation, a shape is an equivalence class of possible datasets, the set of "all the forms that have the same shape" under variation of scale, position, and rotation. Oxnard never mentions position or rotation, and within higher mathematics the notion of an equivalence class comes up mainly in higher geometry (points of projective spaces as equivalence classes), algebraic geometry (equivalence classes of polynomials modulo other polynomials), and other fairly abstruse domains. Still, it is striking that the same author who goes so boldly into nonlinear transformations of his image domains was willing to settle so timidly for merely linear transformations (shears, rotations) of his derived descriptor spaces.

Perhaps the clearest place where Oxnard would have benefited from this technique is in his studies of the "proportions" of single rigid skeletal components like scapula or pelvis. (But any single bone could be profitably studied by this method.) On page 214 of Oxnard (1984), for instance, is a representation of 17 dimensions of the scapula for three subsamples of Old World monkeys. The graphic device used is the Andrews plot, which rotates a vector by a one-parameter list of coefficients that are the first few sines and cosines of the standard one-dimensional Fourier decomposition. There is no corresponding ordination in feature space, although the savvy observer might sometimes sort out clusters of variables after a fashion by inspecting how species are separated when they are separated (C. Oxnard, pers. comm.). A much better display would have been the representation of shape coordinate space after the fashion of the other morphospace 3-D plots in this book. One could thereby see the principal coordinates of these rigid landmark configurations and the relation of both phylogeny and function to the corresponding ordination.

Grids: Thompson, Oxnard, and the thin-plate spline

In the course of Chapter 3 of Oxnard (1984) the discussion turns to the theme of "keeping geometry versus losing it." From the short text here it appears that the topic is keeping the geometrical information that makes it possible to return to the original drawings for interpretation of the multivariate patterns in which algebraic processing culminates. In this context Oxnard mentions his favorite image transformation, Fourier analysis, which, while exactly invertible (the Fourier transform of the Fourier transform is the original image), completely reorganizes the geometrical organization of the dataset, from the spatial domain to the frequency domain or vice versa. He did not consider a less radical approach to descriptor spaces that, so to speak, "leaves the geometrical organization alone." Like the Andrews plot, it replaces vector-valued observations by whole functions; but, in a way not anticipated by Oxnard, the resulting diagram can be viewed explicitly as superimposed over any form of the dataset. We are referring, of course, to the representation of a shape coordinate vector as a transformation grid, the isomorphism of the space of shapes and the space of thin-plate spline interpolants.

We have already quoted Oxnard's comment on Bookstein's original method of transformation grids, the biorthogonal method: it is not easy to comprehend and is limited to considering forms two at a time, whereas Oxnard's applications require observation for forms arbitrarily many at a time (in arbitrarily many systematic subsettings). Oxnard was right, of course, in preferring, among the class of "geometric" methods, those such as Fourier analysis that could be applied to large samples on a routine basis. Had publication been delayed even a couple of years, we are sure Oxnard would have incorporated the elliptic Fourier analysis, which permitted precisely this sort of population implementation. This method, first used in a biometric application by Rohlf and Archie in 1984, appeared in the original image-processing literature only in 1982.

In Oxnard (1984) there is a good, though terse, overview of transformation grid methodology as of the publication date, spanning not only the biorthogonal grid but also Sneath's partial linearization of 1967 (his "trend-surface analysis of transformation grids"). Oxnard was shrewd enough to realize that none of these would yield the sort of representation he needed to proceed with his research program – a representation that would submit to subsequent linear statistical maneuvers and functional interpretation. Today's solution, the thin-plate spline, did not exist even as pure mathematics until 1975, and the borrowing that brought it into morphometrics was dated 1985, well after it could have affected Oxnard's primary research publications and subsequent book-length expositions. It is useful, nevertheless, to see how he "left a place for it": to go over the great variety of graphical displays in his books that would have benefited, in retrospect, if he had realized the need for this device, gone out in active search for it, and reeled it in.

Oxnard has long been aware of the missing technique here, and while waiting invented the remarkably suggestive technique of hand-drawn "Thompsonian deformed coordinates" exemplified in Figures 3.56 and 9.1 through 9.3 of his 1984 volume. In this technique, originally published in 1973, Oxnard selects forms that differ in multivariate morphometric space by a vector aligned with one or another interesting axis, sketches a square grid on one and a plausible homologous grid on the other, and interprets verbally in the language of fabrication, of engineering: in the present examples, a "craniolateral twist," "mediolateral narrowing," or "craniolateral lengthening." These examples are really quite interesting, in that to each of the verbal characterizations corresponds an actual feature from one feature space we now use occasionally, the space of partial warps (Bookstein, 1991): the large-scale features of shape difference that arise as eigenfunctions of the thin-plate spline we are using for representing deformations. The "craniolateral twist" is close to a square-to-trapezoid transformation along the horizontal in Figure 3.56(1) together with an affine shear; the "mediolateral narrowing" is the square-to-kite at large scale aligned with the horizontal in Figure 3.56(2); and the "craniolateral lengthening" of Figure 3.56(3) is a similarly large-scale transformation oblique to the actual axes Oxnard used. He was correct that these are "simple." The language by which they have been simplified is the language of eigenvectors of a now-standard measure for such transformations, the measure called bending energy; in the 1984 discussion Oxnard correctly anticipates the quantity in its alternative sense, that of a transformation's complexity. Figure 4 of Oxnard (2000) explicitly carries out the translation required for the "craniolateral twisting" example.

Producing grids like these is now a routine part of morphometric research. The pairwise analyses are "decompositions into partial warp scores," and the grid visualizations of rotated axes such as principal components or canonical variates are now done automatically in packages such as Rohlf's TPS series of programs.

Comparing ordinations: spanning trees and partial least squares

Of all the domains of biomathematical "dissection" that have emerged since Oxnard wrote, this third theme is the one that shouldn't be here. Oxnard struggled through his whole series of great books with the problem of representing the correspondence between anatomical and functional ordinations, and the solution was right there in front of all of us: a mathematical technique first published in 1872, applied in another branch of statistics since 1972, and explicitly applied in morphometrics (by one of us) already in 1982. This is the method of partial least squares (PLS), the least-squares representation of two or more measurement spaces on the same specimens by singular-value decomposition of their cross-covariance structure.

It is almost endearing to watch Oxnard wrestle with the absence of a suitable tool for this, the last missing rhetorical device in his typical sequence of strategies. At many places in the 1984 book (for instance, Fig. 7.1), we see a fully detailed multivariate ordination annotated by a functional gloss (here, an arch labeled at one end "*Papio*: forelimb less mobile, bearing compression" and at the other "*Hylobates*: forelimb more mobile, bearing tension"). The next figure, of a somewhat star-shaped morphospace configuration, is accompanied by an even more helpful gloss, of a spiral of four different modes of locomotion (leaping, climbing, slow climbing, acrobatic) each with one or two taxa in the corresponding sector of the ordination (*Propithecus*, *Loris*, *Ateles*, *Tarsius*, *Cacajao*). Nearby is a quite dissimilar graphic construction aimed at exactly the same goal. In figures such as 5.25, multiple ordinations are represented by their axes of greatest variability as parallel unidimensional ordinations, and the points corresponding to individual taxa linked by segments down the page. We are shown thereby that "the rank orders of most genera are rather similar."

In the last book of this series, Oxnard *et al.* (1990), the need for a suitable statistical method is even more acute. Figures 6.1 through 6.6 each show a multidimensional ordination of "lifestyle" in one panel and an ordination of anatomy for the same prosimian species in an adjacent panel. The captions instruct us to contemplate these juxtapositions as "comparisons," but there is no algebra of comparison suggested.

Oxnard might not have overlooked this "missing methodology" if his survey of standard multivariate morphometrics had been complete. But in fact it was not. In the list of techniques that were imported from the standard multivariate workup almost verbatim, one candidate is noticeably absent: the technique of canonical correlation analysis, pursuit of pairs of linear combinations, one from each of two separate measurement spaces, that show the greatest correlation (for unit lengths of the two coefficient vectors). One can imagine a great many reasons for not indulging in this particular technique. There are lots of morphometric variables, especially for the outline representations, and sometimes samples are smaller than the variable count, which renders the computation impossible. Even in sufficiently large samples, the notorious pattern of large positive correlation among most measured lengths (recognized in the standard approach of allometric modeling) renders the coefficients of canonical correlates wildly unreliable under conditions of sampling variation, and uninterpretable in most applications. The chapter on canonical correlations in Blackith and Reyment (1971) covers a mere seven pages, and the examples there are not very impressive.

Nevertheless there was no technique closer than canonical correlations for the evident gap in Oxnard's toolkit, and the solution, once announced by Bookstein and others in the 1990s, was simpler than anyone imagined: to compute the normalized linear composites of maximum covariance rather than maximum correlation. This manipulation makes sense when either block of measures has a natural metric, and otherwise when the original variables are normalized as to variance. Computationally, the solutions to this problem are the standard singular vectors of the cross-block covariance matrix, a formalism that was very well-known from the 1960s onward in the closely neighboring field of matrix numerical analysis. Statistically, the singular vectors are the principal component loadings or scores of a specific version of the analysis, uncentered unnormalized analysis, which was always possible in the computer packages, although rarely invoked. Scientifically, the interpretation of a PLS analysis is very close to that of the biplot, which was being introduced into general multivariate practice by Gabriel (1971) at about the same time. In the biplot, variable loadings and specimen scores are scattered simultaneously for one single list of variables; in a PLS analysis, there are two or more sets of variables, each with its own scatterplot, and the loadings pertain to prediction from one set of variables to other sets. The coefficient vectors extracted in PLS, called "latent variables," are geometrically orthogonal within their blocks separately (though, unlike the case in canonical correlations analysis, they are usually correlated as scores). There is no matrix inversion step entailed, and so computation and interpretation go perfectly well when variables are strongly confounded and when there are more variables than specimens.

In both of the books under close consideration here, Oxnard's culminating narratives would have been much stronger if he had realized the need for this one last "borrowing." Partial least squares need not leave canonical axes invariant, but can rotate them to whatever position maximizes the cross-block prediction strengths. The first latent variable of a PLS analysis might align with the first canonical axis of a canonical variates analysis, or the second, or a combination of the first two, or any other combination. But as PLS analyses are invariant under rotation of either block separately, there is no reason to use principal components or canonical variate axes in the first place: the analyses go best when the entire raw data base of one block (the list of measured lengths, or shape coordinates) is analyzed against the entire raw data base of the other (the entire list of locomotive functions, or the entire roster of lifestyle choices). Instead of vaguely asserting a "similarity" between the ordination in morphospace and the prior knowledge of function or lifestyle, Oxnard could have produced the dimensions of morphospace that are best predicted by lifestyle: not only produced them, but also depicted them by thin-plate spline transformation grid. This is now the standard approach for ecophenotypic analysis in contemporary morphometrics (Rohlf and Corti, 2000).

Twenty-first century talk about whole anatomies

We have thus far concentrated on ordination: issues of the arrangement of specimens or taxa in various descriptor spaces and the kinds of biological inferences that follow. This material corresponds, more or less, to Oxnard (1984) up through Chapter 8, "Whole primates – their arrangement by anatomies." This is not the end of the book, however; it is not even the end of the discussion of the extant taxa. In the quite astonishing next chapter, "Whole anatomies – their 'dissection' by primates," Oxnard goes on to grapple with a series of questions to which the field has not turned since his publications, and for which, therefore, we had, until recently, no further answers to offer than those he put forward almost twenty years ago. Oxnard notes that upper limbs fall "in a band-like spectrum" over the order, and lower limbs "in a star-shaped spectrum," in each case correlating well enough with the main functions; and that net morphological dissimilarity, summed over these domains, accords tolerably well with the overall systematics of the order.

But how, Oxnard goes on, might we read all this backward, using these comparisons to reveal "underlying subunits" (p. 285)? At the time Oxnard was writing, this corresponded to a topic that had fallen into desuetude, the concern of "morphological integration." Its statistics had been static since the 1950s, and its graphics for even longer: clearly there was nothing here for Oxnard to borrow. Instead, by an informal combination of his transformation grids and his multivariate axes, he proposes a variety of "localized anatomical regions" that emerge simultaneously from factor analyses (of lengths or proportions), canonical axes, and propinquity. Oxnard declared that arrangements like these "provide impetus for a wholly new set of investigations of the order [Primates]."

In fact, the methods necessary for this further investigation could not be borrowed: they had to be created explicitly by biomathematicians for the purpose of the very analysis Oxnard proposed. Only now, in 2003, are there tools available in which the passage he sought, from ordination back to the study of integration, could go forward. (For instance, the "standard" thin-plate spline toolkit included one tactic, the partial warp decomposition, which was once hoped to produce the long-awaited decomposition, but which actually did not.)

The tools are far more demanding of data than anything that Oxnard would have expected. They mostly require 3-D images of whole organisms, not museum specimens: images produced by whole-body MR or CT scan of both hard and soft tissues. An advantage of recourse to whole solid imagery is that there is no longer any problem of "articulation": the organism can be placed in a standardized posture under conditions of near-lifelike tissue strains. Given that assumption (a rather large one, we admit), the tool for localization was in fact the last one to be produced for the morphometric toolkit. We are referring to the method of creases narrated for the biomathematical audience in Bookstein (2002). A crease is a local extremum of the derivative of a transformation grid. For any transformation, such as the comparison of two mean shapes or a multivariate axis of some larger sample design, creases emerge in an ordered list, discrete in their location on the form and independent of one another in orientation. They are the nearest that morphometrics has yet come to the concept of the "quantitative character."

Separately, also in 2002, the group of morphometrically minded anthropologists at the University of Vienna has updated the classic approach to morphological integration via " ρF-sets" (Olson and Miller, 1958) for application to candidate morphological units as assembled out of landmark points and curving boundaries by the shape-coordinate formalism. The approach rests on an extension of partial least squares from the two-block context (form vs. function or form vs. behavior) to arbitrarily many "blocks," each a contiguous candidate anatomical unit. Different biological processes (e.g., evolution or development) muster these blocks in different ways, which can be studied by the corresponding transformation grids for differences of pattern or localization. The method has thus far been tried out only on a dataset of dried skulls from *Homo* and the hominoid fossil line (Bookstein *et al.*, 2003).

To take this any further will require truly massive commitments of computer resources. At root, what Oxnard invoked in his Chapter 9 of 1984 is the extension of the entire literature of "normal anatomical variation" from its existing context in human medical anatomy outward to the anthropoids and thereafter the entire primate order. But even in medicine the study of normal anatomy has stalled – the last great compendium is Anson's of 1961 – and there is simply no method to be borrowed. Work with the Visible Male and the Visible Female of the

National Library of Medicine at the US National Institutes of Health is just now beginning to put in place the necessary software resources for a specification of variation at the level of the whole organism (Bookstein and Green, 2002). Oxnard (1984) suggested optical methods for this investigation (Figure 3.57), but the current approach is a "hypervariable" approach, the complete estimation of a displacement grid for every voxel of the entire solid image. It is working for brains (Thompson *et al.*, 2000), and may apply to larger subsets of the human anatomy in the near future.

Of course among those who have returned to these largest-scale integrative concerns is Oxnard himself. From a very interesting set of reprints he sent us in response to an earlier draft of this chapter, we highlight a book chapter (Oxnard, 2000) that seems to revisit the discussion of 16 years earlier in light of dozens of subsequent empirical investigations. Writing in 2000, Oxnard focuses on a promising methodological antinomy: whereas species separations, considered morphometrically, tend to relate to the function of anatomical parts severally, analyses of evolutionary relatedness tend to combine variables from several parts at once, and to make sense in terms of dynamic development rather than static adult form. Morphometric analysis, he notes, should always, if only "to some degree," partition the two types of explanation. "Individual studies speak most closely to function and combined studies to evolution," he summarizes; "this thinking relates to a more sophisticated view of what comprises a 'variable' or a 'feature' or a 'character' in morphology" (Oxnard, 2000, p. 260). In terms of the situation of humans within the primates, one of his lifelong foci of work, in terms of the individual functional units humans are generally uniquely different from all other primates. This presumably relates to the totally new functional milieux that humans have come to inhabit. But in terms of combinations of units, the uniqueness of humans comes to be appropriately buried. Instead, there is provided a picture of the relationships of humans that is the same as that evident from molecular investigations.

These recent concerns seem perfectly pre-adapted to the mixed ontogenetic–evolutionary studies of the hominoids just about to emerge from Vienna (e.g., Mitteröcker *et al.*, in press). With this, perhaps, a new cycle of Oxnardian anticipations is beginning for your authors all over again.

Envoi

It must be an odd experience to be the target of a revisionist analytic retrospective that so obviously fails to revise in any substantial aspect. None of the comments here are meant in any way to detract from the magnitude of Oxnard's achievement, as reflected in his own books, recent papers, the careers of his colleagues and the students he has so richly trained, and the other chapters of the present volume. Bookstein relied on Oxnard (1973) for most of the methodological explorations of others he reports in his 1978 volume, and relied on Oxnard, too, for assurance that his (since retracted) method of biorthogonal grids had not been tried by anybody else (for if it had, Oxnard surely would have pounced on it immediately).

Throughout the development of the new morphometrics, Oxnard's soft-spoken demand that morphometrics serve functional anatomy as well as phylogeny always loomed before us as the long-term goal. And it was only when our tools of the 1990s proved obviously more vivid than Oxnard's of the 1980s that we could relax in the belief that we were, indeed, developing a better toolkit for applied morphometrics, not merely a different one. We have listed three tools in the contemporary morphometric toolkit that were not put there by Oxnard: shape coordinates, the thin-plate spline, and partial least squares. For two of these, he was actively looking for such a method, and that one of us found it when he didn't was just a matter of luck. For the third, PLS, he substituted very careful inspections of morphospace for the algebraic shortcuts that we would now apply routinely. For what might or might not emerge as a fourth, the early twenty-first-century Vienna shape-coordinate approach to integration over ontogeny and evolution, the methods have not yet stabilized sufficiently to decide what role Oxnard's own recent experiments in "combined studies" will play out. It is with enormous respect for the insight and integrity of this still-pioneering "mathematical anatomist" that we dedicate this essay, with gratitude, to Charles Oxnard.

References

Anson, B. J. (1961). *An Atlas of Human Anatomy*. 2nd edn. Philadelphia, PA: Saunders.

Blackith, R. B. and Reyment, R. E. (1971). *Multivariate Morphometrics*. London: Academic Press.

Bookstein, F. L. (1978). *The Measurement of Biological Shape and Shape Change*. Lecture Notes in Biomathematics, v. 24. New York, NY: Springer-Verlag.

(1991). *Morphometric Tools for Landmark Data*. New York, NY: Cambridge University Press.

(2002). Creases as morphometric characters. In: *Morphology, Shape, and Phylogeny*, ed. N. MacLeod and P. Forey. Systematics Association Special Volume series, 64. London: Taylor and Francis. pp. 139–174.

Bookstein, F. L. and Green, W. D. K. (2002). *User's Manual, EWSH3.19*. 102 pages. ftp://brainmap.med.umich.edu/pub/fred/ewsh3.19.man

Bookstein, F. L., Gunz, P., Mitteröcker, P., Prossinger, H., Schäfer, K., and Seidler, H. (2003). Cranial integration in *Homo* reassessed: morphometrics of the midsagittal plane in ontogeny and phylogeny. *J. Hum. Evol.*, **44**, 167–187.

Gabriel, K. R. (1971). The bioplot graphical display of matrices with application to principal component analysis. *Biometrika*, **58**, 453–467.

Mitteröcker, P., Gunz, P., Bernhard, M., Schäfer, K., and Bookstein, F. L. (in press). Ontogeny of inter-species differences in hominoid crania. *J. Hum. Evol.*

Olson, E. C. and Miller, R. L. (1958). *Morphological Integration*. Chicago, IL: University of Chicago Press.

Oxnard, C. E. (1973). *Form and Pattern in Human Evolution*. Chicago, IL: University of Chicago Press.

(1984). *The Order of Man*. New Haven, CT: Yale University Press.

(1997). From optical to computational Fourier transforms: the natural history of an investigation of the cancellous structure of bone. In: *Fourier Descriptors and their Applications in Biology*, ed. P. Lestrel. Cambridge: Cambridge University Press. pp. 379–408.

(2000). Morphometrics of the primate skeleton and the functional and developmental underpinnings of species diversity. In: *Development, Growth, and Evolution*, ed. P. O'Higgins and M. Cohn. New York, NY: Academic Press. pp. 235–264.

Oxnard, C. E., Crompton, R. H., and Lieberman, S. S. (1990). *Animal Lifestyles and Anatomies*. Seattle, WA: University of Washington Press.

Rohlf, F. J. and Archie, J. (1984). A comparison of Fourier methods for the description of wing shape in mosquitos (Diptera: Culicidae). *Syst. Zool.*, **33**, 302–317.

Rohlf, F. J. and Corti, M. (2000). The use of two-block of partial least squares to study covariation in shape. *Syst. Biol.* **49**, 740–753.

Thompson, P. M., Mega, M., Narr, K., Sowell, E., Blanton, R., and Toga, A. (2000). Brain image analysis and atlas construction. In: *Handbook of Medical Imaging. Vol. 2, Medical Image Processing and Analysis*, ed. M. Sonka and J. M. Fitzpatrick. Bellingham, WA: SPIE Press. pp. 1061–1129.

19 *Design, level, interface, and complexity: morphometric interpretation revisited*

CHARLES E. OXNARD
University of Western Australia

Origins of morphometrics

The origin of morphometrics lies in the burst of methodological creativity that predates the availability of relevant datasets requiring such methods. A seminal decade saw developments from Hotelling (1931), Wilks (1935), Bartlett (1935), Fisher (1936), Mahalanobis (1936), and others. And the next 20 years provided the extensions of the methods by Rao (1948), Yates (1950), Kendall (1957), and others. Most of these investigations concentrated on developing the methods. The actual data examined during these developments were so few that, though they permitted analysis by manual techniques, they were too restricted to allow examination of real biological problems. One well-known dataset was Anderson's measurements of four variables, the lengths and breadths of sepals and petals, taken on 50 specimens each of three groups of iris. These data were used by Fisher (1936) in the development of the techniques, and by many other investigators since, including myself (e.g., Oxnard, 1973, 1983/84), both for development and for checking.

Of course, the antecedents of morphometrics came from even earlier times (e.g., Galton, 1889; Pearson, 1901) and even the nineteenth and late eighteenth centuries (e.g., Adanson, 1763; Quetelet, 1842). But it was not until the second half of the twentieth century that morphometric methods could be applied to large datasets (many variables, many specimens, many groups) and could be aimed at examining real rather than exemplar anthropological problems (e.g., Trevor, 1955; Ashton *et al.*, 1957, 1975, 1976, 1981; Oxnard, 1967; Howells, 1973). This was almost always done in collaboration and consultation with statisticians (e.g., Mahalanobis, Rao, Healy,

Shaping Primate Evolution, ed. F. Anapol, R. Z. German, and N. G. Jablonski. Published by Cambridge University Press.

Flinn – consultants in prior references). These elaborations also required the coming of the computer.

One earlier ground-breaking book in this area that never received the attention it deserved was Olson and Miller's 1945 book, *Morphological Integration*. The main emphasis of that book was upon the multivariate technology for understanding integration. It was, perhaps, just a little ahead of its time. As a result of this partial neglect, the importance of integration in understanding animal structure became better understood only several decades later. I am delighted to see that it has recently been reprinted. Other ground-breaking books in this area were Blackith and Reyment's (1971) *Multivariate Morphometrics* (now in a second edition: Reyment *et al.*, 1984) and Sneath and Sokal's (1973) *Numerical Taxonomy*. Though aimed at different audiences, the methodologies described in these books are fundamental to morphometric analysis.

The "old" morphometrics and its interpretations

Since the days indicated by this short and incomplete history, the uses of morphometric statistical analyses in biology and anthropology have grown enormously. Their value, however, has been seen by different commentators as lying anywhere along a wide spectrum. One end of that spectrum implied that such studies could give major insights into biological and anthropological problems over and above those of visual assessment (e.g., Oxnard 1973; Stern and Oxnard, 1973). The other end of the spectrum implied that they provided mostly obfuscation and confusion under the guise of mathematical sophistication (e.g., Day, 1977; Herschkovitz, 1977). Happily this latter end of the spectrum has now largely disappeared.

Several reasons exist for such disparity in opinions held at the same time. The data used in analyses were often just that: data, any data, without thought for a biological question. The difference between results and interpretations stemming from results was not always clearly understood. It was not common for publications to include full details of the testing necessary for relying on results. Furthermore, it was not always recognized that multivariate morphometrics was not a single method, but an overall approach that could be used in different ways, and especially because hypothesis-generating and hypothesis-testing usages were not always differentiated. Indeed, the distinction between these two was only then being described (e.g., Tukey, 1977). Finally, the existence of these opposite opinions in anthropology was in large part because many of the results of the morphometric studies of those days implied conclusions about human evolution that were contrary to the then generally accepted views.

For example, morphometric studies of living primates implied, when applied to fossils, that the australopithecines were capable of arboreal activities as well as terrestrial bipedalism. They suggested, in addition, that australopithecine bipedality was not simply "in the manner of man." This indicated that australopithecines might not be direct human ancestors, though, of course, if the results were due to functional mosaicism, they did not necessarily exclude that view.

This controversy is now largely resolved. Most investigators now accept that australopithecines were arboreal as well as capable bipeds though not bipeds in the human manner. For example, many workers nowadays, even those who were originally against the idea, now agree that much australopithecine anatomy speaks to local functional adaptations similar to those of apes, though, of course, not similar to any specific ape today. Thus, relatively new studies of the foot (e.g., Kidd, 1995; Clark and Tobias, 1995; Kidd *et al.*, 1996; Kidd and Oxnard, 2002) now concur with older investigations (Lisowski *et al.*, 1974, 1976; Oxnard, 1975) that the australopithecine foot implies ape-like arboreal activity of some kind.

There are a few investigators who still believe that australopithecines were essentially like human bipeds, and a very few who continue to place them directly on the lineage leading to humans. But the idea of australopithecine bipedality being similar to human bipedality has now been replaced by investigations attempting to find out about different forms of bipedalism. Bipedalism may have arisen more than once. Human bipedalism must now be understood within the framework of various types of bipedal activities.

Even more recent are complex ideas of the precursor or precursors to bipedality. Such a functional precursor may have been the vertical climbing of African apes, an arboreal chimpanzee functional model. It may have been bipedal running as much as walking (e.g., Crompton and Li, 1997; Crompton *et al.*, 1998). It may have been arboreal bipedal movement with or without hand contact on higher branches in the trees (perhaps an orangutan arboreal bipedal functional model: Crompton, work in progress). It may have included completely non-locomotor possibilities (e.g., Jablonski and Chaplin, 1993, and Chapter 13 in this volume).

The "new" morphometrics: obtaining and analyzing the data

Major new developments in morphometrics have occurred in the last two decades. As a result, the new methods are sometimes called "the new morphometrics" or "geometric morphometrics" (see Bookstein and Rohlf, Chapter 18). They have depended upon the work of a large number of investigators including

Bookstein (1989, 1991), Albrecht (1980, 1991), Oxnard (1997)a, Lestrel (1997, 2000), O'Higgins (2000), Richtsmeier *et al.* (2002), and others. Notable amongst these methods is "Morphologika" (O'Higgins, 2000, worked out with the help of statistical collaborators such as Dryden and Mardia, 1998).

One part of the improvement has been the progression from ad hoc anthropological measurements and proportions to three-dimensional coordinate information. It had long been clear that three-dimensional coordinate data would be the most useful. Early attempts (as in stereometrics) were spawned decades ago (e.g., Savara, 1965). In particular, in his sadly shortened life, Norman Creel of Stony Brook obtained coordinate data by adapting map-making technology that used aerial photographs. But in those early days, taking three-dimensional data was time-consuming in the extreme. I had, myself, used the Graf-Pen (Oxnard, 1973). This was an arrangement of three two-foot linear microphones at right angles to one another. They recorded x, y, and z coordinates from a small sound-emitting pointer placed against points marked on (say) a bone. The sonic data were then converted digitally and displayed in a window. The numbers had then to be copied and punched onto IBM cards – scope for error. It worked well for large bones but was not good enough for small bones or teeth. It was also a logistic nightmare in terms of the time and care needed in operating and recording.

Since then, however, the technology has passed through online digital calipers that can put data directly into a computer, through a variety of optical and *x*-ray-imaging methods to, very recently, the MicroScribe for recording three-dimensional field data directly. These are altogether far more powerful data-producing tools for morphometric analysis than ever existed in earlier times.

The "new" morphometrics: interpretations

In addition to carrying out data analysis, the new morphometrics is most valuable for aiding in the interpretation and visualization of results. Thus my almost painful early attempts to display information about variables inherent in statistical axes that separate species (Oxnard, 1973, 1983/84) involved relatively crude inspection of the products of variable loading factors and mean differences. The new morphometrics can render this crystal-clear through computer rotations of axes.

My somewhat more painful attempts to understand what aspects of bone shape were involved in differences between species used D'Arcy Thompson's visual method of deformation of Cartesian coordinates allied to species placement in multivariate axes. However, this was a pencil-and-paper exercise that has been improved out of all recognition by computational developments. These include Bookstein's (1991) biorthogonal grids, Lestrel's Fourier curves (summarized in his 1997 volume), and Bookstein's (1989 and 1991) and Albrecht's (1991) thin-plate splines. Perhaps most of all these include O'Higgins' developments of Morphologika (also including thin-plate splines, as outlined in other chapters of this volume). These methods can visualize differences in structures as the end result of evolution. However they can also visualize changes in structures due to growth, deformations of structures due to biomechanics, and movements of structures due to behavior. They can operate within a single universe of specimens, within groups of specimens, and in age cohorts, geographic communities, ecological guilds, evolutionary clades, and so on.

My especially painful early attempts to understand functional implications of structure (e.g., by theoretical and applied stress/strain analyses such as infinite beam theory, by experimental stress analyses using photoelasticity and moire fringe analyses: Oxnard, 1973) have now been almost completely superseded. (Yet I was delighted to discover that at Stanmore, UK, two-dimensional photoelastic models and three-dimensional photoelastic skins, using the almost identical equipment that I employed so many years ago, are still useful in biomedical applications.) Today, however, direct strain studies using strain gauges (e.g., Hylander *et al.*, 1987) and computational finite elements (e.g., Oxnard *et al.*, 1995) are better. Such studies also involve knowing what the animals are really doing by using electromyography, kinematics, radio-cinematography, kinetics, energetics, and other related modern developments (e.g., Jenkins, 1972, 1975; Jungers and Stern, 1980; Stern *et al.* 1980; German and Franks, 1991; Crompton *et al.*, 1996; etc.). Happily, I have been able to participate in some of these developments through collaborations with colleagues, ex-students, and present students.

Design of observation

In addition to improvements in interpretation that result from the technological advances, there have been improvements that result from asking better biological questions, in particular, from realizing the importance of interactions between biological questions.

One such improvement relates to the "design of observations." I think I first used the phrase in 1976 in a paper prepared for a Wenner-Gren Burg Wartenstein Symposium (and appearing in a more developed form in the published literature in 1979). It exemplified the beginnings of a concept somewhat analogous to Fisher's (1935) "design of experiments," so powerful in the experimental sciences. Though experimentalists often carried out simple studies comparing an experimental group with a control group, Fisher showed years ago how much more powerful was a design of experiments that involved comparing several different experimental groups with the control group. This allowed not only the differences between each experimental and the control to be examined, but also the interactions among the different experimental groups.

A similar concept can be applied to observational studies. Though many observational studies are confined to examining a pair of species differing in a single functional attribute (e.g., Fleagle, 1974, 1976; Mittermeier, 1978; Rodman, 1979; Jungers and Fleagle, 1980), more powerful are designs I first discussed (Oxnard, 1976, 1979) involving several species differing in various ways. This followed the methodology outlined by Fisher (1935) and elaborated over the ensuing years (e.g., Zolman, 1993).

Thus, beyond a pairwise comparison is a simple model that compares several groups of functional attributes differing in a linear manner, e.g., several taxa varying along a locomotor spectrum within the same higher taxon. This is a one-way linear analysis. Any morphological findings consistent with such a linear design are more likely to be related to the functional descriptions with other information segregated (Fig. 19.1, upper frame).

A second level of observational design involves the use of a second attribute, such as several different phylogenetic groups. In this design the two elements, locomotion and taxon, are crossed, that is, each has several levels (types), and each combination of level exists in the design. For example, studies of differences in climbing in Old World monkeys might include comparisons of different degrees of terrestrio-arboreality (one element) over several separate taxa of primates, such as macaques, langurs, and cercopitheques (second element). Such designs may be even more useful in detecting species differences truly referable to the functional hypotheses posited (Fig. 19.1, middle frame).

Finally, this complexity can be extended to multifactorial models, with several elements, such as locomotor spectrum, taxon, and niche (Fig. 19.1, lower frame). Such a design is more powerful than a series of pairwise comparisons, even though, in practice, it might be difficult to fill the full suite of possibilities. They pertain especially to situations where the finer differences in function, behavior, or habitat among closely related species, subspecies, geographic variants, even local communities, are known. Even just thinking about such design possibilities sharpens the mind about the biological questions. Such designs need to be worked out before deciding on the specimens and species, before taking the data, and especially before making analyses. Simply taking data and applying analyses willy-nilly is less useful. These "one- to three-dimensional observational designs" are Fisher's concepts applied to observational data, designs that originated with Fisher's experimental agricultural investigations carried out at Rothamstead Experimental Station.

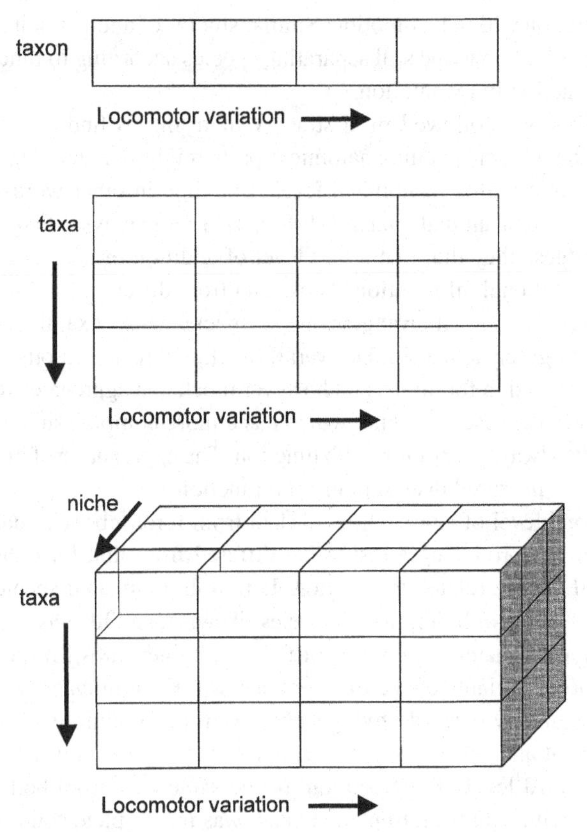

Figure 19.1. Example of observational design for hierarchical studies of primate morphology. The upper frame shows a comparison among four different locomotor types in a single taxon, equivalent to a one-way ANOVA. The middle frame shows a classic two way complete design, with four locomotor types for each of three taxa. The lowest frame indicates a three way design, with four locomotor types, three taxa over three niche groups. This design, as originated by R. A. Fisher (1935) is a more powerful statistical design than the classic pairwise control and experimental, or in the case of many studies of primates, control and observational groups. Perhaps the most significant aspect of this design for any research project is its conception prior to the choice of species and variables, and prior to the collection of data.

Analytical levels: separations of species

A second set of analyses that provided new insights into the separations of species concerned the examination of hierarchical levels of anatomical structure.

Thus, our earliest studies involved a single anatomical unit, the shoulder. This was shown to contain obvious information separating species according to shoulder function within locomotion. Similar studies of another unit, the hip, gave a different result, but one still separating species according to function, in this case hip function in locomotion.

However, this was followed by a strategy of trying to find out what information the data from specific anatomical parts might contain if they were added together into higher anatomical levels (moving, in other words, closer and closer to the whole animal: Oxnard, 1983/84). These analyses first showed, and thank goodness they did, that a first level of addition of anatomical parts still provided functional information. Thus, data from different forelimb parts, all giving similar functional arrangements of species when examined singly, provided, when aggregated, a similar overall functional result. In other words, whatever is measured in forelimbs and however much one aggregates forelimb regions (except, of course, the hand, which has a more complex suite of functions) the results always mirror forelimb function. The aggregation of hind-limb analyses likewise mirrored overall hind-limb function.

Even a second level of aggregation, of data from forelimbs with data from hind limbs, though producing a result that differed from that for either limb alone, was still clearly related to function. Forelimb-dominated species were placed together, hind-limb-dominated species likewise, and likewise too those species with approximately equal uses of fore- and hind limbs irrespective of specific locomotor variants of the different species. Even humans, as unique bipeds but presumably derived phylogenetically from forelimb-dominant antecedents, fitted at an appropriate place within this result (Oxnard, 1983/84).

However, a third level of aggregation, of the same data from both sets of limbs, together with data from trunk and head, was in complete contrast. That is, data with major functional content when examined by localized anatomical part, demonstrated, when examined at a third level of aggregation, groupings of species that seemed most like molecular evolutionary relationships (Oxnard, 1983/84).

For example, in these aggregated studies, Old and New World monkeys could be separately identified, whatever the functional parallels and convergences within and between them. To be fair, however, this distinction melds not only with that from molecular studies, but also with that from a large body of prior morphological investigation. A more unexpected result was that aggregation divided all hominoids into "Asian apes" and "African apes plus humans." These are, again, the same groups produced by molecular studies, but they differ from the classical "great apes" and "humans" of earlier anatomical comparisons. Similar results were found in each major subgroup of the primates. These evolutionary effects were obtained twice, once through aggregation of fine data on many joints, and a second time through aggregation of coarse overall body proportions. Such replication strengthens the idea. It is further strengthened by the fact that it has now been replicated in several other investigations.

Thus a second example of the importance of understanding the hierarchy of anatomical levels is provided by analyses of dental measures in apes and humans. In this case there were only two levels of analysis, individual tooth regions, and the dental battery as a whole. At the first level of analysis it is not surprising that the results mirrored factors like masticatory functions and sexual dimorphisms. At the second level of analysis, however, the results mirrored evolutionary relationships, placing humans with African apes, and Asian apes as separate (Oxnard, 2000).

A third example, though much more restricted, being a study of the single genus *Macaca*, had three anatomical levels of analysis because the variables included many measures of jaws, teeth, and skulls (e.g., Pan and Oxnard, 2002a, 2002b). The first and second levels of analysis examined variables arranged as separate functional elements and aggregated functional units. Both sets of analyses revealed species-size differences and sex-size differences in individual principal component axes in different functional units.

However, a third level of aggregation, into the skull-as-a-whole, produced a different result. Species-size differences and sex-size differences could still be discerned. They were produced, however, by only 35% of the information. The remaining information, contained within several significant higher axes, seemed to be consistent with phylogenetic hypotheses about the macaques as a whole.

Apart from its importance in understanding full biological interpretations of extant anatomies, there is a special importance of these aggregated results for studies of fossils that are often fairly incomplete. This can scarcely be over-emphasized. It suggests that attempting to understand results stemming from fossils with only partial data, a few teeth and a piece of jaw, a partial face, a calvarium alone, absolutely requires an understanding, from complete studies of extant forms, of the effects of aggregation. What may be important is not only what data are missing but also what interpretations are thereby excluded.

Analytical levels: clusters of variables

The questions and datasets described above were all driven from the perspective of a species unit of analysis. However, in a data exploratory paradigm, it is equally valid to shift focus to a variable-unit perspective (Oxnard, 2000). Both the first-level and second-level studies of the postcranium, when examined from this new perspective, provided clusters of variables that seemed to be related to the biomechanical functions of each respective anatomical unit. This is undoubtedly why the separations of species also related to function.

However, in the third-level studies, analysis of these data already known to separate the species according to evolutionary relationship rather than functional adaptation clustered the variables in a new way. These clusters mirrored most obviously the developmental factors now known to be responsible for the form of adult parts. That is, the clusters of variables no longer described individual joint complexes (function), but proximo-distal, cranio-caudal, and serially homologous groups of measures. These are close descriptors of the developmental mechanisms, proximo-distal gradients, cranio-caudal arrays, and serial homologous components, now known to be responsible for the formation of limbs. Does the evolutionary arrangement of species from adult morphometrics occur because adult form overall contains ghosts of the developmental factors that influenced adult diversity?

If this were a single example it would be little more than an anecdote. However, a similar phenomenon has been obtained from the cranio-dental studies (Oxnard, 2000). Thus first- and second-level analyses (which seemed to group species in accord with mastication) clustered the variables in ways that made functional sense within mastication (e.g., all molar areas together, all incisor edge lengths together). In contrast, third-level studies (separating species in relation to evolutionary relationship) clustered the variables into proximo-distal measures along the tooth row, cranio-caudal measures down the jaw, and separate blocks within the jaw. These three clusters resemble the proximo-distal gradients, cranio-caudal trends, and separate islands of neural crest cell aggregation, all of which are involved in the development of the mandible.

Understanding these relationships among variables in living forms in relation to examining the characters exhibited by fossils would certainly impact on cladistic evaluations of the variables.

Interfaces: function, development, and molecules

Overall adult structure, though only a snapshot of structure at a specific time, is nevertheless the complex resultant of development (and growth) of the individual, of functional adaptation in both the individual and the species, and of evolutionary differences among species. The results above imply that a morphometric description of adult structure, perhaps because of its ability to examine groupings of species and clusters of variables in a quantitative manner and to take account of interactions among and between species and variables, may reflect these underlying mechanisms and processes. It may be rather easier to see this at the present time because we know so much more today about developmental biology, functional adaptation, and evolutionary biology.

For example, functional anatomy is much more complex than it used to be. The more recent advances in functional anatomy now include extensions into various aspects of biomechanics, kinematics, kinetics, and energetics of posture and movement, into behavior more broadly defined, into ecology (the biological milieu within which the behavior occurs), and into environment (the physical milieu within which ecology is played out).

Studies in development and growth are also very different from what they used to be. They include the actions of homeobox genes and other molecular factors and mechanisms that produce and control the cascades of processes effecting morphology. They include new ideas about the timing of developmental events and their effects on eventual form. They are providing new insights into the changes that occur from the zygote to the adult in a single species. The differential display of these in different species is what produces, in part, the wide array of structures found within species diversity in evolution.

Finally, advances in evolutionary biology, especially those due to comparisons of molecules, are providing not only bases for accepting or rejecting older hypotheses about the phylogenetic relations of living species but also new information about new phylogenetic possibilities. Yet it is also now beginning to be realized that the molecular studies themselves are subject to some of the same strictures that have long been known to apply to whole-organism studies. That is, the molecular studies not only supply phylogenetic information and therefore hypotheses, but also contain admixtures of other information, and are just as much to be tested as any other technique.

However, it is the complex interface between these three broad disciplines that is the true determinant of the nature of differences among adult organisms. These three biological mechanisms are, in an integrated fashion, responsible for the adult anatomy that we see in animal diversity.

The question, therefore, that has exercised my mind in recent years relates to the degree to which these three – functional adaptation, developmental process, and molecular evolution – can be discerned in, and disentangled through, the morphometrics of adult animals. Can morphometric analyses of static adult form provide interpretations that shine through this complex interface? It ought to be so. For the shadows (ghosts) of these determinants – functioning units, developmental processes, and evolutionary relationships – should be discernible in the final static adult form. After all, the form of the adult at a particular point in time is the overall result of these determinants. Though such shadows may sometimes be discernible through visual examination of anatomies, it is far more likely that they will be visualized, and perhaps partitioned with due allowance for interactions, through integrated quantification and analysis: i.e., morphometrics of one kind or another. The results described above imply that this idea is a distinct possibility.

Of course, direct functional, developmental, and molecular investigations should provide clear information. They are limited in the evolutionary context, however, because they can usually only be carried out on small numbers of experimental extant exemplars. Their enormous strength is that they provide detailed knowledge of the actual underlying mechanisms and processes in particular species. They clearly generate ideas that may well be generalized over wide ranges of species including those (e.g., rare or extinct) that cannot easily be investigated.

In contrast, morphometric investigations give only a single snapshot of static animal form. But morphometrics has the strength that it can be applied to animal diversity very widely distributed, to animal growth in the physiological past (ontogeny), and to animal differences due to the deep past. It also has the strength that it works through already dead materials that (though they may have been collected in the past through mechanisms not in line with today's ethics) at least do not involve current interference with living animals today. Can morphometrics reveal the shadows of the mechanisms and processes that underlie animal form? In other words, the data and analyses of morphometrics, though usually obtained by a particular investigator in the search for an answer to a particular problem, should also contain within themselves information relevant to a wider range of problems.

Complexity: soft tissues, function, behavior, ecology, and evolution

Many of the morphometric studies described above are relatively clear because they examine data from bones and teeth. Indeed, as a result, they have been called "osteometrics" and "odontometrics". Teeth and bones are hard tissues. They can be measured to tenths of a millimeter. The resulting data are "hard."

Soft tissue data: muscle

However, such analyses can also be applied to data from soft tissues. One example of soft tissues is muscle – closely related functionally to bones. The closeness of that relationship allows data from muscles to be used to test studies of bones. For example, morphometric studies of the muscles of the shoulder, arm, and hip carried out a long time ago provided information about function that was similar to that provided by morphometric studies of the bones of the shoulder, arm, and hip (Oxnard, 1973, 1983/84). Such studies enhanced the concept that the form of these bones reflected the function of the muscles that moved them. More recently, quantitative architectural and cytochemical studies of muscle morphology (e.g., Anapol *et al.*, Chapter 6 in this volume; Jouffroy and Medina, Chapter 7) have "fine-tuned" functional analyses in that variability in these physiologically related morphological features now contributes to interpretations of behavior. But muscle studies are inherently more difficult than studies of bones if only because it is more difficult to get quantitative data about muscles in large numbers of specimens and species. Of course, today's non-invasive imaging may yet provide a great deal of soft tissue data for analysis. Apart from that, I am not aware of other morphometric studies of whole-muscle data for a wide range of primate species.

Soft tissue data: brains

Other soft structures, however, can also be examined morphometrically. Thus, our studies have now come to involve the brain. This is somewhat further from animal locomotion and involves functions additional to locomotion, though the brain, of course, is still closely related to the study of bone–joint–muscle and movement units.

The earliest investigations of the sizes of brain parts showed, primarily, a simple relationship with overall brain size and, indeed, also, overall body size. These studies were not multivariate, usually examining variables one at a time or in pairs. They showed an almost complete relationship with size (e.g., Stephan and Andy, 1969; Jerison, 1973, among many other investigators). They led to ideas of increasing degrees of encephalization (in mammals) and increasing degrees of hominization (among primates). They were even sometimes related to increasing degrees of "intelligence."

However, in the last few years, multivariate morphometric method has been applied to brain volumes (e.g., Finlay and Darlington, 1995; Barton and Harvey, 2000; Clark *et al.*, 2001). These authors confirm that size is a most important element of their data that range across several orders of magnitude from the tiniest of insectivores and bats to gorillas and humans. However, they also found that though size (98%, 92%, and 90% of the multivariate information in these studies, respectively) was the main component, there was additional information speaking to the distinguishing of three major taxonomic/phylogenetic groups, and to a special brain organization, the limbic system of the brain.

These kinds of findings led these various workers to suggest that though mammalian brain evolution allowed the distinction of some high-level phylogenetic arrangements (e.g., the separations of insectivores, and strepsirhine and haplorhine primates), the brain was principally under the influence of highly constrained sets of developmental mechanisms and processes. They implied, for example, that the simplest way to evolve a larger cerebellum (e.g., bats) or a larger cortex (e.g., primates) was to evolve a larger brain overall.

Willem de Winter and I were also carrying out multivariate analyses of such data at the same time. Our studies included a greatly extended dataset (three times as many specimens and species) and some of the findings are included in de Winter's contribution to this volume (Chapter 10). They take the story a step further. Thus, in addition to size and evolutionary relationship, and perhaps more important than either, we (de Winter and Oxnard, 2001) have shown, through studies that we call "neurometrics," that additional information is present.

Our preliminary analyses replicated, to the first decimal place, Finlay and Darlington's finding about the supremacy of size. That is not in doubt. And the partial (but statistically significant) separation of strepsirhine and haplorhine primates is also not in doubt. However, our ultimate analyses examined the brain data arranged in association with input/output relationships among the brain parts (i.e., arrangements that reflected, to some degree at least, functional associations). This gave a different result (de Winter, 1997; de Winter and Oxnard, 2001). We found, first, that each of the large orders of mammals had completely different distributions within the data space – insectivores, bats, and primates being almost orthogonal to one another. In contrast to the earlier multivariate studies, this was achieved by organizations of the brain variables that were quite different within each order. We found, second, that within each order the tightest associations of species were not into phylogenetic groups, but into clusters of species that had convergent lifestyles. It was exactly because of these lifestyle convergences (convergences across phylogenetic boundaries) that phylogenetic groups could be only dimly perceived. The information about lifestyle groups within primates is provided in more detail by de Winter (Chapter 10, this volume).

However, yet further studies of these data (Oxnard, in press) relate specifically to the differences between humans and apes (especially chimpanzees). The phylogenetically closest nonhuman primates to humans are chimpanzees. However, in our brain studies, humans are almost equally close to a number of other species, not only chimpanzees, but also the other African ape, gorillas, the Asian great ape, orangutans, the lesser apes, gibbons, and the New World spider monkeys and woolly monkeys. This is a cluster of primates that cuts across several phylogenetic groups. These species have in common, however, many lifestyle similarities (as compared with other monkeys) based upon the dominance of forelimb activities. Such lifestyle features include not only locomotion but also posture, feeding, playing, social interactions, agonistic interactions, attacking, escaping, and so on. However, the nature of the separation of humans from these next-closest primates is of the order of 15 standard deviation units (as revealed by canonical variate analysis). In contrast, the distance encompassing all the nonhuman upper-limb-dominant species is only three standard deviation units, and the distance covering all nonhuman Old World species (from chimpanzees to vervets) a mere eight standard deviation units. The separation of humans does not just involve this major difference. Several higher axes, though statistically significant, contained such a small amount (5–6%) of the information that they are not usually examined. In fact, in three of these axes, almost the entire separation that they performed was a separation of humans from the single group of all other primates. This greatly extends the differences between humans and chimpanzees and especially implies that that difference is complex and not at all just a matter of size.

The particular brain variables contributing to this human/nonhuman species separation were relative increases in humans in neocortex, cerebellum, striatum, and diencephalon as compared with medulla, and independent relative reductions in humans of midbrain, schizocortex, and diencephalon as compared with neocortex. Though not expressible as different brain parts (these nonhuman primates and humans share all the same brain parts) this result reflects differences of input/output relationships among brain parts, and certainly not simply differences in the overall sizes of brain parts themselves. Could such organizational differences relate to expansions of the highest levels of the hierarchy of voluntary activities, and increased capacities to plan and control complex strategies? Perhaps, further, could this be a basis for the major cognitive and other mental differences between chimpanzees and humans? Overall, this result belies the simple quantitative 98.6% DNA similarity of humans and chimpanzees.

We therefore ask, are there any molecular data that are consistent with these results? It turns out that there are. For example, Pääbo (Normile, 2001) noted differences in 165 genes and gene expression factors between humans and chimpanzees. He found that chimpanzees and humans are extraordinarily similar in genes and gene expression factors related to livers, kidneys, bones, and blood. In these tissues, both humans and chimpanzees are markedly different from rhesus monkeys. But in their brains Pääbo found that it was chimpanzees and rhesus monkey that are markedly similar. Humans are so completely different from both chimpanzees and rhesus monkeys that Pääbo estimated that human brains must have undergone at least four times the amount of change in genes and gene expression levels as chimpanzees since the time of their common ancestor. In contrast, the chimpanzee and rhesus monkey brains have hardly changed since their common ancestors. The human brain really seems to be special in having accelerated patterns of gene evolution.

Are there any other molecular differences in the brain? Varki (Normile, 2001) found that the human brain lacks a surface sugar on the cell membrane that is found in the brains of all nonhuman primates. Are there any closely related fossil primates that also lack this sugar? We do not know – fossil brains do not exist. However, this sugar is also found in the bones of all other nonhuman primates examined but not in human bone. It may therefore be of enormous significance that this sugar is lacking in Neanderthal bones. Does this mean it was also lacking in Neanderthal brains? If humans and Neanderthals have been separate for at least 600 000 years (and this is very much a minimum separation; it could easily have been 900 000 years or even 1.5 million years: Tavare *et al.*, 2002) the change in humans may be early indeed.

A final new piece of evidence stems from the work of Britten (2002). The divergence between samples of chimpanzee and human DNA sequences, if inversions and deletions (indels) are counted, is as great as 5% rather than the 1.4% of traditional DNA comparisons. Indeed he finds that the sequence divergence between chimpanzees and humans is in excess of 20% for a few regions. Confirmation will have to wait until comparisons of much larger regions become available.

Nevertheless all these findings do mean that molecular biologists are starting to look for the differences between chimpanzees and humans rather than concentrating on the similarities. Could the neurometric differences between human and chimpanzees found here, differences that render humans neurometrically unique among the primates, be reflecting some of these unique molecular factors?

Soft data: niche-metrics

Morphometric studies can even be applied to data not representing the structure of the animal at all – e.g., data about animal movement, locomotion, behavior, environment, diet: in other words, the niche – and therefore niche-metrics (Oxnard *et al.*, 1990, and see Fleagle and Reed, Chapter 16 in this volume). In those investigations we showed that morphometric analyses of data about locomotor activities, arboreal environments, and dietary choices provided arrangements of the animals (prosimian primates) that made sense in terms of the niches that the animals inhabit and, therefore, also the lifestyles that they display. We also showed that these niche-metric results were extremely similar to parallel morphometric studies of the functional anatomy of these animals.

Thus bushbabies and tarsiers (completely different in an evolutionary sense) demonstrate close relationships in both niche-metrics and morphometrics. Both groups of species have similar locomotor, habitat, and dietary components (niche-metrics) and limb proportions and hip structure (morphometrics). This is also true, separately, for several other groupings of prosimians with parallel functions. Such parallels between niche-metric and morphometric results are entirely at the functional level – like other first-level investigations described in this chapter.

However, another way of examining those same data provides arrangements of the animals in relation to the families into which they are usually placed. For the example of bushbabies and tarsiers cited above, the new statistical analysis separated the bushbabies from the tarsiers (with which they were functionally convergent in the prior analysis) and placed them close to the various lorises (functionally totally different but evolutionarily more closely related). Again, this was also true of a number of other functional groups of parallel species from separate phylogenetic groups. Likewise, *Daubentonia* (see also Groves, Chapter 17 in this volume) was a much greater outlier in this study than it should have been if it were similar to lemurs (as it was in the functional result). Examination of the clusters of variables in these additional studies showed that the achievement of a pattern of evolutionary relationships was through new clusters of variables. These clusters broke niche variable linkages between convergent forms and revealed contrary variable linkages within evolutionary groups (Oxnard *et al.*, 1990, Oxnard, 2000).

Most recently of all, Pan and Oxnard (in press) have attempted other combinations of these strategies. When analyses of the craniodental area of Asian colobines are carried out against an allometric background based on Asian colobines, the arrangement of the species seems to relate most closely to their eco-environmental associations. That is, linked species such as *Pygathrix*, *Presbytis*, and *Nasalis* all inhabit coastal swampy forests: primary and secondary tropical rain forests, evergreen rain forests, and deciduous and semi-deciduous forests. Linked species *Rhinopithecus* and *Semnopithecus* both inhabit coniferous forests. *Trachypithecus* lies between these extremes and inhabits intermediate evergreen moist deciduous forests. But the same data, when examined against the allometric background of Asian macaques, do not show this ecological relationship. Instead the species are linked in ways that approach most closely their evolutionary association (as suggested by molecular studies).

"The softest data of all": theoretical data in modeling studies

Finally we have now entered what is for us a relatively new mode of investigation. That is, whatever we are able to deduce from comparative studies of extant and fossil species, whatever we deduce from molecular investigations, we are dependent upon the accidents of evolution for our data. Is it possible to gain insight into these problems by computational mimicry (perhaps the softest data of all) of evolutionary processes?

For example, most of the species that have ever existed do not appear as fossils. Can we assess the effect of not taking their existence into account? For example, again, we know that interbreeding and migration must have occurred many times in evolution: can we assess the effects of these upon species or subspecies lineages? For example, once more, though we have information from mitochondrial DNA and Y chromosome studies, do we really understand the differences between lineages observed through matrilines, patrilines, and both biological parents? For example, finally, most of the "characters" used in studying fossils are actually complex morphological features that have multiple causations (many of the investigations above have shown this). Do we understand the effects of treating them as though they were truly single characters as in cladistics, rather than mixed complexes of several, perhaps many, underlying characters?

A start has been made in examining questions like these using mathematical mimicry of evolutionary events (Oxnard, 1995, 1997b; Oxnard and Wessen, 2001; Wessen, 2002). Such studies imply, for instance, that whenever we study fossil and living morphologies, we are likely to be two to three times too recent in our assessments of common ancestry. Whenever we make judgments from fossils where interbreeding and migration may intrude, we almost always infer frequencies for these phenomena that are much less – by a factor of two or three – than in fact they were, and also that the phenomena were more recent then in fact was the case – again by a factor of two or three. Whenever we try to look for mothers of mothers, fathers of fathers, and parents of parents, we get different results. In particular, whenever we look for the time depths of individual ancestors of a sample of living individuals, they are much more recent (by hundreds and thousands of generations), than time depths of continuously breeding lineages.

It is obvious that assessments based upon the limited data that investigators have will never be completely correct. That follows inevitably from the partial nature of any investigation where we do not have all the fossils, all the living forms, full knowledge of all the characters, and so on. That is not, however, the problem. The problem is that we need to have estimates of how right or wrong our judgments in such studies may be. Our studies, and the lead investigator here is Ken Wessen (2002), in attempting to mimic evolutionary processes, now allow a wide range of factors to be included: e.g., differential selective advantage, different mating patterns, changing adult sex ratio, population size and structure, and random and catastrophic events. Do we have here the beginnings of a kind of "mimicry metrics"? Certainly such studies are suggesting that some kinds of evolutionary estimates may be far more wrong than we would wish (Oxnard and Wessen, 2001). Combined with new fossil and molecular studies, it is starting to be clear that such modeling is of considerable heuristic value (see also Tavare *et al.*, 2002).

Conclusions

Morphometrics has the potential to aid in the understanding of primate evolution to a degree greater now than at any time in the last 50 years. The new forms of data and the new technologies are particularly important, as are the ways in which they permit or visualize new interpretations. But so too are a number of other factors. A clear definition of the problem, rather than a "let's look at some data, any data" approach, is essential. Taking thought about the design of observations, and the levels of analysis, can both be especially useful. The recognition that the interpretation of morphometric results from whole-organism studies involves understanding the interfaces and interactions between function, ontogeny, and phylogeny brings a welcome synthesis of anatomy, development, and molecular biology that has already been achieved in vertebrate studies. Biological complexity goes beyond hard tissues (bones and teeth – most easily studied using morphometrics) into soft tissues such as muscles, soft organs such as brains, "soft" data such as those of behavior, diet, and environment, and even into the "totally soft" data of modeling (totally soft because they are invented). The application of morphometric methods to such data (e.g., neurometrics, niche-metrics, even "mimicry metrics") is certainly fun and may be important.

Acknowledgments

Many students and colleagues have been involved in these studies and they are recognized throughout the text. My wife Eleanor has helped enormously with my work through her librarianship skills, and also because she has participated with students and colleagues over the years to the extent that they know her as well as they know me.

The work is supported by the Australian Research Council and the (Australian) Health and Medical Research Infrastructure Fund. Special thanks go to the Leverhulme Trust (UK), whose support through a Leverhulme Visiting Research Professorship has permitted my extended consultation and collaboration with Professors R. Crompton, University of Liverpool, and P. O'Higgins, previously University College, London, and now University of York, UK.

References

Adanson, M. (1763). *Famille des plantes*. Vol. 1. Paris: Vincent.

Albrecht, G. H. (1980). Multivariate analysis and the study of form, with special reference to canonical variate analysis. *Amer. Zool.*, **20**, 679–693.

(1991). Thin plate splines and the primate scapula. *Amer. J. Phys. Anthropol.*, **28**, 125–126.

Ashton, E. H., Healy, M. J. R., and Lipton, S. (1957). The descriptive use of discriminant functions in physical anthropology. *Proc. Roy. Soc. Lond. B*, **146**, 555–572.

Ashton, E. H., Flinn, R. M., and Oxnard, C. E. (1975). The taxonomic and functional significance of overall body proportions in primates. *J. Zool. Lond.*, **175**, 73–105.

Ashton, E. H., Flinn, R. M., Oxnard, C. E., and Spence, T. F. (1976). The adaptive and classificatory significance of certain quantitative features of the forelimb in primates. *J. Zool. Lond.*, **163**, 319–350.

Ashton, E. H., Flinn, R. M., Moore, W. J., Oxnard, C. E., and Spence, T. F. (1981). Further quantitative studies of form and function in the primate pelvis with special reference to *Australopithecus*. *Trans. Zool. Soc. Lond.*, **360**, 1–98.

Bartlett, M. S. (1935). Contingency table interactions. *Supp. J. Roy. Stat. Soc.*, **9**, 248–252.

Barton, R. A. and Harvey, P. H. (2000). Mosaic evolution of brain structure in mammals. *Nature*, **405**, 1055–1058.

Blackith, R. E. and Reyment, R. A. (1971) *Multivariate Morphometrics*. London: Academic Press.

Bookstein, F. L. (1989). Principal warps: thin plate splines and the decomposition of deformations. *IEEE Trans. Pattern Anal.*, **11**, 567–585.

(1991). *Morphometric Tools for Landmark Data: Geometry and Biology*. Cambridge: Cambridge University Press.

Britten, R. J. (2002). Divergence between samples of chimpanzee and human DNA sequences is 5%, counting indels. *Proc. Nat. Acad. Sci.*, **99**, 13633–13635.

Clark, D. A., Mitra, P. P., and Wang, S. S.-H. (2001). Scalable architecture in mammalian brains. *Nature*, **411**, 189–193.

Clark, R. J. and Tobias, P. V. (1995). Sterkfontein Member 2 foot bones of the oldest South African hominid. *Science*, **269**, 521–524.

Crompton, R. H. and Li, Y. (1997). Running before they could walk? Locomotor adaptations and bipedalism in early hominids. In: *Archaeological Sciences 1995*, ed. A. Slater, A. Sinclair, and J. A. J. Gowlett. Oxford: Oxford Bow Press. pp. 422–427.

Crompton, R. H., Li, Y. Günther, M. M., and Alexander, R. M. (1996). Segment inertial properties of primates: new techniques for laboratory and field studies of locomotion. *Amer. J. Phys. Anthropol.*, **99**, 547–570.

Crompton, R. H., Li, Y., Wang, W., Günther, M., and Savage, R. (1998). The mechanical effectiveness of erect and 'bent knee, bent hip' bipedal walking in *Australopithecus afarensis*. *J. Hum. Evol.*, **35**, 55–74.

Day, M. H. (1977). *Guide to Fossil Man*. London: Cassell.

de Winter, W. (1997). *Perspectives on Mammalian Brain Evolution*. Ph.D. Thesis, University of Western Australia.

de Winter, W. and Oxnard, C. E. (2001). Evolutionary radiations and convergences in the structural organization of mammalian brains. *Nature*, **409**, 710–714.
Dryden, I. L. and Mardia, K. V. (1998) *Statistical Shape Analysis*. New York, NY: Wiley.
Finlay, B. L. and Darlington, R. B. (1995). Linked regularities in the development and evolution of mammalian brains. *Science*, **268**, 1578–1783.
Fisher, R. A. (1935). *The Design of Experiments*. Edinburgh: Oliver & Boyd.
(1936). The use of multiple measurements in taxonomic problems. *Ann. Eugenics*, **7**, 179–188.
Fleagle, J. G. (1974). Dynamics of brachiating siamang [*Hylobates (Symphalangus) syndactylus*]. *Nature*, **248**, 259–260.
(1976). Locomotor behavior and skeletal anatomy of sympatric Malaysian leaf monkeys (*Presbytis obscura* and *Presbytis melalophos*). *Yrbk. Phys. Anthropol.*, **20**, 440–453.
Galton, F. (1889). *Natural Inheritance*. London: Macmillan.
German, R. Z. and Franks, H. A. (1991). Timing in the movements of jaws, tongue and hyoid during feeding in the hyrax, *Procavia syriacus*. *J. Exp. Zool.*, **257**, 34–42.
Herschkovitz, P. (1977). *The Living New World Monkeys*. Chicago, IL: University of Chicago Press.
Hotelling, H. (1931). The generalisation of "students" ratio. *Ann. Math. Statist.*, **2**, 360–378.
Howells, W. W. (1973). *Cranial Variation in Man: a Study by Multivariate Analysis of Patterns of Differences among Recent Human Populations*. Harvard, MA: Peabody Museum.
Hylander, W. L., Johnson, K. R., and Crompton, A. W. (1987). Loading patterns and jaw movements during mastication in *Macaca fascicularis*: a bone strain, electromyographic and cineradiographic analysis. *Amer. J. Phys. Anthropol.*, **72**, 287–314.
Jablonski, N. G. and Chaplin, G. (1993). Origin of habitual terrestrial bipedalism in the ancestor of the Hominidae. *J. Hum. Evol.*, **24**, 259–280.
Jenkins, F. A. (1972). Chimpanzee bipedalism: cineradiographic analysis and implications for the evolution of gait. *Science*, **178**, 877–879.
(1975) *Primate Locomotion*. New York, NY: Academic Press.
Jerison, H. J. (1973). *Evolution of the Brain and Intelligence*. New York, NY: Academic Press.
Jungers, W. L. and Fleagle, J. G. (1980) Post-natal growth allometry of the extremities in *Cebus albifrons* and *Cebus apella*: a longitudinal and comparative study. *Amer. J. Phys. Anthropol.*, **53**, 471–478.
Jungers, W. L. and Stern, J. T. (1980). Telemetered electromyography of forelimb muscle chains in gibbons (*Hylobates lar*). *Science*, **208**, 617–619.
Kendall, M. G. (1957). *A Course in Multivariate Analysis*. London: Griffin.
Kidd, R. (1995). *An Investigation into the Patterns of Morphological Variation in the Proximal Tarsus of Selected Human Groups, Apes and Fossils*. Ph.D. thesis, University of Western Australia.
Kidd, R., O'Higgins, P., and Oxnard, C. E. (1996). The OH8 foot: a reappraisal of the functional morphology of the hindfoot utilising a multivariate analysis. *J. Hum. Evol.*, **31**, 269–291.

Kidd, R. and Oxnard, C. E. (2002). Patterns of morphological discrimination in selected human tarsal elements. *Amer. J. Phys. Anthropol.*, **117**, 169–181.

Lestrel, P. E. (1997). *Fourier Descriptors and Their Applications in Biology*. Cambridge: Cambridge University Press.

(2000). *Morphometrics for the Life Sciences*. Singapore: World Scientific.

Lisowski, F. P., Albrecht, G. H., and Oxnard, C. E. (1974). The form of the talus in some higher primates. *Amer. J. Phys. Anthropol.*, **41**, 191–215.

(1976). African fossil tali: further multivariate morphometric studies. *Amer. J. Phys. Anthropol.*, **45**, 5–18.

Mahalanobis, P. C. (1936). On the generalized distance in statistics. *Proc. Nat. Inst. Sci. India*, **2**, 49–55.

Mittermeier, R. A. (1978). Locomotion and posture in *Ateles geoffroyi* and *Ateles paniscus*. *Folia Primatol.*, **30**, 161–193.

Normile, D. (2001). Gene expression differs in human and chimp brains. *Science*, **292**, 44–45.

O'Higgins, P. (2000). Quantitative approaches to the study of craniofacial growth and evolution: advances in morphometric techniques. In: *Development, Growth and Evolution: Implications for the Study of Hominid Skeletal Evolution*, ed. P. O'Higgins and M. Cohn. London: Academic Press. pp. 164–185.

Olson, E. C. and Miller, R. L. (1954). *Morphological Integration*. Chicago, IL: University of Chicago Press.

Oxnard, C. E. (1967). The functional morphology of the primate shoulder as revealed by comparative anatomical, osteometric and discriminant function techniques. *Amer. J. Phys. Anthropol.* **26**, 219–240.

(1973). *Form and Pattern in Human Evolution: Some Mathematical, Physical and Engineering Approaches*. Chicago, IL: University of Chicago Press.

(1975). *Uniqueness and Diversity in Human Evolution: Morphometric Studies of Australopithecines*. Chicago, IL: University of Chicago Press.

(1976). Some methodological factors in studying the morphological–behavioral interface. *Burg Wartenstein Symposium*, **71**, 1–65.

(1979). The morphological behavioral interface in extant primates: some implications for systematics and evolution. In: *Environment, Behavior and Morphology: Dynamic Interactions in Primates*, ed. M. E. Morbeck, H. Preuschoft, and N. Gomberg. New York, NY: Gustav Fischer. pp. 209–227.

(1983/84). *The Order of Man: a Biomathematical Anatomy of the Primates*. New Haven, CT: Yale University Press; Hong Kong: Hong Kong University Press.

(1995). The challenge of human origins: morphology and molecules, morphometrics and modelling. In: *The Origin and Past of Modern Humans from DNA*, ed. S. Brenner and K. Hanihara. Singapore: World Scientific Publications. pp. 11–30.

(1997a). From optical to computational Fourier transforms: the natural history of an investigation of the cancellous structure of bone. In: *Fourier Descriptors and their Applications in Biology*, ed. P. Lestrel. Cambridge: Cambridge University Press. pp. 379–408.

(1997b). The time and place of human origins. In: *Conceptual Issues in Modern Human Origins Research*, ed. G. A. Clark and C. M. Willermet. New York, NY: De Gruyter. pp. 369–391.

(2000). Morphometrics of the primate skeleton and the functional and developmental underpinnings of species diversity. In: *Development, Growth and Evolution: Implications for the Study of the Hominoid Skeleton*, ed. P. O'Higgins and M. Cohn. London: Academic Press. pp. 235–264.

(in press). Evolution of the brain: the primate background and the chimpanzee/human comparison. *Int. J. Primatol.*

Oxnard, C. E. and Wessen, K. (2001) Modelling divergence, inter-breeding and migration: species evolution in a changing world. In: *Faunal and Floral Migrations and Evolution in SE Asia–Australasia*, ed. I. Metcalfe, J. M. B. Smith, M. Morwood, and I. Davidson. Netherlands: Swets and Zeitlinger. pp. 373–385.

Oxnard, C. E., Crompton, R., and Lieberman, S. (1990). *Animal Lifestyles and Anatomies: the Case of the Prosimian Primates*. Seattle, WA: Washington University Press.

Oxnard, C. E., Lannigan, F., and O'Higgins, P. (1995). The mechanism of bone adaptation: tension and resorption in the human incus. In: *Bone Structure and Remodelling*, ed. A. Odegard and H. Weinans. Singapore: World Scientific Publications. pp. 105–125.

Pan, R. L. and Oxnard, C. E. (2002a). Metrical dental analysis on golden monkey (*Rhinopithecus roxellana*). *Primates*, **42**, 75–89.

(2002b). Craniodental variation among macaques (*Macaca*): nonhuman primates. *BioMed Central: Evol. Biol.*, **2**, 10–26.

(in press). Functional homology or phylogenetic convergence: a methodological model based on dental variation among Asian colobines. *Proc. Int. Primatol. Soc., Beijing*.

Pearson, K. (1901). On lines and planes of closest fit to systems of points in space. *Phil. Mag.*, **2**, 559–572.

Quetelet, L. A. J. (1842). *A Treatise on Man and the Development of his Faculties*. Paris: Bachelier.

Rao, C. R. (1948). The utilisation of multiple measurements in problems of biological classification. *J. Roy. Statist. Soc.*, **B10**, 159–203.

Reyment, R. A., Blackith, R. E., and Campbell, N. A. (1984). *Multivariate Morphometrics*. 2nd edn. London: Academic Press.

Richtsmeier, J. T., Deleon, S. R., and Lele, S. R. (2002). The promise of geometric morphometrics. *Yrbk. Phys. Anthropol.*, **45**, 63–91.

Rodman, P. S. (1979). Skeletal differentiation of *Macaca fascicularis* and *Macaca nemestrina* in relation to arboreal and terrestrial quadrupedalism. *Amer. J. Phys. Anthropol.* **51**, 51–62.

Savara, B. S. (1965). Applications of photogrammetry for quantitative study of tooth and face morphology. *Amer. J. Phys. Anthropol.*, **23**, 427–434.

Sneath, P. H. A. and Sokal, R. R. (1973). *Numerical Taxonomy*. San Francisco, CA: Freeman.

Stephan, H. and Andy, O. J. (1969). Quantitative comparative neuroanatomy of primates: an attempt at a phylogenetic interpretation. *Ann. N. Y. Acad. Sci.* **167**, 370–387.

Stern, J. T. and Oxnard, C. E. (1973). *Primate Locomotion: some Links with Evolution and Morphology*. Basel: Karger.

Stern, J. T., Wells, J. P., Jungers, W. L., Vangor, A. K., and Fleagle, J. G. (1980). An electromyographic study of the pectoralis major in atelines and *Hylobates*, with special reference to the evolution of a pars clavicularis. *Amer. J. Phys. Anthropol.*, **52**, 13–25.

Tavare, S., Marshall, C. R., Will, O., Soligo, C., and Martin, R. D. (2002). Using the fossil record to estimate the age of the last common ancestor of extant primates. *Nature*, **416**, 726–729.

Trevor, J. C. (1955). The ancient inhabitants of Jebel Moya (Sudan). *Occas. Pubs. Cambridge Univ. Mus. Arch. Ethnol.*, Vol. III. Cambridge: Cambridge University Press.

Tukey, J. W. (1977). *Exploratory Data Analysis*. Reading, MA: Addison-Wesley.

Wessen, K. P. (2002). *Simulating the Origin and Evolution of Ancient and Modern Humans*. Ph.D. Thesis, University of Western Australia.

Wilks, S. S. (1935). On the independence of k sets of normally distributed statistical variables. *Econometrica*, **3**, 309–326.

Yates, F. (1950). The place of statistics in the study of growth and form. *Proc. Roy. Soc. Lond.* B, **137**, 479–489.

Zohlman, J. F. (1993). *Experimental Design and Statistical Inference*. Oxford: Oxford University Press.

20 *Postscript and acknowledgments*

CHARLES E. OXNARD
University of Western Australia

One usually thinks first of the well-known senior individuals who have been responsible for initiating one's career. My initial stimulus came, however, from the enthusiasm of a relatively unknown person. He was the headmaster of the tiny primary school I attended in a small village in Scotland at the outbreak of the Second World War. Knowing that education in Scotland was very classical – English and mathematics, even Latin and Greek in those days, but no science to speak of – he understood, somehow, that this small boy was interested in science. He introduced me to the ideas of Wegener, Goethe, D'Arcy Thompson, and Solly Zuckerman when I was nine years old, two of them, of course, not in the original.

As a result, I may be the only person in the world who knew about the movements of the continents but who did not know that Wegener's ideas were not accepted for almost half a century. By the time I reached university in 1952, the idea was center-stage as plate tectonics. I could not understand what all the excitement was about. I had always known it was so.

Likewise, I understood very well the idea that the skull was simply a series of fused vertebrae. It made sense to me. I did not know that Goethe had it wrong until I later came to read Gavin de Beer's tome on the vertebrate skull. However, it is fascinating that new developmental studies in this last couple of decades show that Goethe was a little bit more right than most of us have thought, though, of course, for the wrong reasons.

I did not understand the mathematical formulae and Greek quotations in D'Arcy Thompson's *On Growth and Form*. But there is much in that book to touch a nine-year-old. The Cartesian coordinate transformation diagrams made so explicit some of the differences in biological forms that Thompson presented. His pictures of the struts in the interior of a bird's wing were so like the struts in the wings of the biplanes of those days. The relationships between stress and architecture of the bridges that he presented were so obvious in the bridge

Shaping Primate Evolution, ed. F. Anapol, R. Z. German, and N. G. Jablonski. Published by Cambridge University Press.

over the Forth that was just down the Firth from my home. Those ideas have been with me all my life. They have figured in my investigations even recently through thin-plate splines, Fourier transforms, and tensegrity towers.

And finally, of course, how could I have known that I would later actually work with Dr. Solly Zuckerman, author of that 1934 book *Functional Affinities of Man, Monkeys and Apes* that my headmaster gave me all those years ago?

I thus also owe much to famous people, especially to Lord Zuckerman himself – I think I was his last full-time student. Eric Ashton and Tom Spence, with whom I worked in Zuckerman's Department of Anatomy at the University of Birmingham, were constant exemplars, colleagues, and friends. Indeed, I owe much to that entire department in Birmingham, even to the wider range of academics with whom Zuckerman was associated in those seminal years. All were critical to my own scientific development. When Zuckerman retired (from the chair in Birmingham – he never retired from problem-solving except at his death in 1995) his efforts at the University of Birmingham had grown Anatomy from about six academics in 1945 to as many as 53 at the time of his retirement in 1966. In those few years, he had increased the department's funding from the widow's mite of a typical anatomy department to one that was third largest in the university, being exceeded only by physics and chemistry. He had moved the department from being almost solely occupied with teaching medical anatomy to medical students to one encompassing many new "anatomies." These anatomies spanned not only humans but also apes, monkeys, mammals, vertebrates, and even many non-vertebrates. They included the anatomies of reproduction, development, growth, and contraception, the anatomies of cells, subcellular organelles, membranes and molecules, and the anatomies of brains, behaviors, cognition, and psychology. What other anatomy department of those days (or of any day) had biochemists, physicists, engineers, psychologists, school masters, and peers of the realm on its staff? And all this without the need to change the name!

Other academics outside the Birmingham department also influenced my career: Professor J. Z. Young who, together with Solly, examined me for my Bachelors degree in 1955.

Professor Young: "What do you see down that microscope, my boy?"

"The substantia nigra, sir," I said.

Covering the microscope eyepiece with his hand: "Is the black stuff inside or outside the cells?" he asked.

Of course, I had no idea – I was not a brain person in those days – but I gave an answer based on standard biological principles. I have never dared go back to see if my answer was right, though I suspect it was. Solly and JZ roared with laughter. They *knew* I had guessed.

And there were others: A. J. E. Cave, Professor of Anatomy at Barts (St. Bartholomew's Medical College of the old days). He of Neanderthal fame. Alec Cave knew that the shambling gait of the Neanderthaler, on its way to modern human walking as it was supposed, was just severe osteoarthritis in a skeleton of an old man. He examined me for my Ph.D.

"Did you write that thesis?" – indicating my doctoral volume on the table.

"Yes."

"Then you pass," he said and immediately signed the examination form and appended: "Highly Commended."

For the next three hours we had a fascinating discussion about the great ideas and great people of anatomy in earlier days (this was the real examination!). And that discussion with him continued sporadically over the years. I sometimes think he and I were almost the last anatomists to dissect the really large animals of this world. He was a wonderful scientist who, though he did not know one end of a statistic from the other, understood intuitively just why we were using multivariate statistical methods to characterize animal form and pattern. He had his telegram from the Queen on his hundredth birthday but died after the commencement of this volume. It is an especial privilege for me to be able to recognize his part in my life.

Zuckerman's departure on retirement from Birmingham in 1966 spawned a diaspora that continued for many years. I was the first emigrant myself in that same year and have enjoyed further seminal decades at the Universities of Chicago, Southern California, and Western Australia. I have found, everywhere in the world, members of that original "Zuckerman Mafia." Each time that I have moved, I have, while keeping the old – old colleagues, old students, old problems, and old techniques – also extended my grasp to the new. All are part of my acknowledgment.

This extension to new fields has especially included disciplines and colleagues very far removed from my original education. This started thus, with a desire to use quantitative methods. Solly sent me to Fisher at Rothamstead, who quickly sent me to his "young man" Yates, who as quickly sent me to his "young man" Michael Healey, who really was almost as young as me. Thus was I introduced to the computer at Rothamstead, to programming in binary code, and to the multivariate statistical approach. This statistical collaboration and consultation was later continued with Paul Meier, David Wallace, and Bill Kruskal when I arrived at the University of Chicago. Further help, in turn, was quickly given by others. Joel Cohen (who almost immediately left for Harvard) helped me with multivariate statistical programmes. Peter Neeley (who almost as quickly left for Kansas) helped me with neighborhood limited classification. D. F. Andrews of Bell Telephone Labs (as it was in those days) provided the method of high-dimensional analysis which he had developed using some of my own prior data. L. A. Zadeh provided early thoughts about the use of fuzzy sets in understanding biological form. Even since then, statistical influences have continued from Norm Campbell (CSIRO, Western Australia) through his extension of the original Blackith and Reyment *Multivariate Morphometrics* text, and from Adrian Baddeley (Applied Mathematics, University of Western Australia). I have always been blessed by having mathematically gifted colleagues with real abilities to understand the biological problems on which we were working.

In the same way, it became obvious, long before the word "biomechanics" had been invented, that I needed to know more about the mechanics that might apply to biological mechanisms. This started with my taking a course on experimental stress and strain analysis at the Royal College of Advanced Technology at Salford (now the University of Salford). The lecturers on that course were partly amused and partly horrified by having one student who turned up with wet bones instead of oily crankshafts for the practical classes. This was quickly followed by a personal course on photoelastic analysis with Ken Sharples of Sharples PhotoLasticity Inc., together with experience in experimental methods and equipment obtained through Westlands Aircraft Ltd. (as it was in those days). Even very recently such collaboration has continued in Australia with colleagues from geomechanics (Chris Windsor and Wayne Robertson from the Geomechanics Unit of the CSIRO at the University of Western Australia). These individuals introduced me to modern elements of finite element analysis performed through FLAC (Fast LaGrangian Analysis of Continua). This collaboration led eventually to the smallest grant that I have ever had (it was a one-year pilot grant) but the one of which I am most proud because it was made to an old-fashioned medical anatomist by the CSIRO of Australia.

Finally, from an interest in form (the external shape of bones) I moved to a desire to understand texture (the internal structure of bones). This led me to look for methods other than multivariate statistics. At first I was introduced to optical data analysis using laser-generated optical Fourier transforms (this was long before the days of computational Fast Fourier Transforms) by John Davis of the University of Kansas and Harry Pincus of the University of Wisconsin at Milwaukee. Both these individuals were in geology departments and interested in such topics as the textures of oil-bearing rocks – topics that were not anywhere as interesting to me as the textures of stress-bearing spongy bones.

It sometimes frightens me that contacts like these were made serendipitously. How many even better techniques have I missed because of the individuals that I did *not* meet?

Perhaps, however, the people to whom I owe most, the people who have contributed most to my researches over the years, indeed, the people who keep me mentally alive even now, are none of the "old greats" or the "accidental colleagues." They are the "new greats." They are students and colleagues, mostly much younger than I, many of them now professors in their turn, in all parts of the world, some of them, however, still students of mine, or of my students, or even of my students' students. Some of them have contributed to this volume. Most of them have shown that many of my original "great" ideas could be "improved upon" – were, in fact, "not quite right" – were, actually, even, "totally wrong." Yet a few of them have found that some of my earlier scientific heresies were not so far from reality. It is through these people that I am still able to be involved in the new developments in our discipline. They keep me alive. They are really the ones honored by this volume. They could not all be present as authors here, but all are present in spirit as I look at my last fifty years in science, and as I contemplate the many research lines with which I will continue to be involved (hopefully) in the next few decades as an honorary investigator.

It is worth recording what has happened to a new group of younger colleagues – honors students, graduate students, and postdoctoral colleagues – in the last few years. Seven of them (all strong students) have opted to leave academia because of the perceived lack of value that our society, our government, our country, even academia itself, places upon a research and teaching career. You would think this would be a worry – yet no one seems willing to do anything about it. This has never happened to me before!

Perhaps the thing of which I am most proud, which all my students and colleagues exemplify in spades, is breadth of learning. Such breadth has required of me membership in a wide range of disciplinary societies: anatomical, anthropological, zoological, human biological, biomechanical, biostatistical, and clinical. These students and colleagues have, equally, moved into disciplines even more widely spread, and in all inhabited continents of the world. It is remarkable that so many of them (not themselves physical anthropologists) could be assembled to attend a symposium organized at the annual meeting of the Association of American Physical Anthropologists. This breadth explains, of course, the breadth in this book.

Index